# Annual Review of IP Communications

IEC
Chicago, Illinois

# About the International Engineering Consortium

The International Engineering Consortium (IEC) is a non-profit organization dedicated to catalyzing technology and business progress worldwide in a range of high technology industries and their university communities. Since 1944, the IEC has provided high-quality educational opportunities for industry professionals, academics, and students. In conjunction with industry-leading companies, the IEC has developed an extensive, free on-line educational program. The IEC conducts industry-university programs that have substantial impact on curricula. It also conducts research and develops publications, conferences, and technological exhibits that address major opportunities and challenges of the information age. More than 70 leading high-technology universities are IEC affiliates, and the IEC handles the affairs of the Electrical and Computer Engineering Department Heads Association and Eta Kappa Nu, the honor society for electrical and computer engineers. The IEC also manages the activities of the Enterprise Communications Consortium.

# Other Quality Publications from the International Engineering Consortium

- *Achieving the Triple Play: Technologies and Business Models for Success*
- *Business Models and Drivers for Next-Generation IMS Services*
- *Delivering the Promise of IPTV*
- *Evolving the Access Network*
- *The Basics of IPTV*
- *The Basics of Satellite Communications, Second Edition*
- *The Basics of Telecommunications, Fifth Edition*

For more information on any of these titles, please contact the IEC publications department at +1-312-559-3730 (phone), +1-312-559-4111 (fax), *publications@iec.org*, or via our Web site (http://www.iec.org).

ISBN: 978-1-931695-70-1

International Engineering Consortium
300 West Adams Street, Suite 1210
Chicago, Illinois 60606-5114 USA
+1-312-559-3730 phone
+1-312-559-4111 fax

# Contents

## Academic Perspectives

## Corporate Perspectives

## Internet Protocol Television (IPTV)

## Internet Protocol Multimedia Subsystem (IMS)

## Network Architectures and Design

# Quality of Service

# Contents by Author

# University Program Sponsors

The IEC's University Program, which provides grants for full-time faculty members and their students to attend IEC Forums, is made possible through the generous contributions of its Corporate Members. For more information on Corporate Membership or the University Program, please call +1-312-559-4625 or send an e-mail to *cmp@iec.org*.

Based on knowledge gained at IEC Forums, professors create and update university courses and improve laboratories. Students directly benefit from these advances in university curricula. Since its inception in 1984, the University Program has enhanced the education of more than 500,000 students worldwide.

## IEC–Affiliated Universities

The University of Arizona
Arizona State University
Auburn University
University of California at Berkeley
University of California, Davis
University of California, Santa Barbara
Carnegie Mellon University
Case Western Reserve University
Clemson University
University of Colorado at Boulder
Columbia University
Cornell University
Drexel University
École Nationale Supérieure des Télécommunications de Bretagne
École Nationale Supérieure des Télécommunications de Paris
École Supérieure d'Électricité
University of Edinburgh
University of Florida
Georgia Institute of Technology

University of Glasgow
Howard University
Illinois Institute of Technology
University of Illinois at Chicago
University of Illinois at Urbana-Champaign
Imperial College of Science, Technology and Medicine
Institut National Polytechnique de Grenoble
Instituto Tecnológico y de Estudios Superiores de Monterrey
Iowa State University
KAIST
The University of Kansas
University of Kentucky
Lehigh University
University College London
Marquette University
University of Maryland at College Park
Massachusetts Institute of Technology
University of Massachusetts

McGill University
Michigan State University
The University of Michigan
University of Minnesota
Mississippi State University
The University of Mississippi
University of Missouri-Columbia
University of Missouri-Rolla
Technische Universität München
Universidad Nacional Autónoma de México
North Carolina State University at Raleigh
Northwestern University
University of Notre Dame
The Ohio State University
Oklahoma State University
The University of Oklahoma
Oregon State University
Université d'Ottawa
The Pennsylvania State University

University of Pennsylvania
University of Pittsburgh
Polytechnic University
Purdue University
The Queen's University of Belfast
Rensselaer Polytechnic Institute
University of Southampton
University of Southern California
Stanford University
Syracuse University
University of Tennessee, Knoxville
Texas A&M University
The University of Texas at Austin
University of Toronto
VA Polytechnic Institute and State University
University of Virginia
University of Washington
University of Wisconsin-Madison
Worcester Polytechnic Institute

# Academic Perspectives

# An Empirical Evaluation of IPv6 in Linux and Windows Environments

## Sanjay P. Ahuja

*Professor, Computer and Information Sciences Department*
University of North Florida

## Krishna Dendukuri

*Professor, Computer and Information Sciences Department*
University of North Florida

## Abstract

The Internet has revolutionized the world and continues to grow exponentially. The current widely used Internet protocol version 4 (IPv4) is straining to keep up with the demands of today's Internet. The Internet Engineering Task Force (IETF) developed Internet protocol version 6 (IPv6) to provide support for growth and address the shortcomings of IPv4. While IPv6 is becoming more widespread, more empirical research on its performance needs to be carried out. This paper addresses this need and presents an evaluation study of IPv6 protocol stack implementations in Linux and Windows 2000 operating system environments. Given the wide deployment of Windows 2000 and the even more widespread popularity enjoyed by Linux, the impact of IPv6 on end-to-end applications on these platforms must be adequately assessed. We investigate the impact of IPv6 overhead on end-to-end applications using transmission control protocol (TCP) and user datagram protocol (UDP) with different packet-size messages. Several experiments to measure and compare a set of performance metrics were performed on a test bed of two end stations connected through direct point-to-point link running Linux and Windows 2000 in dual-boot mode loaded with IPv4/IPv6 stacks. This paper also explores the work done in IPv6 implementations and discusses some performance evaluation results for IPv6 implementations in routers and end host applications.

## Introduction

Growth is the basic issue that created need for IPv6 [3]. The fixed 32-bit address space in IPv4 [4] is inadequate for exponential growth of computer networks. With the changing nature of the Internet, business networks, and the growing demand for new multimedia applications and mobile communications, IPv4 is straining to keep up with its demands. The Internet Engineering Task Force (IETF) has designed the business and technical case for IPv6 and is working on the deployment of IPv6. The new IPv6 has been designed to provide 128-bit address space and to enable high performance and scalable internetworks that should operate as needed for decades. Part of the IETF design process involved correcting the inadequacies of IPv4 and adding a number of enhanced features [1, 2].

As many computer corporations started developing IPv6 applications and providing support for IPv6 implementations, performance measurements of IPv6 support in routers and mobile devices were largely published [5, 6, and 7]. As IPv6 is maturing through its development process, various projects have started evaluating its functionality and support features. However, end-to-end user application performance comparisons in multiple operating system environments have not been widely available. The main goal of this paper is to explore the impact of IPv6 protocol stack on TCP and UDP applications running on Linux and Windows 2000 operating systems using a set of performance metrics that include throughput, latency, host central processing unit (CPU) utilization, TCP connection time, and the number of client/server interactions per second.

The rest of the paper is organized as follows: Section 2 discusses background information on IPv6 and covers the various aspects of research performed in the IPv6 arena related to the transitioning mechanisms, porting applications to IPv6, and the deployment of IPv6; Section 3 investigates related work in the literature on IPv6 performance evaluations in different projects; Section 4 describes the project, test-bed configuration, and performance metrics used for this comparison study; Section 5 focuses on the experimental results and details the metrics captured in the evaluation study; and Section 6 provides conclusions and discusses future work.

## IPv6 Background

In the early 1990s, it became apparent that IPv4 would run out Internet address space. IPv6 was developed to restore the ability to provide a globally unique address for every device on the Internet and to permit the hierarchical addressing necessary as the Internet grew. This section briefly describes the IPv6 architecture, mechanisms to transition from IPv4, and deployment in the IPv6 arena covered in the literature [3, 9, 10, 11, 12, and 18].

### IPv6 Architecture

IPv6 was designed to take an evolutionary step from IPv4 and with future applications in mind. Working functions in IPv4 were kept in IPv6, and non-working functions were removed. IPv6 increases the IP address size from 32 bits to 128 bits to support more levels of addressing hierarchy, a much greater number of addressable nodes, and simpler auto-configuration of addresses. IPv6 addresses are assigned to individual interfaces on nodes. IPv6 allows three types of addresses—unicast, anycast, and multicast addresses. In a unicast address, a packet is delivered to the interface identified by that address. In a new type of address called an anycast address, a packet is delivered to any one of a group node. In multicast addresses, a packet is delivered to all the interfaces identified by that address. The scalability of multicast routing is improved by adding a "scope" field to multicast addresses [3, 9].

Although the address space in IPv6 is four times larger than that of IPv4, the header in IPv6 increases by only 20 bytes. The IPv6 header is simplified by excluding checksum and variable options fields and introduces extension headers. IPv6 requires no manual configuration or dynamic host configuration protocol (DHCP) setup and provides better network management. Fragmentation support has been moved from routers to the sending hosts. The flow-label field in IPv6 tells a router that a flow wants special handling, doesn't reveal information about the service being used, and improves quality of service (QoS) [3, 9].

### IPv6 Transition Mechanisms

The transition phase from IPv4 to IPv6 networks will need a set of mechanisms to ensure a transparent communication environment bearing two protocols. The key to a successful IPv6 transition is compatibility with the large installed base of IPv4 hosts and routers. Maintaining compatibility with IPv4 while deploying IPv6 will streamline the transition of the Internet to IPv6 [10, 11, 12].

#### Dual Stack Host

This technique provides complete support for both IPv4 and IPv6 in hosts and routers. The most straightforward way for IPv6 nodes to remain compatible with IPv4 only nodes is by providing a complete IPv4 implementation. Each host has both IPv4 and IPv6 stacks and addresses. The domain naming system (DNS) is used in both IPv4 and IPv6 to map between hostnames and IP addresses. DNS resolver libraries on IPv6/IPv4 nodes may return both IPv4 and IPv6 addresses, only IPv4 addresses, and only IPv6 addresses to the application. A host will either send packets in IPv4 or IPv6 depending on the protocol used by the destination. Dual stacks are easiest to implement, but complexity increases at the hosts because of the support for both stacks, which in turn creates higher cost [10, 11, and 12].

#### Tunneling IPv6 in IPv4

Tunneling is the most widely used transition mechanism. Tunneling uses the existing IPv4 routing infrastructure while the IPv6 infrastructure is deployed over time. Tunneling provides a way to use an existing IPv4 routing infrastructure to carry IPv6 traffic. Encapsulation of IPv6 packets within IPv4 packets allows two IPv6 hosts/networks to be connected with each other while running on existing IPv4 networks. IPv6 packets are encapsulated in IPv4 packets and then are transmitted over IPv4 networks like ordinary IPv4 packets. At the destination, these packets are decapsulated to the original IPv6 packets. Tunneling can be implemented in a variety of ways, including the following:

- *Router to router*: IPv6/IPv4 routers interconnected by an IPv4 infrastructure can tunnel IPv6 packets between the routers.

- *Host to router*: IPv6/IPv4 hosts can tunnel IPv6 packets to an intermediary IPv6/IPv4 router that is reachable via an IPv4 infrastructure.

- *Host to host*: IPv6/IPv4 hosts that are interconnected by an IPv4 infrastructure can tunnel IPv6 packets between the hosts.

- *Router to host*: IPv6/IPv4 routers can tunnel IPv6 packets to their final-destination IPv6/IPv4 host.

Tunneling techniques are classified according to the mechanism through which the encapsulating node determines the node address at the end of tunnel. Two major tunneling techniques, automatic and configured, differ primarily in how tunnel endpoint address is determined. In configured tunneling, for each tunnel, the encapsulating node must store the tunnel endpoint address. Encapsulating node explicitly configures the tunnel endpoint address. Endpoints of tunnel must be dual-stack nodes. When an IPv6 packet is transmitted over a tunnel, the configured tunnel endpoint address is used as the destination address for the encapsulating IPv4 header. The exit node of the tunnel (decapsulating node) receives the encapsulated packet, reassembles the packet if needed, removes the IPv4 header, updates the IPv6 header, and processes the received IPv6 packet. In automatic tunneling, the tunnel endpoint address is determined by the IPv4 compatible destination address of the IPv6 packet being tunneled. An implementation will have a special static routing table entry to direct automatic tunneling. Packets that match with routing table entry prefix are sent to a pseudo-interface driver that performs automatic tunneling. The encapsulation and decapsulation procedures of IPv6 packets remain the same as configured tunneling. Automatic tunneling allows IPv6/IPv4 nodes to communicate over IPv4 routing infrastructures without preconfiguring tunnels [10].

#### Stateless IP/ICMP Translation (SIIT)

This mechanism allows IPv6–only hosts to interoperate with IPv4 hosts. As the number of available unique IPv4 addresses decreases, there will be a desire to take advantage of the large IPv6 address, and every node does not require having a permanently assigned IPv4 address. The SIIT proposal assumes that IPv6 nodes are assigned a temporary IPv4 address for communicating with IPv4 nodes and does not specify a mechanism for the assignment of these addresses. This temporary IPv4 address will be used as an IPv4–translated IPv6 address, and the packets will travel through an SIIT that will translate the packet headers between IPv4 and IPv6 and translate the header addresses between IPv4 addresses on one side and IPv6–translated IPv6 addresses on the other side [11, 12, and 13].

## Network Address Translation – Protocol Translation (NAT–PT)

This mechanism allows IPv6–only hosts to interoperate with IPv4 hosts and vice versa. Its process is similar to the network address translator (NAT) in an IPv4 box. A pool of globally unique IPv4 addresses is assigned to IPv6 hosts dynamically in response to any request for packets leaving one of the boundaries. The DNS is used and mandates that that DNS queries should be translated from IPv4 to IPv6 and vice versa [11, 12, and 14].

## Bump in the Stack (BIS)

This mechanism allows the hosts to communicate with other IPv6 hosts using existing IPv4 applications. The technique inserts new modules to local IPv4 stacks and translates IPv4 into IPv6 and vice versa, and makes them self-translators. On the BIS host, the IPv6 address is mapped to the local internal IPv4 address and acts as a dual-stack host [15, 11, and 12].

## Dual Stack Transition Mechanism (DSTM)

In this mechanism, an IPv4 address is allocated to a dual-stack node if the connection cannot be established using IPv6. This allows either IPv6 nodes to communicate with IPv4–only nodes, or IPv4–only applications to run on an IPv6 node without modification. The DSTM architecture includes a DSTM address server and DSTM–capable nodes. Protocols such as the dynamic host configuration protocol version 6 (DHCPv6) server can be used as a DSTM server to assign a temporary IPv4 address to the node if connection fails to establish using an IPv6 address. Once IPv6 nodes have obtained IPv4 addresses, dynamic tunneling is used to encapsulate the IPv4 packet within IPv6 and then forward that packet to an IPv6 border router, where the packet will be decapsulated and forwarded using IPv4. This allocation mechanism is coupled with the ability to perform IPv4–over–IPv6 tunneling, hiding IPv4 packets inside the native IPv6 domain to simplify the network management [16].

## IPv6 Deployment

IPv4 is an extraordinary success and cannot be replaced overnight. Incremental deployment has started with additional care to avoid affecting IPv4 traffic. IPv6 currently has low value because it has a much smaller number of users than IPv4, which has a huge population of users. Network managers are reluctant to make the transition to a new protocol. The missing infrastructure required for realistic production-level deployment is another obstacle for IPv6 deployment. Hardware support, operating systems, middleware, applications, management tools, and trained technical staff are needed to complete the deployment. These will require a significant investment in time, resources, and money [18].

Many major vendors have announced or introduced IPv6–compatible products. Sun Microsystems shipped Solaris 8 with IPv6 support, and Microsoft released a "technology preview" that prefigures IPv6 protocol stack in the next major release of Windows. Cisco also published a three-phase road map for delivering IPv6 services. All Linux latest kernels come with the in-built USAGI IPv6 protocol stack support. Many organizations started releasing their host and router implementations for testing [19]. Some of

these implementations are in beta or developer releases, while others are fully supported, production-ready products. In some significant key industries like mobile communications such as cellular phone industry, strong undercurrent of deployment activity is already taking place. The Mobile Wireless Internet Forum has mandated IPv6 support in its architecture [17, 18].

The IETF ngtrans working group has designed a set of IPv4–to–IPv6 transition tools to address the various needs of different networks. The 6bone is an IPv6 test bed for IPv6 deployment and a worldwide informal collaborative project. The experimental activities carried out inside the 6bone are coordinated by the IETF to provide feedback to IETF IPv6–related activities [20].

## Related Work in the Literature on IPv6 Evaluations

As many computer organizations started releasing their IPv6 products for testing and IPv6 is maturing through its development process, various projects have started evaluating the functionality and its support features. In this paper, we present the evaluation of end-to-end performance as noticed by TCP and UDP applications on Linux and Windows environments. A survey of the literature has revealed no evaluations on end-user application performance on the Linux operating system for the IPv6 protocol stack and no performance comparisons between Linux and Windows environments for end-user applications. The following section briefly describes some of the related work in the literature on IPv6 evaluations and performance comparisons in router and end-user implementations.

P. Xie et al. in [22] have evaluated the Microsoft Research (MSR) IPv6 beta protocol stack for Windows NT 4.0. The performance of protocol stack was measured by analyzing the network latency, throughput, processing overheads, and CPU utilization at host for TCP and UDP applications. This work only evaluated IPv6 and did not compare it with IPv4.

S. Ariga et al. in [23] have evaluated the performance of data transmission for digital video application in TCP and UDP approaches over the IPv6 network using IP security (IPSec). The evaluation system consist of two end hosts with a FreeBSD 2.2.8 and KAME IPv6 protocol stack and a router implemented in a PC platform also running a FreeBSD 2.2.8 and KAME IPv6 protocol stack.

K. K. Ettikan et al. in two papers [24, 25] have presented an evaluation of IPv6 compared to IPv4 using the dual-stack implementation of KAME over FreeBSD OS using the ping utility and a FTP application. The performance metrics used for comparison were latency and file transfer throughput. FTP application has been used to capture the throughput value and ping utility to calculate the latency. In [25] two end hosts were connected via a hub. In [24] two software-based routers running FreeBSD were used to connect the two end hosts. In their experiments, they have shown more than the theoretically expected performance loss in IPv6 compared to IPv4 application.

Ioan Raicu et al. in [8] presented evaluation for IPv6 protocol stack performance with IPv4 stack in Solaris 8 and

Windows environments. The network throughput, latency and percentage of CPU utilization were used as metrics and study was conducted in both TCP and UDP applications. In [21], the same authors evaluate the IPv6 protocol stack performance through similar experiments. In this project, two workstations were connected through routers in the evaluation study. Two dual stack IBM and Ericsson routers were used in three test bed configurations to connect the workstations.

C. Bouras et al in [26] have evaluated the IPv4 and IPv6 protocol stacks' performance for IPv6 ported OpenH323 library. OpenH323 library is used in the rapid applications development that uses the H.323 protocol for multimedia communications over packet-based networks. The experiments were conducted on a real IPv6 test-bed network so that communication from one endpoint to the other had to pass through two hops connected with a 10 Mbps link. In this study, application evaluation was done through performance and bandwidth consumption characteristics to measure the quality and usefulness of the ported application.

Tim Chown et al in [27] have presented an overview to the different IPv6 projects being undertaken in the European Commission's Fifth Framework Information Society Technologies (IST) Program. The program's coverage includes a broad range of topics across the IPv6 networking spectrum. GÉANT is the name given to both the project and the network that emerged from it. The GÉANT project offers a network backbone that interconnects more than 25 national research and education networks. GÉANT is planning an introduction of IPv6 services to the existing IPv4 backbone, 6NET, which has deployed a pan-European IPv6–only academic network, and Euro6IX, which is building a network of IPv6 exchange points. This paper gives a clear idea for various collaborative works in IPv6 developments in European countries.

Yukiji Mikamo et al in [5] have described an evaluation of seven types of IPv6 multicast routers in Japan Gigabyte Network (JGN) IPv6 project. The JGN IPv6 project has performed various evaluations and verifications of IPv6

routers. The evaluations include source-specific multicast interoperability. All the tests were performed on ATM and Ethernet connections. In the study, no problem was revealed in the various evaluation tests for unicast addressed target routers. However, some routers were observed to have performance problems in IPv6 multicast transactions.

Mobile communication is increasingly oriented toward the use of all–IP networks as fixed-network components. Steffen Sroka et al in [29] and Marc Torrent-Moreno et al. in [7] have evaluated the quality of service features in mobile communications, specifically, IPv6 versions mobile IPv6 (MIPv6), hierarchical mobile IPv6 (HMIPv6), and fast handover mobile IPv6 (FMIPv6) through measurements such as hand-off latency, packet-loss rate and bandwidth per station. Hong-Sun Jun et al in [28] have evaluated multicast-based localized mobility support scheme is proposed in IPv6 networks. Through simulation, the performance of the proposed scheme is analyzed in terms of packet loss and throughput.

## Project Description

As IPv6 is becoming more widespread, given the wide popularity and deployment for Windows and Linux operating systems, impact of IPv6 on end-to-end applications on these platforms must be adequately assessed. Application programs use IP in combination with either TCP or UDP. This research presents results of an empirical evaluation of IPv6 stack implementations in Gentoo Linux and Windows 2000 environments through application programs developed using both TCP and UDP approaches.

### Test Bed

The test bed used is shown in *Figure 1*. Two identical workstations are directly connected using a point-to-point link. This enabled us to eliminate most variables from the experiments, including router processing. One workstation acts as the sender, which connects to the receiver for all experiments. The second workstation, acting as a receiver, responds to sender requests. Each workstation was configured with Intel X86 665 MHz processor, 256 megabytes of

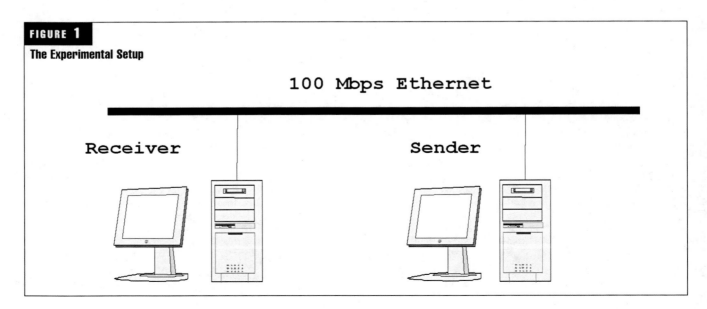

**FIGURE 1**

**The Experimental Setup**

100 Mbps Ethernet

Receiver

Sender

RAM, two identical 3.9-gigabyte hard drives and 100 Mbps Ethernet network adapters. The workstations were loaded with two different operating systems, Gentoo Linux with 2.4.26 – r9 kernel version and Windows2000 SP4 version in dual bootable mode. These workstations were also loaded with an add-on IPv6 protocol stack package available via Microsoft IPv6 Technology Preview release [30] for Windows 2000 operating system and the inbuilt Universal playground for IPv6 (USAGI) protocol stack [31] for the Linux operating system.

### Performance Metrics

The following metrics were used to compare the protocol stacks as they directly represent the performance perceived by the end user. Each of the following metrics is captured by sending different packet-size messages—ranging from 128 bytes, a very small size, to 64 kilobytes, the maximum possible—transmitted between host and target workstations.

- *Latency* (measured in microseconds): This metric represents the time it takes to send a packet from sender to receiver and back.

- *Throughput* (measured in Mbps): This metric represents the data transfer rate from the sender to the receiver.

- *CPU utilization* (measured as a percentage): This metric represents the amount of available CPU resources consumed at the sending workstation.

- *Socket creation time* (measured in microseconds): This metric represents the time taken by the API–provided call to create a socket in the application program.

- *TCP connection time* (measured in microseconds): This metric represents the time it takes to connect with the receiver workstation by the TCP application program running at the sender workstation.

- *Number of connections*: This metric represents the possible number of new connections between the sender and receiver workstations in a one-second time period.

### Measurement Methodology

To compare the protocol stack performance and calculate metric values, sample TCP and UDP application programs were developed in C++. The latest versions of network-socket API calls were used in the application programs as explained in Winsock2 reference programs available on the Microsoft Developer Network (MSDN) [32] and the latest Unix network programming samples for IPv6/IPv4 stacks [33]. Each metric was measured for various data lengths transmitted between the sender and receiver workstations by changing the packet sizes.

In the latency measurement test, round-trip time (RTT) was calculated by sending a packet from sender to receiver and back to the sender. The same process was repeated for 20,000 iterations. The processing time for 20,000 iterations was measured in the application program through timing functions, and the RTT for a single iteration was calculated. The same test was repeated twice and the average was used as the end result to avoid any inconsistencies.

In the throughput measurement test, the data transfer rate was calculated by sending a packet from sender to receiver. This process was repeated for 50,000 iterations. The processing time for 50,000 iterations was measured in the application program through timing functions and the number of bits transferred per second was calculated. The same test was repeated twice and the average used as end result to avoid any inconsistencies.

For the CPU utilization measurement test, the percentage of CPU utilization was measured by performance monitoring tools and commands provided in the Windows and Linux operating systems at the sender workstation during the throughput measurement experiments.

In the number of connections measurement test, the sender sent a 1-byte message to the target. After receiving the message, the receiver sent a confirmation message. Upon confirmation message receipt, the sender workstation took down the connection and terminated the socket in the TCP application. These measurements were repeated for 5,000 iterations. The processing time for all iterations was measured in the application program through timing functions and the number of possible new connections calculated.

For the TCP connection time measurement test, the processing time for socket API calls to connect with the receiver was measured in the sender TCP application program. In the socket creation time measurement test, the processing time for socket API calls to create a socket for network communication was measured in the host TCP application program. Since UDP does not use a connection mechanism, the connection time was not measured.

## Experimental Results

The packet header format in IPv6 is 20 bytes larger than IPv6. IPv6 header length increases to 40 bytes from 20 bytes in IPv4. Since Ethernet's maximum transmission unit (MTU) size is 1,514 bytes, theoretically, increased header overhead in IPv6 should be minimal and limited to 1.3 percent. However, experimental results from this study reveal a larger-than-expected difference in performance results and larger impact in small packet-size experiments.

### Latency

Latency, also known as round-trip delay, is very important for network applications sensitive to delay. Transactional applications using request-reply operations and media applications such as real time audio are very sensitive to latency value. The lower value represents the better performance delivered to end users.

From TCP and UDP, application latency results shown in *Figures 2, 3, 4,* and *5,* round-trip delay in Windows 2000 was higher than in Linux. The percent increase in IPv6 latency decreased with larger packet-size messages in both the Windows 2000 and Linux environments. UDP results illustrated in *Figures 4* and *5* for IPv6 compared to IPv4 indicate that Windows 2000 showed more than a 10 percent increase for messages with packet sizes of up to 384 bytes. The 9 percent increase in IPv6 latency for packet sizes of 512 bytes gradually dropped to 2 percent with packet sizes of 64K bytes in the Windows 2000 environment. With Linux, more than a 15 percent increase in latency for messages up to 512 bytes was noticed. The IPv6 latency increase of 9 percent for packet sizes of 512 bytes gradually dropped to 3 percent with packet sizes of 64K bytes in the Linux environment.

TCP application results shown in *Figures 2* and *3* for IPv6 compared to IPv4. Windows 2000 showed more than a 10 percent increase for messages with packet sizes of up to 512 bytes. The 8 percent increase in IPv6 latency for packet sizes at 640 bytes gradually dropped to 3 percent with packet sizes of 64K bytes in the Windows 2000 environment. With Linux, latency values also increased in IPv6 compared to IPv4. In TCP and UDP comparisons of both environments, TCP application shows a higher delay, which can be attributed to the additional processing overhead of the TCP protocol. However, with increasing message sizes, the latency difference between IPv4 and IPv6 packets decreases on both operating systems for both TCP and UDP applications.

*Throughput*
Throughput is a direct measurement to indicate the end-to-end performance delivered to the end users. Higher throughput means better performance of an application. TCP and UDP application throughput results are shown in *Figures 6, 7, 8,* and *9.* Throughput in Windows 2000 was lower than in Linux for both IPv4 and IPv6. The differences decrease as message sizes get larger. In Linux, throughput values remained very close with IPv6 and IPv4 for message sizes greater than 1,024 bytes. UDP application results are illustrated in *Figures 8* and *9* for IPv6 compared to IPv4. These figures show that Windows 2000 exhibited more than a 10 percent decrease for IPv6 messages with packet sizes up to 640 bytes. The 8 percent decrease in IPv6 throughput for packet sizes of 728 bytes gradually dropped to 3 percent with packet sizes of 16kilobytes and remained constant for packet sizes larger than 16 kilobytes in the Windows 2000 environment. With Linux, more than a 10 percent decrease in throughput was

noticed for IPv6 messages up to 384 bytes. The 6 percent decrease in IPv6 throughput at 512 bytes gradually dropped and a 2 percent drop was noticed with packet sizes larger than 32 kilobytes.

TCP application results are shown in *Figures 6* and *7* for IPv6 compared to IPv4 values. Windows 2000 showed more than a 10 percent decrease for IPv6 messages with packet sizes up to 1,408 bytes. The 6 percent decrease in IPv6 throughput at 2-kilobyte packet sizes gradually dropped and a 3 percent decrease was noticed for packet sizes larger than 16 kilobytes. With Linux, more than a 10 percent decrease in throughput for IPv6 messages up to 512 bytes was noticed. The 6 percent decrease in IPv6 throughput at packet sizes of 640 bytes gradually dropped to 3 percent with packet sizes of 32 kilobytes and remained constant for packet sizes larger than 32 kilobytes in the Linux environment. In both environments, the TCP application showed lower throughput values than UDP. This is due to the higher TCP overheads.

*CPU Utilization*
CPU utilization was calculated at the sender workstation in throughput experiments through the performance monitor tool in Windows and performance commands available in Linux. CPU utilization represents the amount of consumed processing resourced by the application programs and also availability for other processors. From multiple experiments using varying packet sizes, the most frequent difference in utilization was chosen for comparisons.

In TCP and UDP application comparisons of both environments, TCP application was observed to consume more CPU resources for similar throughput results. UDP IPv6 application used 10 percent more resources in Windows and 8 percent in Linux than IPv4. TCP IPv6 application used 12 percent more resources in Windows and 10 percent more in Linux than IPv4.

---

**TABLE 1**

**Socket Creation Time and TCP Connection Time for IPv6 and IPv4**

| Operating system | Protocol | IP version | Socket creation time (microseconds) | Connection time (microseconds) |
|---|---|---|---|---|
| Linux | TCP | IPv4 | 18 | 329 |
| | TCP | IPv6 | 33 | 390 |
| | UDP | IPv4 | 16 | N/A |
| | UDP | IPv6 | 32 | N/A |
| W2K | TCP | IPv4 | 1,946 | 753 |
| | TCP | IPv6 | 2,102 | 871 |
| | UDP | IPv4 | 1,944 | N/A |
| | UDP | IPv6 | 2,097 | N/A |

**FIGURE 2**

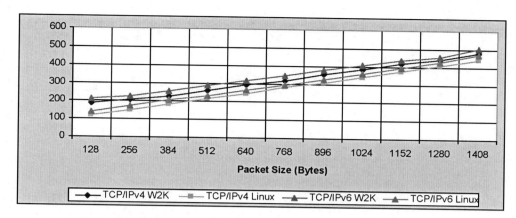

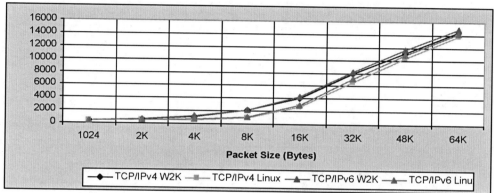

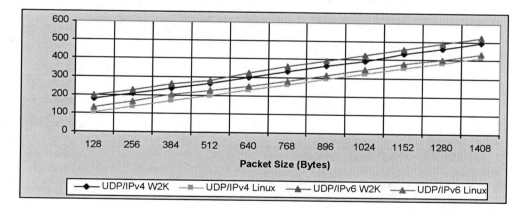

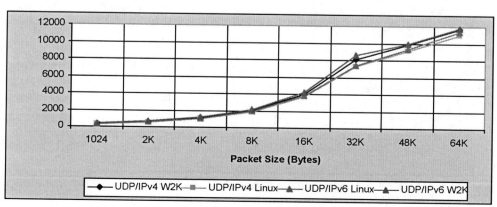

**FIGURE 2**
**(continued)**

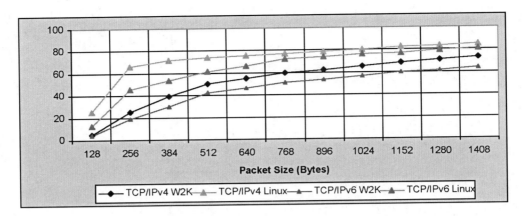

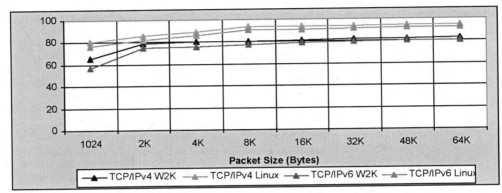

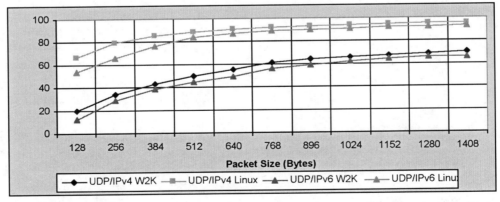

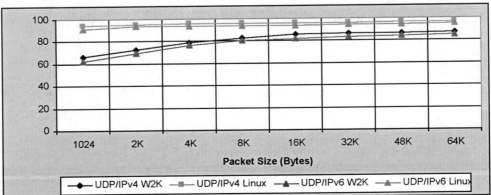

*Socket Creation Time and TCP Connection Time*
Sockets are used to exchange information between two workstations by network applications. Socket creation time and TCP connection time are partially responsible for performance obtained in end-user applications.

Socket creation time and TCP connection time are significantly lower in Linux compared to Windows. Since the IPv6 socket address size is constant, we attribute the difference in Windows 2000 and Linux to a more efficient socket layer in Linux than Windows 2000 and improved protocol stack implementations. As shown in *Table 1*, Linux socket creation time increased approximately 100 percent and TCP connection time increased 18 percent from IPv4 to IPv6. Windows socket creation time increased approximately 8 percent and TCP connection time increased 16 percent from IPv4 to IPv6. The increased connection times with IPv6 is likely due to the increased overhead resulting from the larger header and address sizes. As UDP is not a connection-oriented protocol, connection time is not applicable.

*Number of Connections*
Transaction-oriented network applications involve request-reply operations. In client-server architectures, the server that accepts more number of new connections per second will process more requests and generally offer better end performance obtained at the client application.

In our TCP experiments, Linux environment has shown higher possible new connections count compared to Windows 2000 for both IPv6 and IPv4. For TCP, new possible connections count decreased to 823 from 978 in Linux IPv6 from IPv4. On Windows 2000, count decreased to 690 from 850, a 19 percent drop compared to 16 percent in Linux. The higher socket creation and TCP connection times have an impact on the possible new connections count in both the Linux and Windows environments, but more so in the Windows environment. This data illustrates the performance penalty that IPv6 clients would incur compared to IPv4 clients in a situation where the server supports dual stacks.

## Conclusions and Future Work

In this paper, we have presented an empirical evaluation for IPv6 protocol stack implementations in Linux and Windows environments through data transfer applications. In similar conditions, Linux has shown better performance than Windows environment. Higher UDP application performance was obtained compared to TCP applications. The IPv4 version yielded better performance than the respective IPv6 protocol stack. This variation in performance has decreased as the transmitted message size increased. We hope that our study on end-to-end performance will offer insights to protocol and application designers as they seek to deploy IPv6–based applications.

All the experiments in this project were performed on two directly connected workstations. As the hardware cost of routers with IPv6/IPv4 dual-stack support becomes affordable, we would like to continue these experiments using routers. The primary focus of this project has been protocol stack end performance as seen in data transfer applications. In the future, we would like to evaluate the other IPv6 features such as quality of support in multimedia applications and security-support features provided in the IPv6 protocol.

## References

1. Steve King, et al., "The Case for IPv6," Internet Draft draft-ietf-iab-case-for-ipv6-06.txt, Internet Architecture Board of Internet Engineering Task Force, www.6bone.net/misc/case-for-ipv6.html, December 1999.
2. Robert M. Hinden "IP Next Generation Overview," Communications of the ACM archive, June 1996.
3. Steve Deering et al., "Internet Protocol, Version 6 (IPv6) Specification," Request for Comments 2460, Internet Engineering Task Force, December 1998.
4. Information Sciences Institute, University of Southern California, "Internet Protocol," Request for Comments 791, Internet Engineering Task Force, September 1981.
5. Yukiji Mikamo et al., "An Evaluation of IPv6 Multicast Router in JGN IPv6 Networks," 2004 Symposium on Applications and the Internet Workshops, January 2004.
6. T. Yazaki et al., "High-Speed IPv6 Router/Switch Architecture," 2004 Symposium on Applications and the Internet Workshops, January 2004.
7. Marc Torrent-Moreno et al., "A Performance Study of Fast Handovers for Mobile IPv6," 28th Annual IEEE International Conference on Local Computer Networks, October 2003.
8. Sherali Zeadally, Ioan Raicu, "Evaluating IPv6 on Windows and Solaris," IEEE Internet Computing , May 2003.
9. Robert M. Hinden "IP Next Generation Overview," Communications of the ACM archive, June 1996.
10. R. Gilligan, E. Nordmark, "Transition Mechanisms for IPv6 Hosts and Routers," Request for Comments 2893, Internet Engineering Task Force, August 2000.
11. Hossam Afifi, Laurent Toutain, "Methods for IPv4-IPv6 Transition," The Fourth IEEE Symposium on Computers and Communications , July 1999.
12. Viagénie IT consultant firm, IPv6 presentations, www.viagenie.qc.ca/en/ipv6/presentations/IPv6-transition-mechanisms_v1.pdf.
13. E. Nordmark, "Stateless IP/ICMP Translation Algorithm (SIIT)" Request for Comments 2765, Internet Engineering Task Force, February 2000.
14. G. Tsarists, P. Srisuresh, "Network Address Translation – Protocol Translation (NAT–PT)," Request for Comments 2766, Internet Engineering Task Force, February 2000.
15. Tsuchiya, et al., "Dual Stack Hosts using BIS" Request for Comments 2767, Internet Engineering Task Force, February 2000.
16. Jim Bound, "Dual Stack Transition Mechanism," Internet Draft draft-bound-dstm-exp-01.txt, www3.ietf.org/proceedings/02mar/I-D/draft-ietf-ngtrans-dstm-07.txt, April 2004.
17. Jun-ichiro itojun Hagino, "Implementing IPv6: experiences at KAME project," 2003 Symposium on Applications and the Internet Workshops, January 2003.
18. Alain Durand, "Deploying IPv6," IEEE Internet Computing, January 2001.
19. IPv6 implementations, playground.sun.com/pub/ipng/html/ipng-implementations.html.
20. IPv6 testbed for deployment, www.6bone.net.
21. Ioan Raicu, Sherali Zeadally, "Impact of IPv6 on end-user applications," 10th International Conference On Telecommunications, March 2003.
22. P. Xie et al., "Profiling the Performance of TCP/IP on Windows NT," 4th International Computer Performance and Dependability Symposium, March 2000.
23. S. Ariga et al., "Performance Evaluation of Data Transmission Using IPSec over IPv6 Networks," The 10th Annual Internet Society Conference, July 2000.
24. K. K. Ettikan, "IPv6 Dual Stack Transition Technique Performance Analysis: KAME on FreeBSD as the Case," Faculty of Information Technology, Multimedia University, Jalan Multimedia, October 2000.
25. K. K. Ettikan et al., "Application Performance Analysis in Transition Mechanism from IPv4 to IPv6," Faculty of Information Technology, Multimedia University, Jalan Multimedia, June 2001.
26. C. Bouras et al., "Performance Evaluation of an IPv6–capable H323 Application," 18th International Conference on Advanced Information Networking and Applications, March 2004.

27. Tim Chown, Jordi Palet, "Results and plans of the IPv6 test-bed initiatives within the European Commission IST Programme," 2003 Symposium on Applications and the Internet Workshops, January 2003.

28. Hong-Sun Jun, Miae Woo, "Performance Analysis of Multicast-Based Localized Mobility Support Scheme in IPv6 Networks," Second Annual Conference on Communication Networks and Services Research, May 2004.

29. Steffen Sroka, Holger Karl, "Performance Evaluation of a QoS-Aware Handover Mechanism," Eighth IEEE International Symposium on Computers and Communications, July 2003.

30. Microsoft IPv6 Technology Preview for Windows 2000, msdn.microsoft.com/downloads/sdks/platform/tpipv6.asp.

31. Linux IPv6 Development Project, Universal playground for Ipv6 (USAGI) downloads, www.linux-ipv6.org.

32. Windows socket programming reference, Windows Socket 2, msdn.microsoft.com/library/default.asp?url=/library/en-us/winsock/winsock/winsock_reference.asp.

33. Richard Stevens et al., "Unix Network Programming, Vol. 1: The Sockets Networking API," Third Edition , Addison-Wesley Pearson Education , October 2003.

# An XML–Based Approach for Network Protocol Description and Implementation

## Fabio Baroncelli

*Researcher, National Laboratory of Photonic Networks*
National Inter-University Consortium for Telecommunications (CNIT), Pisa, Italy

## Barbara Martini

*Researcher, National Laboratory of Photonic Networks*
National Inter-University Consortium for Telecommunications (CNIT), Pisa, Italy

## Piero Castoldi

*Associate Professor, Center of Excellence for Communication Networks Engineering (CEIRC)*
Scuola Superiore Sant'Anna, Pisa, Italy

## Abstract

The specifications of most transport and signaling protocols used in the Internet—for example, those reported in the requests for comment (RFCs) of the Internet Engineering Task Force (IETF)—are expressed using informal and textual-based languages. This format is far from the rigorous format needed for the practical implementation of a protocol in a network element. Indeed, formal specifications for network protocols are needed during the implementation process to guarantee full interworking among protocol software from different vendors. Extensible markup language (XML) is emerging as a good supporting technology for the definition of a formal language to document protocol specifications thanks to its hierarchical structure, its inherent support for validation data, and to the large software library availability over the principal software platforms.

This work proposes the XML–based multiprotocol language (XMPL), a new language suitable for a protocol description in terms of protocol logic, protocol message set, and protocol data structures. XMPL can be used not only for documenting a protocol, but also as a configuration language for protocol software implementations. Accordingly, we also propose the XML–based multiprotocol framework (XMPF), a new software protocol engine that we have designed and developed to be configured by protocol specifications con-tained in native XMPL files. As a result, XMPF is a software tool that can concurrently support different protocols without the need to write a specific code for each of them.

A test bed implementing the generalized multiprotocol label switching (GMPLS) protocol suite is used to validate the adopted XML approach and to evaluate the relevant performance. At the current stage of development, XMPF enjoys several benefits when used to support network protocols as a competitor of coded software modules, and it appears suitable to efficiently support signaling protocols with a low-medium messaging rate.

## Introduction

Documents containing current network protocol specifications (e.g., RFCs developed by the IETF) are mainly written using natural languages for the description of the protocol behavior. They also exploit graphical representation to describe message fields and data structures. Thanks to this pragmatic approach, RFCs have become the de facto standard for protocol specifications being widely taken as a reference by the world of research and manufacturers. Nevertheless, RFCs sometimes fail in thoroughly documenting all possible aspects and details relevant to a protocol operation, as it may happen using an informal (e.g., wording-based) representation of state transitions and conditions under which they occur. Indeed, the process of translating RFCs specifications into a working protocol software often ends up with protocol implementations that are unintentionally (or, sometimes, deliberately) different from vendor to vendor, eventually preventing complete interworking among devices of different vendors.

Two main motivations have driven the work reported in this paper and correspond to the two non-independent goals outlined below. The first motivation is suggested by software developers and also by a part of the research world: Protocol specifications should be expressed in a formal language, in a way to have a rigorous protocol description that is directly usable for software development. The second motivation reflects the need to speed up protocol deployment in network nodes: If a network node could have the possibility to process via software a file containing the protocol specifications, this facility would allow to quickly provide for upgrade to any new protocol version or for the support of a new one.

The first goal of this work deals with the representation of protocol specifications and aims at bridging the gap between the formats required at the dissemination and at implementation level. Several protocol languages have been developed to facilitate the protocol software development, including the following:

- Languages specific for a determined class of protocols (e.g., Web services description language [WSDL]) [1]
- Languages intended for the specification of a generic protocol behavior (e.g., the SDL) [2]
- Languages specific for describing a data-structure syntax of a protocol (e.g., the abstract syntax notation 1 [ASN.1]) [3]

Nevertheless, to best of our knowledge, a single implementation-oriented language able to comprehensively represent the specifications of a protocol in terms of protocol behavior, protocol message set, and data structures is missing. To this purpose, we designed the XMPL, a formal language based on XML [4]. On the one side, XML allows an efficient description of protocol specification thanks to the XML Schema technology and it permits an efficient processing of software thanks to the large XML software library availability, (i.e., it is implementation-oriented). On the other side, XML, in spite of its rigorous statement, is simple to write and easy to understand because it is based on textual tags [5].

The second goal of this paper is to show that XMPL has been designed in such a way that files written in this language can be directly used by software tools realizing an automatic processing of the protocol specifications. To this purpose, we designed and implemented the XML–based multiprotocol framework (XMPF), a protocol "engine" that, once configured with XMPL–compliant files, is able to run as a specific protocol application. Therefore, no specific code is needed for the protocol implementation. This reduces the deployment time and likewise improves the software stability. In addition, no extra code is needed for data validation because of the intrinsic checks offered by the XML library.

A complete demonstration of the solution is provided by a simple test bed consisting of an IP network representing the control plane of a GMPLS–based transport network. XMPF is used in the test bed to support the three protocols of the GMPLS protocol suite: OSPF, RSVP, and LMP whose specifications are contained in XMPL files.

The rest of the paper is organized as follows: Section 2 presents the XMPL formalism, Section 3 describes a use case showing XMPL utilization, Section 4 introduces the XMPF and the relevant software architecture, Section 5 illustrates the test bed used for the XMPF validation, and Section 6 draws some conclusions.

## The XML–Based Multiprotocol Language

We first introduce the concept of protocol language, the requirements we have within the scope of this paper, and the advantages of choosing XML as supporting technology.

By protocol language, we mean any specification language that is used to document the protocol operation. A protocol language should describe the following:

- The format and the syntax of the set of messages (message set) handled by a protocol
- The finite state machine (FSM) related to the protocol behavior (logic)
- The organization of the information supporting the operation of the protocol (data structure)

In addition, to be implementation-oriented, a protocol language should be suitable to be interpreted at run time by a dedicated protocol software. The selected technology able to satisfy the above requirements is the XML [4]. Originally designed to meet the challenges of large-scale electronic publishing, XML is playing an increasingly important role in the exchange of a wide variety of data on the Web and elsewhere. According to The World Wide Web Consortium (W3C), "The extensible markup language (XML) is the universal format for structured documents and data on the Web." The use of XML technology in protocol specification is already suggested in IETF RFC 3470 *Guidelines for the Use of Extensible Markup Language (XML) within IETF Protocols* [5]. The strength of XML consists in a large library available over all operating systems and programming languages that permits to easily parse, format, navigate, and validate the XML–based information.

The language that we created to meet the above requirements is called XMPL; this allows the generic description of any type length value (TLV) network protocol, after its specifications have been organized in terms of message set, behavior, and data structure. XMPL consists of the following parts:

- XMPL message-set specification (XMPL–MS)
- XMPL logic specification (XMPL–LS)
- XMPL data-structure specification (XMPL–DS)

Each part consists of a specific XML Schema file. An XML Schema [6] defines the structure, content, and semantics of a set of XML documents. It can be used with validation software to ascertain whether a particular XML document is conformed to the set of defined rules.

### The XMPL Message-Set Specification (XMPL–MS)

XMPL–MS is an XML Schema file that represents the basis for defining the protocol-specific message set. In particular, the XMPL–MS allows the following:

- Expression of the structure of the protocol message that satisfies the TLV paradigm
- Specification of the mapping between the XML–based format (where the message is encoded as an XML document) and the binary format (where the message is encoded as a string of bytes) of the protocol messages and vice versa
- Validation of the syntax of a protocol message

In the XMPL–MS, each field of a protocol message is characterized by the XML tag's name and type. The name identifies a specific field of the message. The type specifies the number of bits that should be used to represent a specific field value in the binary-format and it is useful for executing the binary/XML–format conversions. The types that we have defined are reported in *Table 1*.

### The XMPL Logic Specification (XMPL–LS)

The XMPL logic specification (XMPL–LS) is the XML Schema defined to express the logic of a generic network protocol. It is based on the virtual finite state machine (VFSM) theory, used to describe the FSM peculiar to a certain protocol. We briefly review both concepts in this subsection.

An FSM is a model of computation consisting of a set of states, inputs, and actions and a transition function that maps inputs and current state and indicates the set of actions that should be executed. The VFSM [7] is a specification method derived from the FSM theory that separates the definition of the inputs and actions from the definition of the states and the transition function. In fact, all the inputs that affect a determined state are represented by an Input-Name and the actions, consequence of specific events, are represented by an Action-Name. The mapping function between the Input and Action-Names is separately defined from the specification of the states and relevant conditions and transitions. This separation has the following advantages:

- It is possible to create a library of inputs and actions definitions that can be reused by different FSMs.
- It is possible to change the definition of an input or an action without changing the FSM work flow.

### TABLE 1
**XMPL–MS Types**

| Type | Number of bits |
|------|---------------|
| byte | 8 |
| short | 16 |
| int | 32 |
| longInt | 64 |
| IPv4 | 32 |

In a VFSM, the conditions that control the changes of state or the execution of a set of action are strictly expressed in terms of boolean AND and OR operators only applied to Input-Names. The boolean NOT operator is not allowed, because the result of a negation of an input is usually not defined. The VFSM workflow is depicted in *Figure 1*. When a new state is entered, a set of actions (entry actions) is executed, then the system that implements the VFSM checks for new Inputs. If a set of conditions (input conditions) is satisfied over the inputs, a set of actions (input actions) is executed. Finally, if another set of conditions (transition conditions) is satisfied over the same Inputs, a set of action (exit actions) is executed and a change of state occurs. Otherwise, the system checks for new Inputs.

Accordingly, XMPL–LS has three parts: the input section, the action section, and the state section, designed as follows:

- *Input Section*: It defines the Input-Names and describes the mapping function between them and the value of the data stored in the protocol data structures and/or in the messages received. Each Input-Name is characterized by a name and a set of conditions. The conditions are expressed using the XPath grammar and built using the positive logic algebra. The XPath [8] is a language for addressing parts of an XML document and for the manipulation of strings, numbers, and Boolean. XPath operates on the abstract, logical structure of an XML document, rather than the relevant syntax. The XPath is used in XMPL–LS for retrieving values from two types of XML objects: the received messages and the data structure supporting the protocol logics.

### FIGURE 1
**VFSM Flow Chart**

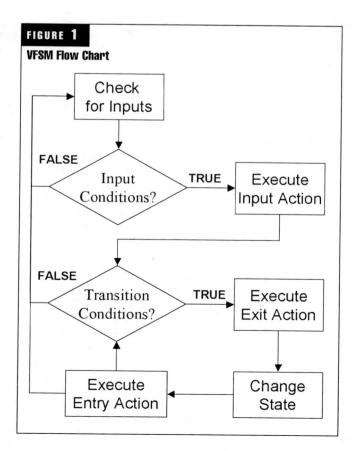

- *Action Section*: It defines the Action-Names and describes the set of actions that is executed for each Action-Name. Each Action-Name is characterized by a name, an identifier of the target of the action, and a string that expresses directives over the protocol data structure (e.g., the update of a specific field) or the creation of specific messages (e.g., the creation of a Hello Message with a given expiration time).

- *State Section*: It defines the State and describes for each State the conditions over the Input-Names, the transitions among States, and the production of the Action Names. For the State Section, the VFSM XML Schema defined in [7] is adopted because it is the de facto standard for expressing a generic VFSM State name-space using the XML syntax.

### The XMPL Data-Structure Specification (XMPL–DS)

The XMPL data specification (XMPL–DS) is the XML Schema that expresses the data structures of a specific protocol. Data structures are used by network protocols for storing the permanent or volatile information supporting the protocol operation. The XMPL–DS specifies the hierarchical structure and the relevant fields of a protocol data structure. In particular, the name, the type defined in Section 2.A, and the value range is specified for each field.

XML, thanks to its hierarchical structure, can express protocol data structures in a simple and powerful manner. Firstly, the use of XPath technology for storing, retrieving, updating, and deleting attributes allows to avoid proprietary solutions with significant code saving. Secondly, the ability to convert XML to any other arbitrary format (including code generation) via extensible style-sheet language transformation (XSLT) [9] permits to express the data structures using different syntax, thus improving the interworking with different protocol applications. Thirdly, an XML–based data structure can be easily validated using the XML Schema technology, overcoming the well-known problem of invalid data structures often present in commercial devices.

### XMPL Advantages

The adoption of XMPL for network protocol description permits to achieve several advantages:

- XMPL is portable, because XML technologies are widely supported over the principal platforms.
- XMPL is easily understandable by both human and machine.
- XMPL permits to easily validate protocol descriptions thanks to the XML Schema technology.
- XMPL does not require the use of intermediate code, unlike other protocol specification languages in the technical literature (e.g., SDL [2] and LOTOS [10]).
- Tools able to interpret XMPL can be easily written thanks to the large XML software libraries available nowadays.

### XMPL Drawbacks

The principal drawbacks of the XMPL can be summarized as follows:

- The number of bytes needed for representing a protocol message is considerably larger using the XML format with respect to the binary format.

- The time needed for describing a protocol provided with complex behavior is longer using the XMPL respect using a natural and informal language.
- The XML technology is evolving and, consequently, the XMPL syntax could change in the future.

### The OSPF Hello Use Case

The RFC 2428 regarding the open shortest path first (OSPF) protocol specification [11] is taken as use-case to show the shortcoming of a textual and informal protocol behavior specification and to explain how the XMPL can be applied to the description of real protocol. In particular, we focus on the specifications of the OSPF Hello Message that is responsible for the establishment and the maintenance of the adjacency relationships among neighboring network nodes.

The FSM of the OSPF Hello Message, named "neighbor state changes" in the RFC, is seen in *Figure* 2.

This protocol specification fragment is an example of non-exhaustive description that often afflicts the current protocol recommendations (in particular, the IETF specifications). In fact, the considered FSM description reports only the states and the events that trigger the state transitions. A comprehensive description of the relative FSM is missing, because the associated actions carried out upon each state transition are not indicated. For example, it the arrival of a Timer-Expired event within the Attempt state could cause the dispatch of a Hello Message, but it is not explicitly reported in the RFC.

The OSPF fragment related to the Hello Message has been represented in XML syntax, using the XMPL paradigm, as detailed in the following sections. Consequently, three XML Schema documents were defined: the XMPL–MS, representing the message set; XMPL–LS, representing the behavior; and XMPL–DS, representing the data structure.

### OSPF Hello Message

The structure of an OSPF Hello Message as defined in [11] is reported in *Figure* 3.

According to the XMPL–MS definition, the XML Schema file with constraints over the type and the value range allowed for each field is derived. The structure of an OSPF Hello Message expressed in XML format and compliant with the XMPL–MS specifications is explained as follows:

```
<?xml version="1.0" encoding="UTF-8"?> <protocol
xmlns:xsi="http://www.w3.org/2001/XMLSchema-
instance"
xsi:noNamespaceSchemaLocation="XMPL-MS.xsd"
name="OSPF" ID="204" IPdest="217.9.70.87">
<object name="header" ID="0" length="48">
<field name="Version" type="byte">2</field>
<field name="Type" type="byte">1</field>
<field name="Length" type="short">48</field>
<field name="RouterID" type="int">23</field>
<field name="AreaID" type="int">67</field>
<field name="Checksum" type="short">873</field>
<field name="AuType" type="short">0</field>
<field name="Authentication" type="longInt">0</field>
</object>
<object name="Hello" ID="3" length="20">
```

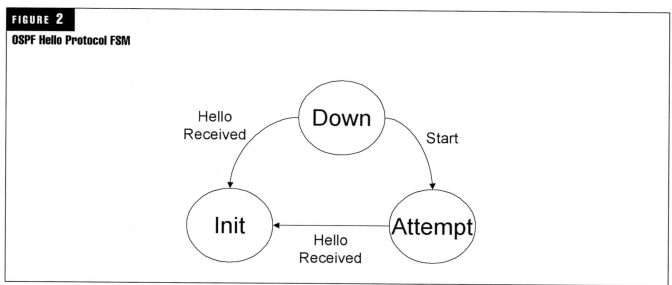

**FIGURE 2**

**OSPF Hello Protocol FSM**

**FIGURE 3**

**OSPF Hello Message Structure**

| 0        7 | 8        15 | 16        23 24        31 |
|---|---|---|
| Version | Type | Packet Length |
| Router ID | | |
| Area ID | | |
| Checksum | | AuType |
| Authentication | | |
| Authentication | | |
| Network Mask | | |
| HelloInterval | Options | Rtr Pri |
| RouterDeadInterval | | |
| Designated Router | | |
| Backup Designated Router | | |
| Neighbor | | |

<field name="NetworkMask" type="IPv4">14</field>
<field name="HelloInterval" type="byte">10</field>
<field name="Options" type="byte">0</field>
<field name="RtrPri" type="byte">0</field>
<field name="RouterDeadInterval" type="int">40</field>
<field name="DesignatedRouter" type="IPv4">0</field>
<field name="BackupDesignatedRouter" type="IPv4">0</field>
<field name="Neighbor" type="int">0</field>

</object>
</protocol>

*OSPF Hello Behavior*
To define the OSPF Hello behavior in terms of XMPLS–LS Schema, the relevant FSM must be first described. *Table 2* represents the States, Input-Names and Action-Names entities of the FSM Hello Protocol obtained from the analysis of the OSPF RFC.

| TABLE 2 |
| --- |

**OSPF Hello Protocol VFS Entities**

| States | Down, Attempt, Init |
| --- | --- |
| Inputs | start, helloTimerExpired, helloReceived |
| Actions | sendHelloMessage, startHelloTimer, updateAdjacency |

From the entities defined in *Table 2*, it is possible to derive the relevant XMPL–LS Schema fragment as follows:

```
<vfsmml projec="true">
<Name>Hello FSM Example</Name>
<VFSM>
<State>
<Name>Down</Name>
<Transition>
<Condition> <ci>start</ci></Condition>
<StateName>Attempt</StateName>
<Action>sendHelloMessage</Action>
</Transition>
</State>
<State>
<Name>Attempt</Name>
<EntryAction>startHelloTimer</EntryAction>
<InputAction>
<Condition><ci>helloTimerExpired</ci></Condition>
<Action>sendHelloMessage</Action>
<Action>startHelloTimer</Action>
</InputAction>
<Transition>
<Condition><ci>helloReceived</ci></Condition>
<StateName>Init</StateName>
</Transition>
</State>
<State>
<Name>Init</Name>
<EntryAction>updateAdjacency</EntryAction>
</State>
</VFSM>
</vfsmml>
```

### OSPF Hello Data Structure

To define the XMPL–DS fragment, we take into consideration the control structures involved in the OSPF neighbor discovery process. We identified the following two control data structures supporting the OSPF Hello protocol operations:

- *Hello data*: It is the structure containing the noteworthy fields related to the Hello Messages that the network nodes periodically send on all interfaces to establish and maintain neighbor relationships.

- *Neighbor Conversation*: It is the structure containing all the information pertinent to the forming or formed adjacency between the two neighbors. It reports data related to preliminary contacts between two routers that, after a negotiation phase, become officially adjacent. An adjacency can be viewed as a highly developed conversation between two routers.

The specific XMPL–DS fragment for the data structure of the OSPF Hello protocol just described is as follows:

```
<?xml version="1.0" encoding="UTF-8"?> <protocol
xmlns:xsi="http://www.w3.org/2001/XMLSchema-
instance"
xsi:noNamespaceSchemaLocation="XMPL-DS.xsd"
name="OSPF Hello Control Data Structure">
<HelloData>
<AttachedNetworkConfiguration>
<NetworkMask type="int">14</field>
<HelloInterval type="byte">10</field>
<RouterDeadInterval type="int">40</field>
</AttachedNetworkConfiguration>
<NeighborID type="ipv4">217.9.70.87<\record>
</HelloData>
<NeighborConversation>
<InactiveTime type="int">10</field>
<NeighborData>
<ID type="int"></field>
<IPaddress type="ipv4">217.9.70.67</field>
<\NeighborData>
</NeighborConversation>
</protocol>
```

In both Hello Data and Neighbor Conversation structures, information about the neighbors is contained. The NeighborID of the HelloData structure is an identifier that is learned when Hello messages are received from the neighbor. The IPaddress of the NeighborConversation structure is the IP address of the neighboring router's interface that is learned during the routing table building process.

### The XML–based Multiprotocol Framework

#### Overview

The XMPF is the software "engine" that was developed for implementing network protocols. It has been designed so that it can be totally configured "at runtime" using XMPL files. XMPF is able to manage several protocols simultaneously sharing the same network interfaces. Each protocol runs in a dedicated process. The communication between protocols is achieved thanks to the shared memory spaces. This approach guarantees better performance instead of other inter-process communication techniques (e.g., semaphores, queues). Several XMPF instances can coexist over the same machine.

XMPF is composed by a set of software modules. XMPF modularity is achieved by means of an object-oriented approach that uses the Design Pattern [12] principles and "Conduits+" [13] as a starting model. XMPF is written in

ANSI C++ and it makes use of open-source libraries like the standard template library (STL), the LibPCAP and the LibNet for massage handling, the Xerces-C for XML files handling. XMPF architecture, depicted in *Figure 4*, consists of the following building blocks:

- *Network Controller*: It collects incoming messages from a set of network interfaces and it passes them to the Message Parser. It receives outgoing messages from the Message Formatter and it delivers them to a specific network interface based on the information stored in the messages (e.g., IP address, IP port, Protocol ID).

- *Message Parser*: It handles messages received from the Network Controller and converts them from the binary format to the XML format. The Message Parser permits validation of the transformed XML format messages using the XML Schema of the same XMPL–MS file used for the conversion. This allows the discarding of malformed messages (i.e., messages with errors or that are not compliant with the protocol specification) without writing specific code.

- *Message Formatter*: It handles XML–format messages received from the Message Creator and, thanks to the information stored in the XMPL–MS file, converts them in a binary format before passing them to the Network Controller.

- *Message Queue*: It acts as a message buffer to ensure that no message will be lost. In particular, it waits for messages arriving from the Message Parser and stores them

in a Queue, giving one message at a time to the Input Processor that is unable to handle multiple messages.

- *Input Processor*: According to the XMPL–LS file, it creates the set of Input-Names from the information stored in the Data Controller and from the fields of the message obtained from the Message Queue. Then, the Input-Names are passed to the VFSM Executor.

- *VFSM Executor*: It is the core block of the framework. It reads the set of Input-Names managed by the Input Processor and it executes the VFSM logic according to the XMPL–LS file. Finally, it produces a list of Action-Names that is interpreted by the Action Executor. Because of this mechanism, the set of actions to be performed depends on the current protocol state, the content of the received message, and the occurrence of some conditions determined by the value of the information stored in the Data Controller.

- *Action Executor*: It analyzes the set of Action-Names given by the VFSM Executor and performs the actions on the Data Controller and the Message Creator according to the directives declared in the XMPL–CS file. Examples of actions are retrieving and/or setting the protocol data on the Data Controller and creating and/or sending a message to another network node via the Message Creator.

- *Message Creator*: It creates XML–format messages that adhere to the directives received from the Action Executor and comply with the XMPL–MS file.

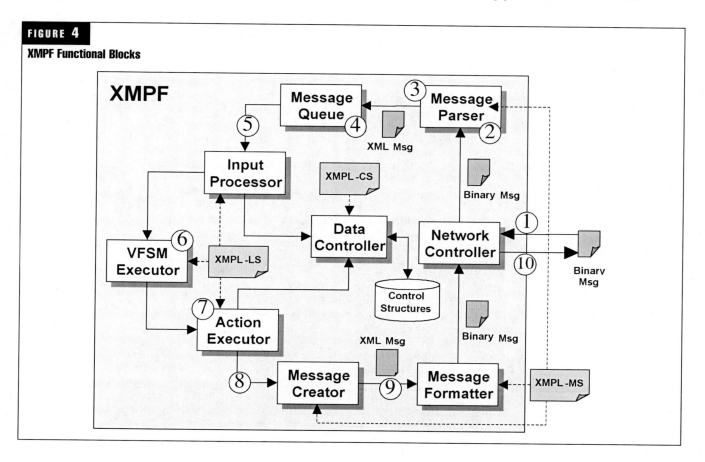

**FIGURE 4**

**XMPF Functional Blocks**

- *Data Controller*: It handles the information supporting the protocol operation. This module is configured by the XMPL–CS file that defines the hierarchical control structure needed for storing the protocol information. In particular, it implements the set of functionalities that permits the Input Processor and the Action Executor to store, update, and delete the information describing the protocol behavior.

With reference to the numbering reported in *Figure 4*, we describe the following possible sequence of steps related to generic message processing:

1. The Network Controller collects the protocol message coming from the network interfaces and passes them to the Message Parser.
2. The Message Parser converts the binary-format message to an XML–format message.
3. The Message Parser validates the message.
4. The Message Queue stores the message in a queue.
5. The Input Processor takes the message from the Message Queue and, according to the protocol logic expressed in the XMPL–LS file, obtains a set of Input-Names for the VFSM Executor.
6. The VFSM Executor elaborates the set of Input-Names obtained from the Input Processor and obtains a set of Action-Names according to the protocol logic and the current state of the protocol.
7. The Action Executor converts the set of Action-Names in a set of actions over the Data Controller and the Message Creator according to the directives expressed by the XMPL–LS file.
8. If an action requires the creation of a message, the Action Executor sends directives to the Message Creator to obtain an XML–format message compliant with the XMPL–MS file.
9. The obtained message is passed to the Message Formatter, which converts the XML format to a binary format according to the XMPL–MS file.
10. The Network Controller sends the message to the network interfaces.

### XMPF Advantages
A protocol implementation that comes as a combination of XMPF and its configuration files written in XMPL has the following benefits:

- *Reduction of implementation time*: The protocol developers are required to implement only the protocol logic as a set of XML files compliant with the XMPL specifications without writing any code. Commercial software libraries are available for supporting the manipulation and the validation of XML structures.

- *Reduction of deployment time*: The time-consuming procedures needed by administrators of complex networks to update and manage new versions of protocol software in commercial devices are significantly shortened. In fact, the support of new protocols can be made available by furnishing the set of XMPL files that represent the protocol logic and relevant message set.

- *Improvement of software stability*: The execution of the protocol, implemented by the XMPF code, is separated

from the protocol description given in terms of a set of XMPL files. This approach allows improvement of the software stability because no additional code is required for new protocols and a bug fixed in the framework is a bug fixed for any protocol implementation. In addition, the use of the XML library, available on most common operating systems, reduces and simplifies the framework code. The choice of converting binary-format messages into XML–format messages allows for easy validation of incoming messages thanks to the XML Schema technologies without writing specific code for data checking.

- *Software portability*: The XMPF presents great portability because it is written in ANSI C++, every employed library is available over a large variety of operating systems, and the XML text file is a platform-independent format.

### XMPF Drawbacks
The combination of the XMPF "engine" and its configuration files written in XMPL has been shown to have several architectural advantages. Nevertheless, some drawbacks exist in terms of performance, especially processing time as outlined in the section dedicated to the performance analysis. The main two concerns are the following:

- *Processing time*: Protocol messages expressed in binary format allows software tools to directly access specific message fields, thus achieving fast protocol processing. But XML–format messages require protocol tools to scan the entire XML file until they find the field of interest. This considerably increases the number of instructions needed for generating, processing, and interpreting the protocol messages, which in turn increases relative latency.

- *Memory space*: A second fundamental drawback, which pertains to XMPL as a language but affects XMPF as protocol framework, is the size overhead brought by XML files with respect to the actual information they carry. Because of this, XMPF should be run on machines with adequate memory space.

## Experimental Activity

A test bed was set up to validate the designed protocol software supported by XMPF and XMPL files and to evaluate its performance. But the prototype implementation was not designed with the primary goal of performance optimization, rather to validate the approach and assess the XMPF's ability to manage several protocols at the same time.

### Test-bed description
The test bed consists of three hosts inter-connected by a switch as shown in *Figure 5*. Each host is a 2.4 GHz PC running Linux operating system (Fedora Core 4 distribution) and equipped with a 100 Mbps Ethernet interface. Each host runs an instance of XMPF and was configured by means of XMPL files to support concurrently the OSPF, the Reservation Protocol (RSVP) [14], and the LMP [15]. Each PC also runs an instance of a protocol analyzer software tool (Ethereal) that was used for capturing and analyzing the packets created and sent by the XMPF instances.

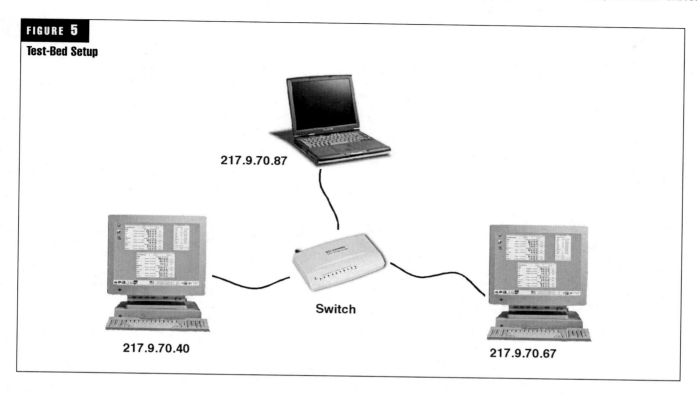

**FIGURE 5**
**Test-Bed Setup**

217.9.70.87

Switch

217.9.70.40

217.9.70.67

The scenario reproduces a control plane network based on a simplified version of the GMPLS [16] protocol suite, by implementing a subset of the OSPF, RSVP, and LMP messages.

The implemented set of GMPLS messages is presented in *Table 3*.

*Validation and Performance Evaluation*
We defined packet processing time (PPT) as the time latency that occurs for creating an XML-format message, translating it into binary-format, and finally for sending it to the network interface. The throughput T expressed in messages per second has been calculated as [1/PPT]. It represents the maximum number of messages that XMPF is able to process without message losses. We measured the average PPT as a function of the message length. The PPT is calculated by inserting a time-stamp in the XMPF code immediately before the execution of the procedure that creates an XML–format message and immediately after the procedure that sends the binary-format message to the network interface. The experimental data we obtained, represented by square dots, and their relevant best-fit line is presented in

the graph shown in *Figure 6* where the the average PPT is plot as a function of the length of a generic message expressed in byte.

For the given test-bed setting, some simple conclusions about the performance can be drawn. First, PPT is directly proportional to the length of the message. PPT for a 1-byte message is about 85 ms, so the maximum throughput that the XMPF is able to achieve is 11 messages per second (i.e., 660 messages per minute).

*Figure 7* shows a snapshot taken from a PC of the test bed during the trial. It is related to the node 217.9.70.87 that processes LMP, OSPF and RSVP messages coming from the node 217.9.70.40. For each type of message, the counting and the seconds elapsed from the last message sent are reported, while specific message fields are not displayed. Each XMPF instance was set for sending up to 300 messages per minute and, consequently, able to process up to 600 messages coming from the other two nodes without losing messages.

Of course, the XMPF is conceived to be only a prototype. Nevertheless, thanks to the performance measured, XMPF can be used for the implementation of network protocols employed in simple real networks (e.g., GMPLS control plane with a moderate signaling rate).

The XMPF performance is mainly determined by the capability of the XMPF to process messages in XML format. For this reason, our next step will be to improve the software algorithm for processing the XML messages and to replace the Xerces-C XML library with other open-source libraries (e.g., LibXML) to reduce the PPT and then increase throughput.

**TABLE 3**

**Implemented GMPLS message set**

| Protocol | Messages implemented |
|----------|---------------------|
| OSPF | Hello, Link-State |
| RSVP | Hello, Request |
| LMP | Hello, Config, Test |

## FIGURE 6

**PPT Plot as a Function of the Message Length**

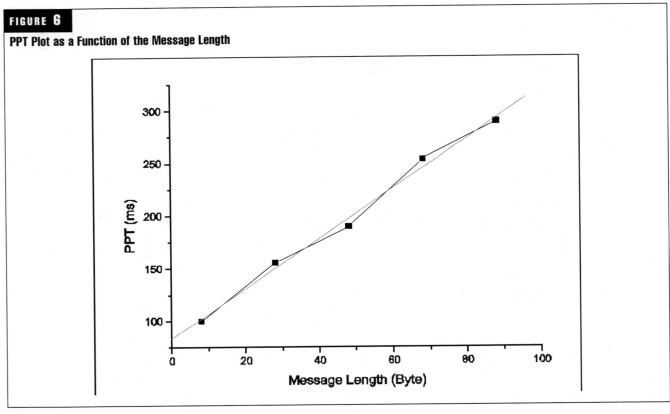

## FIGURE 7

**Test-Bed Node Snapshot**

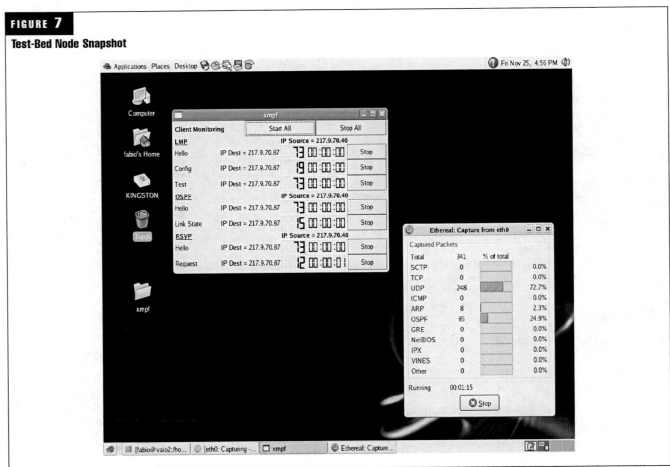

## Conclusion

This work presents two main achievements: (i) XMPL definition, that is, a general purpose XML–based protocol language for representing protocol specification in an implementation-oriented manner. Specifically, using XMPL, it is possible to describe a protocol in a comprehensive way in terms of message set (XMPL–MS), behavior (XMPL–LS), and data structure (XMPL–DS); (ii) XMPF development, that is, an engine that implements the concurrent support of protocol application instances whenever provided with the three previously mentioned external configuration files written in XMPL.

The proposed approach, consisting of the combination of XMPL and XMPF, significantly decreases the protocol software development time and, at the same time, increases its robustness thanks to data checking provided by the XML Schema technology. In addition, the usage of the XML makes both XMPL and XMPF portable and scalable due to the wide support over all the principal software platforms.

Finally, a test bed for validating the XMPL language and the XMPF tool was presented. In particular, it has demonstrated that the XMPF is capable of managing network protocol suites (e.g., GMPLS) even if, at the current stage of development, it may not be the best solution when very large message rates or real-time protocol processing represents the primary requirements of the protocol operation.

## Acknowledgments

Part of this work has been supported by EU through the NOBEL (FP6-IST 506760) project.

## References

[1] W3C. Web Services Description Language (WSDL) 1.1, March 2001.

[2] ITU–T. Rec. Z.100: Specification and description language (SDL), March 1993.

[3] ITU–T. Rec. X.680: Abstract Syntax Notation One (ASN.1): Specification of basic notation, July 2002.

[4] Extensible Markup Language (XML) 1.0 (Third Edition). www.w3.org/XML, 4 February 2004.

[5] S. Hollenbeck. RFC 3470 – Guidelines for the Use of Extensible Markup Language (XML) within IETF Protocols. Technical report, IETF, January 2003.

[6] XML Schema Part 1: Structures Second Edition. www.w3.org/TR/xmlschema-1, 28 October 2004.

[7] James O'Grady Mark Staskaukas Anal Flora-Holmquist, Edward Morton. The Virtual Finite-State Machine Design and Implementation Paradigm. *Bell Labs Technical Journal*, Winter 1997.

[8] XML Path Language (XPath) Version 1.0. www.w3.org/TR/xpath, 16 November 1999.

[9] XSL Transformations (XSLT) Version 1.0. www.w3.org/TR/xslt, 16 November 1999.

[10] Language of Temporal Ordering Specifications (LOTOS), 1989. ISO 8807.

[11] J. Moy. RFC 2328 – OSPF Version 2. Technical report, IETF, April 1998.

[12] Hany H. Ammar and Sherif M. Yacoub. *Patter-Oriented Analysis and Design*. Addison Wesley, 2003.

[13] Robert Engel, Hermann Hüni, and Ralph Johnson. A Framework for Network Protocol Software. *OOPSLA*, 1995.

[14] R. Braden. RFC 2205: Resource ReSerVation Protocol (RSVP) – Version 1 Functional Specification. Technical report, IETF, September 1997.

[15] J. Lang. RFC 4204: Link Management Protocol (LMP). Technical report, IETF, October 2005.

[16] E. Mannie. RFC 3945: Generalized Multi-Protocol Label Switching (GMPLS) Architecture. Technical report, IETF, October 2004.

# An Epidemiological View of Worms and Viruses

Thomas M. Chen

*Associate Professor, Department of Electrical Engineering*
Southern Methodist University

## Introduction

The communal nature of the Internet exposes organizations and home computer users to a multitude of worms, viruses, and other malicious software (malware) threats such as spyware and Trojan horses. Viruses are program fragments attached to normal programs or files that hijack the execution control of the host program to reproduce copies of the virus. Worms are automated self-replicating programs that seek out and copy themselves to vulnerable new targets over the Internet. In the same way that germs are quickly shared among people, worms can spread rapidly among networked computers. In the second half of 2004, Symantec reported 7,360 new Windows worms and viruses, an increase of 63 percent over the number of new worms and viruses in the first half of 2004 [1]. The most prevalent worms were variants of Netsky, MyDoom, Beagle, and Sober. In the 2005 CSI/FBI Computer Crime and Security Survey, 75 percent of the surveyed organizations reported being hit by worm and virus attacks [2]. Worms and viruses were the most frequent and costly type of attack, despite the use of antivirus software and firewalls by 96 percent of the surveyed organizations.

Biologists tackle infectious diseases at both microscopic and macroscopic levels. However, very little effort is spent to treat worms and viruses at the macroscopic or epidemiological level. Today, the security industry focuses on the treatment of worms and viruses exclusively at the "microscopic" level, analogous to the microbiological approach to infectious diseases. Antivirus companies collect samples of worms and viruses through donations and honeypots. The malicious code is disassembled into a more human readable format to study its programmatic structure and develop a new antivirus signature. The new signatures are downloaded to update antivirus software programs.

Epidemiology is more interested in the dynamics of diseases spreading through populations than their biochemical mechanism. In the long history of medicine, epidemiology has been a relatively recent development. The foundations of epidemiology are often traced to Dr. John Snow, who studied an outbreak of cholera in London in 1848 [3]. In treating patients, he became convinced that the disease was spread by ingesting germs from polluted water. At the time, many physicians did not believe in germs as the cause of infectious diseases. To avoid controversy, Snow described the cause of cholera as a "poison" that had the ability to "multiply itself" within cholera victims, before being spread to new victims through polluted water. He came across a district supplied by two private water companies. Snow collected a vast amount of statistical evidence that linked a high mortality rate to people supplied by one of the water companies, and a much lower mortality rate to the other water company. Unfortunately, because Snow was the first person to use a survey of statistical incidence and distribution of an epidemic in an effort to determine its cause, his evidence was not believed by other doctors.

In 1853, another outbreak of cholera occurred in a neighborhood close to Snow's home in the London district of Soho. He traced the water supplied to cholera victims to a water pump on Broad Street. Snow was able to convince the Board of Guardians to turn off the pump, and the local cholera outbreak quickly ended. When Snow died in 1858, his theory about the spread of cholera still had not been accepted. The germ theory of disease did not gain acceptance until the 1860s, after it was demonstrated by the chemist Louis Pasteur. In historical perspective, Snow's important contribution was his persistent efforts to determine how cholera was spread by means of statistical and mapping methods, which have become standard methods in epidemiology.

## Successes of Epidemiology

The practical usefulness of epidemiology was demonstrated by the successful eradication of smallpox. Smallpox is an acute contagious disease caused by the variola virus. It is believed to have originated more than 3,000 years ago in India or Egypt. For centuries, devastating epidemics have swept across continents, decimating populations. In the absence of vaccination, humans are universally susceptible to infection. No effective treatment has ever been developed, and the mortality rate is about 30 percent. Survivors are often left with scars or blindness.

The mathematician Daniel Bernoulli made a major contribution to epidemiology by mathematically proving that variolation (inoculation with a live virus obtained from a victim with a mild case of smallpox) was beneficial. Variolation usually resulted in immunity from smallpox. Bernoulli was able to formulate differential equations to show that variolation could reduce the death rate [4].

In 1798, Edward Jenner demonstrated inoculation with cowpox. The smallpox vaccine contains live vaccinia virus, which is closely related to the variola virus. Vaccine administered up to four days after exposure to the virus, and before the rash appears, provides protective immunity that can prevent infection for many years or at least reduce the severity of an infection.

In the 1950s, there were an estimated 50 million cases of smallpox in the world each year. Smallpox vaccination became part of the mission of the Center for Disease Control and Prevention (CDC), originally established in 1946 as the Communicable Disease Center led by Dr. Joseph Mountin in the U.S. Department of Health and Human Services [5]. Its broad mission is to monitor the prevalence of infectious diseases, develop public health policies, enact strategies for disease prevention, and investigate problems of public health. Dr. Mountin envisioned the CDC as a center for epidemiology responsible for all infectious diseases. Dr. Alexander Langmuir joined the CDC in 1950, when the Korean War posed the threat of biological warfare, to establish the CDC's Epidemic Intelligence Service (EIS). Medical epidemiologists were rare at the time, and the EIS was instrumental in training epidemiologists. In the 1950s, the CDC was instrumental in overseeing the polio inoculation program and developing a national vaccination program for a major influenza epidemic in 1957.

The CDC established a smallpox surveillance unit in 1962. It worked to refine a smallpox vaccination and introduce the vaccine to millions of people in central and West Africa. The CDC established the application of scientific principles of surveillance to the problem. In 1967, the World Health Organization followed the success of the CDC and resolved to intensify their plan to eradicate smallpox. The WHO had passed an earlier resolution for global eradication of smallpox in 1959 but had not dedicated many resources. The intensified program consisted of a combination of mass smallpox vaccination campaigns and surveillance and containment of outbreaks. Through the success of the global eradication campaign, smallpox was finally pushed back to the Horn of Africa and then to a single last natural case in Somalia in 1977. The global eradication of smallpox was certified by the WHO in 1980.

## Role of an Epidemiology Center for Worm Control

Today, no counterpart of the CDC exists for worm and virus control or prevention. Although analogies can be drawn between worm outbreaks in the Internet and disease outbreaks in the human population, there is no national level organization responsible for coordinating and responding to worm outbreaks. Given the success of the CDC for human diseases, an argument could be made by analogy for the need for a national "center for worm control." The establishment of a national center for worm control could have several benefits to network security.

First, the prominence of a national center would elevate the worm problem to a national priority. Although the importance of infectious diseases affecting public health is obviously a national priority, the health of the Internet is not seen as a problem concerning the federal government. It might be argued that the Internet has evolved to the point of becoming a critical infrastructure essential for national productivity, and even national security. However, the Internet is generally viewed as a commercial enterprise, although it began as a Defense Advanced Research Projects Agency (DARPA)–funded project. It is somewhat loosely administered by the Internet Society (ISOC) a professional membership society with more than 100 organization and 20,000 individual members in 180 countries [6]. It includes the Internet Architecture Board (IAB) and the Internet Engineering Task Force (IETF) responsible for Internet infrastructure standards. The ISOC is really a facilitator to coordinate the efforts of various stakeholders in the Internet. The Internet is really administered by the many companies and organizations that own parts of the Internet. Worms and network security problems in general are viewed as problems of the separately administered networks.

Second, a national center for worm control could be instrumental in developing an Internet-wide "health policy" to maintain the security and integrity of the Internet, in the same way that the CDC devises public health policies. Health policies could include standard practices for software patching, antivirus software updates, sharing worm information among companies and organizations, and coordination of local responses to new worm outbreaks. Today, worms are not treated as a single Internet-wide problem. Instead, individual networks are responsible for their own protection and defense. By design, the Internet is highly distributed and decentralized. Consequently, worm protection and defense is carried out in a piecemeal manner. However, worm infections of one network obviously affect other networks. An infected network not only increases the chances of infecting another network, but could also substantially increase the level of congestion with worm traffic. Therefore, it is not difficult to see the advantage of a national network security health policy that enforces consistency among security practices for the benefit of all networks.

Third, a national center for worm control could facilitate the collection and sharing of worm samples and information. Today, antivirus companies collect their own worm samples through donations and honeypots and informally share samples with each other in a limited way. They publish their own libraries of worm information. In addition, there are informal vendor-neutral groups such as the Anti-Virus Information Exchange Network (AVIEN) for exchanging worm and virus information among security specialists [7]. However, there is no centralized repository, which makes it difficult for anyone else to obtain worm samples or detailed information without subscribing to a proprietary service. Obviously, security researchers depend on access to real worm code, and the lack of data availability is a hindrance to further research. In addition to making worm samples available for research, a central repository could have additional benefits such as consistency in worm/virus names and terminology, pooling of information about specific worms, and consistent and safe practices for worm code sharing.

Fourth, the idea of information sharing could be taken further to propose that a national center could provide an early warning for new worm outbreaks. Current approaches to early warning, like the approaches for information sharing, are either proprietary or grassroots. A well-known example of a proprietary approach is Symantec's DeepSight Threat Management System [8]. It collects log data from 24,000 sensors (firewalls, intrusion detection systems, honeypots, and hosts running Symantec antivirus) distributed throughout 180 countries, in addition to 2 million decoy e-mail accounts. The log data is correlated and analyzed for signs of attacks, including worm outbreaks. The wide geographic coverage of the DeepSight System enables it to theoretically detect a new worm outbreak that might originate anywhere in the world. Another example is AT&T's Internet Protect Service, which monitors traffic going through AT&T IP backbone routers. These backbone routers handle a considerable fraction of the total Internet traffic. The traffic data is correlated and analyzed for signs of worms, viruses, and denial of service (DoS) attacks. An example of a grassroots early-warning system is the Anti-Virus Information and Early Warning System (AVIEWS), an outgrowth of the AVIEN information sharing network.

Fifth, a national center for worm control could coordinate real-time responses to new worm outbreaks. Due to the decentralized nature of the Internet, responses today are piecemeal and ad hoc. System administrators are generally responsible for protecting their own networks. When a new worm outbreak is discovered, they respond in a variety of ways, such as configuring firewalls, patching systems, updating antivirus programs, and taking systems off-line. Unfortunately, there is little coordination among system administrators of different networks.

Lastly, a national center for worm control could promote the scientific principles of epidemiology that have been successful for human diseases and apply them to worms. Little epidemic theory has been developed for worms. The idea of epidemiology for worms was suggested as early as 1993 but has not been pursued far [9].

## Goals of Worm Epidemiology

How can epidemiology apply to worms, and what can be learned? The so-called "simple epidemic model" fits random scanning worms fairly well [4, 10]. The vulnerable hosts in the Internet are viewed as a fixed size population, all initially in a "susceptible" (vulnerable but not infected) state. A small number of infected hosts are introduced. After contact with a worm from an infected host, susceptible hosts will change state to "infected" and subsequently remain permanently in the infected state. An infected host makes contacts with susceptible hosts at a certain "infectious contact rate" that depends on the scanning rate of the worm and the likelihood that any scan will reach a susceptible (and not already infected) host.

A more complicated "general epidemic model" adds another "removed" state to factor in the possibility of worm disinfection—that is, system administrators are assumed to be removing the worm from infected hosts by patching software or running an antivirus. Infected hosts may change state to "removed" and subsequently remain permanently in the removed state, immune from future re-infection. The transitions from infected to removed state occur at a certain "removal rate."

One of the obvious goals of epidemiology is to predict how far a worm outbreak can spread as a function of time. This is important knowledge because it always takes some time to detect and respond to a new worm outbreak. In the meantime, a new worm might spread without any constraint. Containment of the outbreak to a given infection level would require a response time that can be calculated.

Another goal of epidemiology is to quantify the effectiveness of immunization. Hosts can be protected against infection by keeping software patches and antivirus software up to date. In practice, however, it is difficult to keep up patching and antivirus updates on all hosts in a network. Epidemiology can predict how a given level of immunization can slow down a worm outbreak.

Still another goal of epidemiology is modeling of active responses such as quarantining [11]. Quarantine of worms works in the same way as quarantine of human diseases. The idea is to prevent infected hosts from making contacts with susceptible hosts. Epidemic models can be used to evaluate quarantine strategies by proper selection of infectious contact rates between pairs of hosts.

## Conclusions

We have made a case arguing for the success of biological epidemiology and the need to further develop a similar body of theory for worms. A national-level center for worm control, analogous to the CDC for human diseases, could be instrumental in fostering and applying this theory.

## References

[1]   D. Turner, et al., "Symantec Internet security threat report: trends for July 2004–December 2004," available at www.symantec.com.

[2]   L. Gordon, et al., "2005 CSI/FBI Computer crime and security survey," available at www.goscsi.com.

[3]   W. Winterton, "The Soho cholera epidemic of 1854," *History of Medicine*, vol. 8, 1980, pp. 11–20.

[4]   D. Daley, J. Gani, *Epidemic Modeling: An Introduction*, Cambridge University Press, 1999.

[5]   Centers for Disease Control and Prevention home page, available at www.cdc.org.

[6]   Internet Society (ISOC) home page, available at www.isoc.org.

[7]   AVIEN home page, available at www.avien.org.

[8]   Symantec DeepSight Threat Management System, available at tms.symantec.com.

[9]   J. Kephart, D. Chess, S. White, "Computers and epidemiology," *IEEE Spectrum*, vol. 30, May 1993, pp. 20–26.

[10]  D. Moore, C. Shannon, J. Brown, "Code-Red: a case study on the spread and victims of an Internet worm," *ACM Internet Measurement Workshop*, Nov. 6–8, 2002, pp. 273–284.

[11]  D. Moore, et al., "Internet quarantine: requirements for containing self propagating code," *IEEE Infocom 2003*, pp. 1901–1910.

# Corporate Perspectives

# Services over IP: Delivering New Value through Next-Generation Networks

## Jeanette Carlsson

*Global Communications Sector Leader*
IBM Institute for Business Value

## *Executive Summary*

The communications industry is facing unparalleled changes that are breaking down traditional industry boundaries. Providers from different technologies and backgrounds all focus on the same future—integrated service offerings for next-generation customers. Telecom service providers (SPs) are challenged as never before to defend market share and increase revenues.

On one hand, peaking broadband penetration and customer demand for multimedia services and content provide new revenue opportunities. At the same time, intensifying competition from existing industry players and new market entrants, including rejuvenated cable multiple-system operators (MSOs), Internet service providers (ISPs), and consumer brands, bring renewed demands for innovation. Increasingly discerning customers expect a seamless, multichannel experience, regardless of technology. At a time of rapid technology change, these factors add to existing pressures on telecom infrastructures and market expectations for growth.

Internet protocol (IP) technology is the catalyst for change. To telecom SPs, IP brings unprecedented changes beyond what is considered an ongoing evolution, in particular to the network: Network convergence brings together previously parallel networks (cellular, fixed, and enterprise) onto a single IP–based infrastructure. Service convergence enables integrated service propositions (e.g., "triple" and "quadruple" plays [1]). Network convergence links with service convergence to enable virtually "anytime, anywhere, anyhow" service delivery. Such changes present major challenges and demand renewed SP attention to their network strategies.

SPs operating traditional circuit switched (CS) networks are increasingly constrained to deliver against the convergence vision and extract additional value from their legacy environments. Parallel, service-specific networks, based on proprietary technologies, are typically difficult to integrate and inflexible in terms of what services they can deliver and how. Many proprietary technologies are also nearing the end of their lives and are costly to maintain, affecting network performance and the economics of service delivery. Finally, the emergence of multiple new access technologies and devices sparks demands for service and technology interoperability, requiring an IP–based network infrastructure.

To position themselves for leadership in the converging communications world, SPs around the world need to follow the example of today's industry leaders and build a transformation path to next-generation networks and service delivery capabilities, based on the following:

- An IP/multiprotocol label switching (MPLS)–based core network strategy for revenue and cost optimization
- Open-standard technologies and platforms for cost-efficiency, flexibility, and interoperability
- Reusable, commercial off-the-shelf (COTS) solution components for lower infrastructure costs
- A service creation environment to drive and support a large ecosystem of applications and, therefore, revenue-generation capabilities
- A common, standards-based service delivery platform that can be easily integrated with existing and third-party assets for multimedia service delivery

The urgency of migration will vary by operator type and competitive pressures in local markets. However, transition to all–IP networks is under way. How individual SPs respond to the reality of IP will create the foundations of the industry of tomorrow and determine winners and losers.

## Meeting the Challenge While Seizing the Opportunity

### Responding to the Growth Challenge

Telecom providers around the world face tremendous pressures to sustain core revenues. Fixed-line voice is being "squeezed" by substitution of broadband (cable and digital subscriber line [DSL]), mobile, and IP. According to Forrester Research, mobile's popularity among European consumers continues to grow at the expense of fixed-line telecom services. About a quarter of mobile users have switched at least a portion of their fixed-line use to mobile. More intend to follow: Six percent of mobile users plan to cancel their fixed connections in the future [2]. In addition, with reinvigorated cable companies invading traditional telecom territory with voice over IP (VoIP) "pure plays" and digital phone rollouts, the assault on the fixed voice business is more acute than ever. Industry analysts Frost & Sullivan expect that, by 2009, there will be nearly 20 million VoIP subscribers in the United States alone [3].

Meanwhile, increasing operating expenses are further eroding margins and overall growth. With network and network development costs typically accounting for more than 35 percent of fixed-line operating costs, telecom SPs are under pressure to reduce their network-related expenses to sustain margins [4].

In mobile, while some markets still have opportunities to grow voice usage (minutes of use [MoUs]) and revenues, the double-digit revenue growth and healthy profit margins once enjoyed have been lost to market saturation, intensifying competition and low average revenue per unit (ARPU) net additions in emerging markets. According to Ovum, global mobile voice revenue growth of 23.1 percent from 1999 to 2003 will decline to 8.0 percent from 2004 to 2009 [5]. To compensate, mobile operators have looked to the financial management service (FMS) to boost top-line growth. However, in some markets, the scope for further substitution is starting to slow.

Moreover, mobile operators cannot be complacent about VoIP. With fixed-line players increasingly offering IP–based telephony, they are able to compete more effectively on price, potentially threatening mobile voice—including lucrative international roaming—revenues. In addition, although commercial deployment of mobile VoIP (e.g., voice over wireless area network [VoWLAN]) may be immature,

mobile SPs need to consider the potential risk to future voice revenues against new, potential service opportunities enabled by IP–based wireless networks (e.g., WLAN, WiMax) and advanced devices. In the short term, pervasive instant messaging (IM) and e-mail may start to make highly profitable service management system (SMS) look like an expensive alternative, impacting data revenues. Although accelerating third-generation (3G) adoption brings renewed hope, new emerging technologies such as high-speed downlink packet access (HSDPA) offer double/triple download speeds, enabling TV streaming. Finally, with the increasing importance of content, SPs may be braced for a revenue battle with content providers, making the future growth path for mobile far from clear.

## Seizing the Convergence Opportunity

As growth moderates, the convergence opportunity is increasingly important as a source of new revenues.

The global broadband market is forecast to pass 190 million subscribers in 2005, growing to more than 430 million subscribers in 2010 (see *Figure 1*), and customers are increasingly demanding multimedia services and content such as digital music, IM, TV/video on demand (VoD), games, and customized content.

According to Multimedia Research Group Inc., global IPTV service revenues are likely to hit US$880 million in 2005, growing to $9.9 billion in 2009, a compound annual growth rate (CAGR) of 83 percent [8]. Europe and North America are generating the majority of this revenue (see *Figure 2*).

IPTV services enable telecom companies to compete with rejuvenated cable providers and start-ups, while generating new revenues. However, aggressive competition is leading to intense price pressures, putting telecom returns on investment at risk unless they can significantly reduce their cost base.

Differentiated triple- and quadruple-play bundles around the world are beginning to earn higher ARPUs and reduce

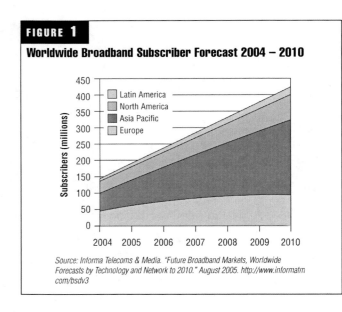

**FIGURE 1**

**Worldwide Broadband Subscriber Forecast 2004 – 2010**

*Source: Informa Telecoms & Media. "Future Broadband Markets, Worldwide Forecasts by Technology and Network to 2010." August 2005. http://www.informatm.com/bsdv3*

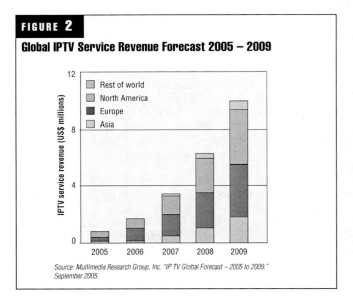

**FIGURE 2**

**Global IPTV Service Revenue Forecast 2005 – 2009**

*Source: Multimedia Research Group, Inc. "IP TV Global Forecast – 2005 to 2009." September 2005.*

churn. Korea Telecom has led the way with its state-of-the-art "Digital Home" package that offers multimedia/telecom services over its seamlessly integrated fixed-mobile networks [11]. FastWeb in Italy has generated some of the highest annual residential ARPUs in the world—more than $1,000—by offering both telephony and video services [12]. In the United States, SBC Communications is jumping on the quadruple-play bandwagon with its plans for a full IP network, incorporating fixed and mobile voice, high-speed data, TV, and video services.

As convergence breaks down traditional industry barriers, Sprint in the United States is teaming up with four cable TV companies—Comcast, Time Warner, Cox, and Advance/Newhouse Communications—to launch a $200 million joint venture to create a quadruple play of voice, video, Internet, and wireless services [13].

In the enterprise market, IP is opening up major new VoIP–based opportunities. Forrester Research predicts that by 2015, 95 percent of enterprise voice calls will be VoIP–based [14]. However, the fastest growing VoIP market is "hosted IP voice" or "IP Centrex," which is expected to expand from about $60 million in 2004 to more than $7.6 billion by 2010, representing a CAGR of 282 percent [15]. By that time, VoIP technologies are expected to be handling 45 percent of the total voice-telephony market [16].

Fixed-mobile convergence (FMC) is another significant new opportunity in the consumer and enterprise markets: Pyramid Research expects FMC revenues to reach $80 billion in 2009, or 6 percent of total communications spending worldwide [17]. With original WLAN/Wi-Fi offerings aimed at providing ubiquitous broadband access—now largely a commodity—SPs aim to use FMC to defend their share of voice minutes and drive new service revenues from target customers. For example, in mid-2005, British Telecom (BT) Group in the United Kingdom launched its FMC service offering called Fusion—a cellular handset using Bluetooth technology to switch seamlessly from the cellular network outside the home to a fixed-line service through a broadband DSL hub in the home, all billed to the end user as one service [18].

However, telecom companies are challenged to match the innovative service propositions of powerful cable giants, ISPs, and consumer brands such as America Online (AOL), Yahoo, Microsoft Network (MSN), and Google. For example, Google is offering Google Talk (IM and VoIP service) along with streaming video of prime-time TV [19]. To compete, telecom companies need to improve their understanding of evolving user needs, and match these with the "right" services and content targeted at specific market segments at the "right" price. Soon, next-generation services, based on presence management (information about a user's online status), will mix real-time multimedia components with legacy services within one "call," enabled by the information management system (IMS), an IP–based architecture framework enabling multimedia and converged service delivery (see section entitled "Delivering Next-Generation Services"). Among these services are push-to-talk/message and other "push-to" services; combinational (voice/picture) and location-based services; group chat; and online, multiplayer games. IMS–based enterprise services include IP–based virtual private network (VPN), IP Centrex/hosted private branch exchange (PBX), or IP Centrex combined with Web-based conferencing and messaging services. Ultimately, we anticipate that all services will become next-generation network (NGN)– or NGN–IP/MPLS–enabled.

## Mastering the Infrastructure Challenge

Turning out content-rich services is operationally complex, requiring integrated infrastructure and converged multimedia service delivery capabilities. However, most carriers today operate distinct CS and packet-switched (data) service networks, based on different proprietary protocols, which are often difficult to integrate, inflexible in terms of service delivery, and costly to maintain, which drags down earnings. With aggressive competition and increasing consumer choice, the ability to bring new offerings to market quickly is a critical competitive differentiator. Finally, legacy infrastructures typically have no means to address increasing user fatigue of today's fragmented communications experience, involving multiple services, devices, technologies, and user names/passwords.

With competition heating up, leading SPs have begun the shift from legacy environments to NGNs. While traditional wireline capital expenses are forecast to grow at 3 percent CAGR between 2002 and 2007, global spending on IP infrastructure is forecast to grow at 12 percent CAGR during the same time period [20]. We expect the migration to IP networks to intensify over the next three to five years with the urgency of migration varying by operator type and local competitive pressures.

## The NGN Environment

### From CS to IP Technology
NGN is a general term used to describe networks characterized by the use of IP, fiber optics, and software-based platforms for service delivery. One of the most dominant trends in the evolution of the IP–based NGN is the move from CS to IP technology, which helps reduce the cost of delivering existing services and enhances SPs' capabilities to deliver new, integrated services more rapidly. IP–based NGNs use quality of service technology to support all types of customer traffic—voice, data, and video—on a single, common network, "marrying" different networks within one ubiquitous platform or "single piece of wire." The term NGN refers to the network core (IP/MPLS in *Figure 3*) and the interface of the core with the local access ("last mile") part of the network (the "network edge").

In the IP environment, SPs can increase network efficiency using optimized IP transport and coding solutions. By reducing overall capacity requirements and other cost factors, SPs help reduce the expense and inefficiencies of isolated legacy networks supporting different service types. Freed from proprietary constraints by open standards-based IP networks, SPs can shop for the technologies they need instead of building proprietary solutions, or counting on one legacy vendor to supply them, providing significant scope for cost reductions.

Key themes in the NGN evolution include the following:

- Convergence of multiple, parallel networks onto a single, IP–based infrastructure

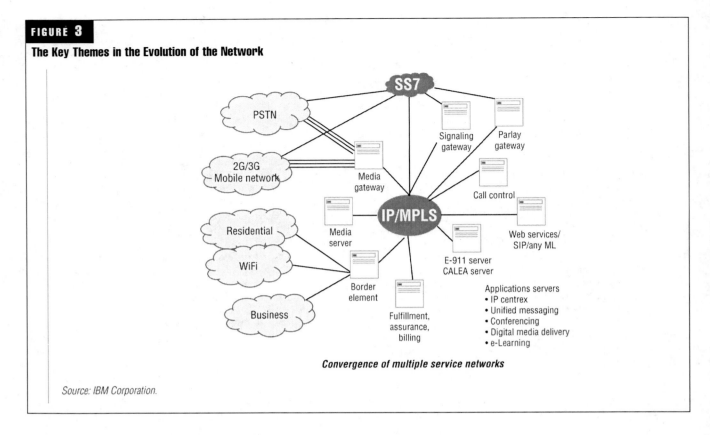

**FIGURE 3**

**The Key Themes in the Evolution of the Network**

*Convergence of multiple service networks*

Source: IBM Corporation.

- Convergence of the telecom network with the IT operating environment
- Convergence of service offerings (VoIP, FMC, triple/quadruple play)
- Network and service convergence designed to enable "anytime, anywhere, anyhow" service delivery

Network convergence is the enabler—the means by which network operators facilitate better access to converged, value-added services and applications across different forms of IP connections and devices (see *Figure 3*).

The benefits of network convergence stem from SPs' ability to deliver current and new value-added services more effectively at reduced cost, provide seamless user access to services/content regardless of broadband connection and device, reduce cost, and enhance flexibility.

Other key themes in the NGN evolution include the following:

- Replacing technology "stovepipes" with horizontal network layers
- Replacing time-division multiplexed, narrowband transmission with open, broadband, carrier-grade IP services
- Replacing expensive proprietary hardware and operating systems with open software-based computing platforms, operating systems and interfaces
- Replacing monolithic applications with offerings from hundreds of competitive firms, including applications, middleware, operating systems, and other technologies
- Replacing many small, distributed elements with fewer, more centralized servers

- Replacing closed standards taking 15 to 25 years to evolve with open standards advancing with timescales closer to five to ten years

*New Sources of Value*

With the shift from proprietary, commoditizing hardware to IP– and software-based platforms, SP value shifts from enabling voice traffic to delivering software-based services and supporting infrastructure to global operations on common IT hardware and software platforms. This environment enables SPs to do the following:

- Use common Internet protocols such as hypertext transfer protocol (HTTP), secure socket layer (SSL), session initiation protocol (SIP), and extensible markup language (XML)
- Move to open operating systems such as Linux
- Reduce capital expenses through COTS solutions that usually can be developed and deployed for less than the cost of bringing legacy network equipment up to speed
- Build an environment for the creation of new services and service packages, based on multiple "basic services" or capabilities, using industry-standard tools and technologies
- Build portfolios around hundreds of horizontally integrated services and supporting infrastructures at potentially lower costs
- Shift to open, integrated business models, enabling SPs to draw on the capabilities of a wider ecosystem, and use the competitiveness of multiple software vendors to lead the way to major savings and flexibility in service delivery

## The Architectures of the NGN

The IP/MPLS core network forms the underpinning of the IP NGN architecture. With key architectures and standards still being defined, there are multiple sources of architecture for an NGN. However, some individual technologies are widely accepted, including IP, SIP, and H.248.

Built on open standards, with COTS hardware and software, the NGN divides into three planes (see *Figure 4*). The transport plane manages traffic flows (movement and routing across the network) and separates traffic from the service and control planes, enabling user access to multimedia services and content and service interoperability, independent of access network or device. Applications and services (and therefore revenue generation) reside in the service plane, which supports service orchestration and application logic. Call session functions and subscriber-related information are handled by the control plane. For example, the control plane supports advanced SDPs (including IMS) and provides gateway management, selection, and control. The control plane is designed to enable the convergence of technology stovepipes, improve operating efficiency, and connect the service and transport planes.

Leading European carriers such as BT and Koninlijke PTT Nederland (KPN) are at the forefront of the move to all–IP networks. Through the rollout of its 21st Century Network project—expected to cost more than $17 billion—BT aims to become the world's first multiservice, all–IP network by 2009. Annual cost reductions are forecast at $1.7 billion [21]. Similarly, KPN is aiming to become all–IP by 2009, projecting operating expense savings of approximately $1 billion by end of Phase 2 in 2009—a significant proportion from a staff reduction of 45 percent [22].

### Delivering Next-Generation Services

To drive value, the NGN needs services. One of the weaknesses of legacy environments is that end-user services and functions driving those services are vertically integrated, with parallel silos combining a particular service, its enablers, the operations support systems (OSS)/business support systems (BSS), and the network. For each new service, vertical integration leads to extensive and costly replication of functions such as subscriber management, which is common to multiple end-user services but typically not reused for another end-user service that requires the same function.

In the NGN, multiple services share a common set of enablers (the SDP), and common OSS/ BSS. Integration and operating costs are paid once and amortized across all new services. Enablement of a common SDP is necessary to exploit the multimedia content services opportunity fully, independent of access. SPs that do not have an SDP will likely struggle to deliver NGN services rapidly, flexibly, and cost-effectively. The economics of the software industry show that when a common platform is introduced, it generates significant added value.

## The IMS Advantage

The framework of choice for multimedia and convergent service delivery is IMS. IMS is a standards-based solution defined by third-generation partnership project (3GPP) and 3GPP2, as well as by organizations using Internet Engineering Task Force (IETF) protocols, to enable IP services for mobile and fixed-line operators. IMS specifications define five planes: transport, control, service, access, and enablers (*Figure 5* shows a subset view of IMS, excluding the access part of the network and the enablers). The transport plane moves the bits back and forth and interconnects them to the legacy network, while the control plane controls the session state, the subscription and service information, and the selection and state of the gateways involved in the sessions. The service plane, where the application logic resides, is composed of a service-enabling plane (supporting service

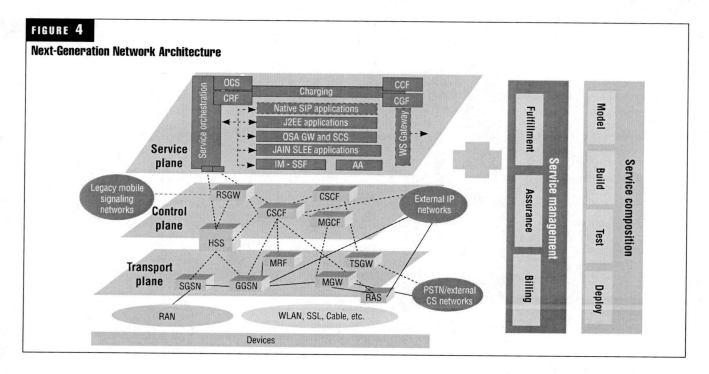

**FIGURE 4**

**Next-Generation Network Architecture**

orchestration, charging, etc.) and the applications. The access part of the network comprises the various fixed, mobile and wireless access technologies (including DSL, global system for mobile communications [GSM], general packet radio service [GPRS], WLAN, universal mobile telecommunications system [UMTS], and cable). Finally, the enablers comprise the IP network, the public switched telephone network (PSTN), etc.

Originally defined by 3GPP as an application-enabling technology for mobile Internet, IMS is now generally accepted as the future convergence platform for fixed and mobile carriers; no major alternative architecture exists, even though alternative "point solutions" are available. In a recent survey of the telecommunications industry by IDC, all respondents indicated an interest in IMS, with most listing a timeline for deployment. With the overall market still in its infancy, leading SPs are not expected to begin commercial IMS deployments until 2006, with spending ramp-up not likely until 2008. Worldwide IMS revenues are expected to grow to $14.1 billion by 2010, with Western Europe driving adoption [24]. The bulk of revenues are expected to come from the service plane.

IMS supports multiple services and access types and is designed to enable interoperability of IP services and applications, and service interoperability between subscribers. IMS is distinct from but complementary to the NGN and any existing SDPs. Unlike NGN, IMS is specifically optimized for SIP and multimedia applications.

*Benefits of IMS*
From an operator perspective—mobile or fixed—IMS delivers a framework and common capabilities to enable converged and multimedia services to meet user demands for personalized services and content. New, attractive services and service bundles typically improve ARPU and customer stickiness (loyalty), reducing churn. The ability to roll out services rapidly and flexibly gives a company an edge over competitors. As a single, horizontal infrastructure for fixed and mobile platforms, IMS can potentially enable significant cost savings.

To the end user, life is not about technology but what the user wants and how well services fulfill those needs. IMS allows the simultaneous use of real-time voice and data with non-real-time data, based on presence management, including push-to-x, combinational and VoIP–based services, in virtually any setting (e.g., stationary, in motion) and to suit individual preferences and allow multi-user activities (e.g., collaborative working or multiplayer games). IMS also enables an intuitive, integrated user interface and consistency of service across technologies and devices, helping meet user demands for a simplified, improved experience.

*Mapping Your Transformation*
Consensus is beginning to emerge about how the evolution to NGN will happen, even as the industry debate continues on the detailed architecture of the NGN and the best deployment schedules (especially for replacing legacy networks and OSS). Transition to NGNs affects the following key areas:

- The access network
- The core network
- The service plane
- The control plane
- Systems
- New services
- The device

**FIGURE 5**

**IMS Is Defined by Standards 3GPP and 3GPP2 and by Organizations Utilizing IETF Internet protocols.**

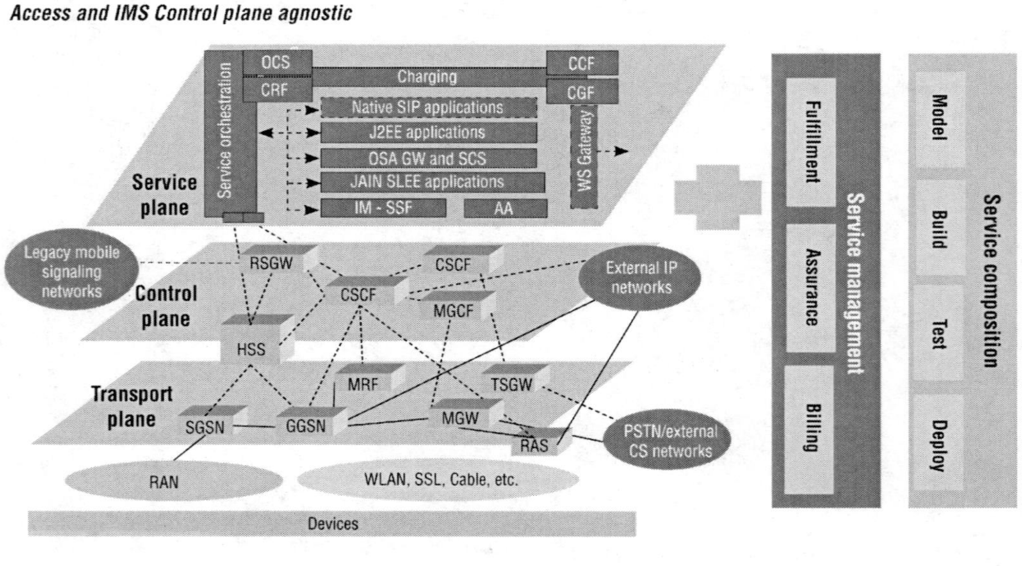

Source: IBM Corporation.

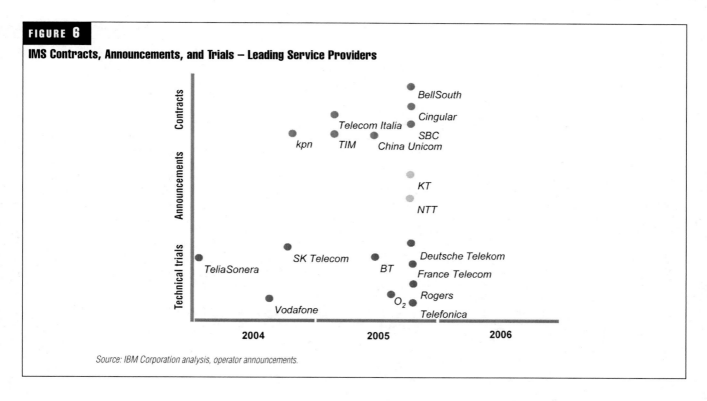

**FIGURE 6**

**IMS Contracts, Announcements, and Trials – Leading Service Providers**

Source: IBM Corporation analysis, operator announcements.

Each SP's NGN migration path will depend on its strategic and commercial objectives, unique network technology, maturity, and depreciation schedules. However, there are dependencies between the seven areas, which lead to some obvious staging of the steps. These principles apply to both fixed and mobile networks, with more areas of similarity than difference.

Generally, with a broadband core network in place, the service plane can begin evolving to a set of application services that use open interfaces that link to the control plane and modern software technology, such as service-oriented architecture (SOA) and application programming interfaces for service creation. SPs can also deploy SDPs before the control plane of the network has been migrated, enabling interaction with third-party services.

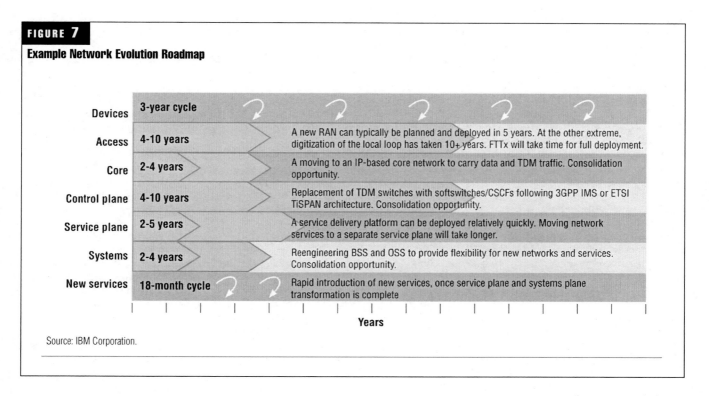

**FIGURE 7**

**Example Network Evolution Roadmap**

| | | |
|---|---|---|
| Devices | 3-year cycle | |
| Access | 4-10 years | A new RAN can typically be planned and deployed in 5 years. At the other extreme, digitization of the local loop has taken 10+ years. FTTx will take time for full deployment. |
| Core | 2-4 years | A moving to an IP-based core network to carry data and TDM traffic. Consolidation opportunity. |
| Control plane | 4-10 years | Replacement of TDM switches with softswitches/CSCFs following 3GPP IMS or ETSI TiSPAN architecture. Consolidation opportunity. |
| Service plane | 2-5 years | A service delivery platform can be deployed relatively quickly. Moving network services to a separate service plane will take longer. |
| Systems | 2-4 years | Reengineering BSS and OSS to provide flexibility for new networks and services. Consolidation opportunity. |
| New services | 18-month cycle | Rapid introduction of new services, once service plane and systems plane transformation is complete |

Years

Source: IBM Corporation.

Despite the differences in migration paths and the ongoing evolution of the network planes, several conclusions can be drawn regarding the transformation to an NGN. Of key importance is the transition to a single-core IP network. Once broadband is in place, this will lead to the relatively independent evolutions of the access network and devices. To design an on-demand, cost-effective network, services must be access-agnostic. The systems plane should evolve toward an NGN architecture that is network plan–agnostic, based on SOA principles, and supportive of NGOSS. Services must be separated from the network and able to support open interfaces with service enablers. The service plane integrates SOA principles and the IMS architecture to deliver new services rapidly. Open standards underpin the entire evolution, aiming to produce greater flexibility, economy, and investment returns.

### Questions of Strategy

Network transformation carries a range of strategic implications for SPs and the industry at large. First, IP and convergence are forcing players to rethink their roles in the expanding telecom value chain, as well as how they add value and with which core competencies. "Walled gardens" have given way to open standards and business models to meet customer demand for the services/content of their choice, producing an industry shift away from a single, one-size-fits-all model to multiple business models and industry ecosystems with which SPs must partner to create new value.

Separation of transport from service delivery and control, as well as the shift to software and services, are changing industry dynamics. Linkages between service delivery and control allow for service innovation. Separation of transport and control from network operations improves service focus. Open standards and a shared software environment support multiple third-party interfaces.

These trends have caused the emergence of a complex mix of new network, software, and IT players. In areas beyond their core competencies, SPs will increasingly use the skills of these players, focusing on their wider strategic supplier ecosystem. NGN equipment and SP consortia of telecom equipment manufacturers, network equipment providers, and system integrators are emerging as dominant players.

To support next-generation service offerings, telecom providers need to build next-generation operations support system (NGOSS) and BSS capabilities. Advanced service delivery requires enhanced network and service management. Next-generation customers are also demanding a seamless, superior customer experience to move to convergence, including single sign-on, one bill, and real-time charging, which usually requires upgrades to telecom business support, including billing systems. In the transformed NGN environment, the value proposition of OSS and BSS is elevated from business support to business enablement.

Finally, in a horizontal network environment, service providers must flatten their vertical, single product–based structures to enable horizontal service delivery to better serve customer needs and reduce IT, general operating, and network costs. For instance, Telecom Italia and Telecom Italia Mobile (TIM) have announced the complete integration of their fixed and mobile assets [29]. IMS, in particular, enables data center consolidation and the virtualization of network infrastructure, helping reduce costs and improve flexibility.

### Higher Revenues, Lower Costs: Proving the NGN Business Case

The rationale for moving to NGNs rests on capturing two major opportunities—increasing revenue and reducing costs (both capital and operating expenses). Combined, these opportunities offer very significant margin potential to companies that may have been struggling to increase their bottom line. An August 2005 article in Dataweek reports that IP networks may save carriers up to 38 percent in the long run [30].

Similarly, detailed IBM business case analysis shows that IMS has the potential to generate returns on the incremental costs, with related network renewal investments generating further savings.

IMS is expected to produce revenues from IMS–enabled services from the outset, while the magnitude of revenues and costs depends on the exact service portfolio mix. Based on conservative assumptions, scenarios show that for a mobile carrier with approximately 15 million customers, a positive cash flow on the full IMS investment was reached early in the second year. This estimated return requires a relatively low upfront investment of $27.5 million to get the services up and running. The analysis showed an ongoing incremental cash flow of approximately $63.50 per IMS–enabled user per year, net of the full incremental costs and revenues.

Wider NGN investments—including introduction of an optical IP core, migration of subscribers from home location register (HLR) to home subscriber server (HSS), softswitching and consolidation of sites, and server and data center consolidation—provide the operator with the potential to reduce costs further, by an estimated $5.20 per user per year.

The picture for fixed-line carriers is similar, depending on service portfolio mix and associated costs. Core network savings from introducing a VoIP core network are estimated to provide a 50 percent reduction in operating expenses from $1.40 to $0.72 per subscriber per year. The analysis also shows access network costs halved from $6.6 to $3.2 per line per year.

IMS can also yield important indirect benefits, each with its own financial rewards and further value to the business case for NGN transformation. The anticipated benefits include the following:

- More rapid service delivery, reduced time to market, and faster investment returns
- Greater flexibility in building solutions, either through internal resources or by combining and/or modifying applications to create differentiated services
- Shared functions across applications, leading to higher end-user loyalty
- Seamless views of customers and services and, therefore, new understandings of customer preferences, habits, and interests

- Simplified provisioning architecture, as well as streamlined maintenance, new deployments, and technical recruitment.

## Capitalizing on NGN Opportunities

The fixed-line business is under attack as never before. Mobile revenue growth is slowing as developed markets saturate and emerging market net additions come at low ARPU. NGNs offer service providers revolutionary new capabilities to drive dramatic improvements in revenues by enabling new, advanced services across multiple media, including rich voice applications, music, IPTV, VoD, location-based services, push-to-x, video messaging, and interactive games. NGNs also offer significant potential for cost cutting and capital efficiency, hence scope for margin improvements. Heightened competition is driving increased transformational capital expense, and the list of SPs embarking on the path to NGN services is growing rapidly.

By moving to open, IP–based network technologies supported by flexible, on-demand SDPs, SPs will become better equipped to respond to new opportunities quickly. The NGN environment enables SPs to link service creation and delivery to the device, thereby strengthening their abilities to customize and enhance the user experience, attract new subscribers, and reduce churn. Transition to a COTS software-based environment creates potential for significant cost savings. Clearly, the NGN demands new investment and expertise. However, its foundation in advanced IT and software-based technologies opens the way for SPs to collaborate with an extended ecosystem of partners and suppliers, enhancing their responsiveness to rapidly evolving user demand. IP transformation poses challenges and requires change, but ultimately we believe that the implementation of integrated, on-demand service delivery can place telecom companies at the heart of the real-time economy, enable innovation, and create new value.

## References

[1] Triple play refers to the delivery of voice, video and data services to a single user over an integrated IP platform. Quadruple play adds wireless (or mobility) to triple play.

[2] "Countering Fixed-Mobile Substitution Threats." Forrester Research, Inc., Michelle De Lussanet. September 7, 2004.

[3] McElgunn, Tim. "The Year Ahead: Cable Outlook 2005." Frost & Sullivan. January 19, 2005; updated by Mr. McElgunn 3Q2005, as yet unpublished.

[4] Network operating expenses include those for plant and network operations. IBM Institute for Business Value analysis and Thankappan, Jeevan M. "The Great Telecom Scale-Up." GTL. February 1, 2004. www.gtllimited.com/NewsReports_27.asp.

[5] MacKinzie, Michele, Marta Munoz Mendez-Villamil, Carrie Pansey, Eden Zoller and Pauline Trotter. "Ovum forecasts global wireless markets: 2004-2008." Ovum. December 2004 and Mendez-Villamil, Marta Munoz. "Ovum Mobile Forecasts." Ovum. October 2005.

[6] Informa Telecoms & Media. "Future Broadband Markets, Worldwide Forecasts by Technology and Network to 2010." August 2005. www.informatm.com/bsdv3.

[7] "IBM and Cirpack attract Magnet Networks to next generation voice infrastructure." IBM press release. February 9, 2005. www.ibm.com/news/ie/en/2005/02/ie_en_news_20050214_1.html.

[8] Multimedia Research Group, Inc. "IP TV Global Forecast – 2005 to 2009." September 2005. www.mrgco.com/TOC_Global_Forecast_0805.html.

[9] "Telecom '05: SBC Says IPTV Will Be Cool." Converge! October 25, 2005. www.convergedigest.com/Bandwidth/newnetworksarticle.asp?ID=16388&ctgy.

[10] "SBC CIO Confirms Project Lightspeed Timing, Milestones at Analyst Conference." SBC press release. November 3, 2005. www.sbc.com/gen/press-room?pid=4800&cdvn=news&newsarticleid=21874.

[11] "Ember Supplies ZigBee for SK Telecom 'digital smart home' Service." Converge! October 3, 2005. www.convergedigest.com/WiFi/wlanarticle.asp?ID=16149.

[12] "FASTWEB: revenues at 258.5 million Euro(+39%) in the third quarter 2005." FastWeb press release. November 11, 2005. http://company.fastweb.it/index.php?sid=93&idc=1028 and IBM Institute for Business Value analysis.

[13] "Sprint Nextel, Comcast, Time Warner Cable, Cox and Advance/Newhouse to form Landmark Cable and Wireless Joint Venture." Sprint press release. November 2, 2005. www2.sprint.com/mr/news_dtl.do?id=8961.

[14] "European Incumbent Telcos VoIP Road Map." Forrester Research, Inc., Lars Godell. October 2003.

[15] Muraskin, Ellen. "Report: Hosted IP Voice Will Storm the Enterprise." eWeek. September 29, 2004. www.eweek.com/article2/0,1759,1662501,00.asp.

[16] "European Incumbent Telcos VoIP Road Map." Forrester Research, Inc., Lars Godell. October 2003.

[17] Issaeva, Svetlana. "Fixed-Mobile Convergence: Creating value with successful business models," Pyramid Research. June 2005. www.pyramidresearch.com/documents/EXFIXMOBCONV.pdf

[18] "BT Fusion." www.btbusinessshop.com/page/btfusion_hub and "BT Fusion Fact Sheet." BT. June 14, 2005. www.downloads.bt.com/business/btfusion/bt_fusion_factsheet.pdf.

[19] Ward, Mark. "Google sets tongues wagging with Talk." BBC News. August 24, 2005. news.bbc.co.uk/1/hi/technology/4180182.stm.

[20] "Fixed CAPEX." Pyramid Research, Inc., January 2003. www.researchandmarkets.com/reportinfo.asp?report_id=45610.

[21] Walko, John. "Ericsson To Pay $2.1 Billion For Assets Of U.K.'s Marconi." The Networking Pipeline, courtesy of EE Times. October 25, 2005. nwc.networkingpipeline.com/172900146; Walko, John. "BT plans pure IP-based network by 2009." CommsDesign. June 9, 2004. www.commsdesign.com/showArticle.jhtml?articleID=21600074 and IBM Institute for Business Value analysis.

[22] Le Maistre, Ray. "KPN Lays Out IP Migration Plan." Light Reading. March 7, 2005. www.lightreading.com/document.asp?doc_id=69419&page_number=1&site= and IBM Institute for Business Value analysis.

[23] Springham, Justin. "Sprint Chases SDP Success." Unstrung. June 1, 2005. www.unstrung.com/document.asp?doc_id=74850.

[24] Doyle, Lee, Shiv Bakhshi, Elizabeth Rainge, Tom Valovic. "IDC Multiclient Study, IMS Status Report: Demand and Supply of a Key Enabling Technology." IDC. July 2005.

[25] "NTT unveils FTTH Kit." Light Reading. November 8, 2005, www.lightreading.com/document.asp?doc_id=83803&WT.svl=wire1_.

[26] "MCI'S Private IP Service Now Delivers Even More Reach, Access Choices and Features." VoIP NEWS. October 28, 2005. www.voipnews.com/art/10k.html.

[27] Geitner, Thomas. "Vodafone Group Technology." September 19, 2005. www.vodafone.com/assets/files/en/Vodafone_Group_Technology_Thomas_Geitner.pdf.

[28] "T-Mobile implemented an integral surveillance system of telecommunication network in cooperation with IBM." IBM press release. February 18, 2004.

[29] "TI, TIM Complete Merger." Light Reading. June 20, 2005. www.lightreading.com/document.asp?doc_id=76128 and "Re-structuring of the Telecom Italia Group." Telecom Italia press release. October 5, 2005. www.telecomitalia.it/cgi-bin/tiportale/TIPortale/ep/contentView.do?channelId=-9793&LANG=EN&contentId=19285&programId=9596&programPage=%2Fep%2FTImedia%2FTISearch.jsp&tabId=3&pageTypeId=-8663&contentType=EDITORIAL.

[30] "Calculating ROI for centralized IP-network analysis in large enterprise systems." DATAWEEK. August 27, 2005. www.dataweek.co.za/news.asp?pklNewsID=11794&pklIssueID=348&pklCategoryID=.

# A Layered Approach to Evaluating Processor and System Performance for VoP Applications

## Markus Levy
*President, Embedded Microprocessor*
Benchmark Consortium (EEMBC)

## Danny Wilson
*Principal Solutions Architect, DSP Products Division*
LSI Logic

## Abstract

Voice over packet (VoP) at the customer premises is delivered in a variety of ways. An analog telephone adapter (ATA) allows a plain old telephone to be connected directly to an Internet connection. Internet protocol (IP) phones connect via an Ethernet cable or Wi-Fi interface to an Ethernet router. A personal computer can become a VoP portal by using a broadband Internet connection. An IP–based private branch exchange (PBX) allows a business to interoperate with the public switched telephone network (PSTN) by incorporating an IP gateway with PBX functionality.

These systems use a combination of digital system processor (DSP), reduced instruction set computing (RISC), and complex instruction set computing (CISC) processors to share the processing load of a VoP system. VoP systems vary widely in their complexity but share many common components, thereby making it possible to develop a common set of benchmarks to allow engineers to analyze and appropriately design these systems. Using case studies, this paper presents a layered approach to evaluating processor and system performance that could be used for both ends of the VoP spectrum.

## Introduction

VoP is a term encompassing many packet voice transport technologies, including voice over Internet protocol (VoIP), digital subscriber line (DSL), cable, asynchronous transfer mode (ATM), and frame relay. Regardless of the transport method, the underlying technology, particularly the DSP–related functions, are the same. By evaluating the processors and systems using this base technology, you can use that effort in many VoP markets.

VoP telephony is transforming communications at enterprises and customer premises worldwide, as it replaces legacy circuit-switched equipment. Emerging applications for VoP are embedded in products ranging from high-density carrier gateways to high-volume VoP handsets. Using VoP technology, enterprises can reduce overall telecommunications spending, address workforce demands for flexible hours and decentralized virtual offices, and provide a platform for new telephony services. Home offices benefit from value-added features such as video calling. Consumers benefit from the convergence of voice and data with services such as click to dial/call logging, which integrates voice with contact lists on home PCs.

### Voice Quality: What Matters?

Voice quality in a packet-based system is primarily determined by network delay and variation in delay (jitter), echo control, speech codec quality, and packet loss.

### Delay

End-to-end delay, or latency, is the time between generation of sound at one end and reception at the other end. ITU G.114 recommends that a total one-way delay of 0 to 150 ms is acceptable for most applications. As delay increases above this limit, echoes become more noticeable and the dynamics of conversational speech become disrupted by simultaneous starts and awkward silences. Managed packet networks use the most direct path through the network to minimize delays associated with processing and buffering.

As shown in *Table 1*, delays in a packet-switched network fall into three categories: propagation delay, processing delay, and buffering delay. (1) Propagation delay for a fiber-optic trunk is about 5 Is per kilometer. Network failures can increase propagation delay by forcing a less direct path. (2) Processing delay includes the time associated with voice codecs, packetization, echo cancellation, noise reduction, and packet loss concealment. Processing delay can be compounded by transcoding—multiple conversions from packet

**TABLE 1**

**Type and Causes of One-Way Delay**

| Delay Source | Range (ms) | Comments |
|---|---|---|
| Propagation | 1–100 | Longest terrestrial delay through fiber: ~20,000 km * ~5_s/km = ~100 ms |
| Processing | | |
| Codec | 20–100 | Encoding and packetization for a single hop with one frame per packet |
| Other DSP | 0–30 | PLC/noise suppression/silence suppression/echo cancellation |
| Buffering | 1–20 | Variation in network delay determines delay caused by jitter buffers |

to time division multiplex (TDM) and back. (3) Buffering delay comes from queuing at routers and jitter buffers at the decoder. Voice must be played back at a constant rate, but individual packets move through the network at varying rates. A jitter buffer removes the network variation and presents packets at a constant rate. Jitter buffer delay is dependent on the variation in delay across the network. If packets are delayed beyond the length of the jitter buffer, they are not available for processing and are considered lost.

*Echo*
Echo effects result from a coupling of the transmit path with the receive path, causing outgoing speech to be sent back to the talker. Echo tolerance is a function of both the amplitude

and delay of the echo. Amplitude of an echo is measured relative to the outgoing speech signal, and called echo path loss. For a constant echo path loss, a longer delay will cause the echo to seem louder and more annoying (see *Figure 1*).

Although a fully digital network has no network echo paths, it is still subject to acoustic coupling between the receiver and the microphone and crosstalk in analog circuits at end devices. When the packet network has to interconnect to a circuit-switched network, echoes are caused by reflections from two-to-four-wire hybrids in the analog lines.

DSPs running echo canceller algorithms can reduce the echo by 26 to 30 dB. Echo cancellers are adaptive digital filters

**FIGURE 1**

**Talker Echo Tolerance as a Function of Delay**

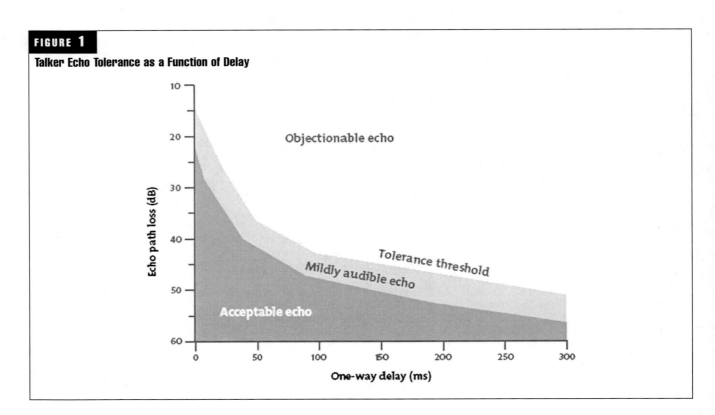

that look for delayed signals on the return path that are highly correlated with a signal on the incoming path. Residual echo is further removed using a non-linear process.

## Codec

Popular codecs in packet networks include G.711, G.726, G.729, G.729A/B, G.723.1, and Global System for Mobile Communications (GSM)–enhanced full rate (EFR) (see *Table 2*). Bit rates vary from 64kbps of G.711 to the 5.3kpbs for low-rate G.723.1. Total delay is the delay for an ordinary implementation of the codec on a DSP. Total delay is twice the frame size plus the "look-ahead," assuming one frame per packet. The most commonly requested codec for non-wireless VoP is G.729A, since it provides a low-complexity codec at 8 kbps with only 25 ms total delay.

The bandwidth required for a voice call can be significantly reduced through voice activity detection (VAD) techniques. VAD removes silence, which accounts for as much as 40 percent of the voice information that is transmitted. VAD processing at the encoder is combined with comfort noise generation (CNG) at the decoder. G.729A/B is a combination of G.729A with an integrated G.729B VAD/CNG algorithm for silence compression.

## Packet Loss

Compressed voice information is broken into small samples or packets for transmission through the network. Travel time varies with individual packets and, sometimes, packets fail to arrive at the destination. These lost packets can cause gaps in voice communications, which can cause clicks or unintelligible speech at the decoder.

Quality of service (QoS) protocol and call admission control are two methods of minimizing packet loss in the network. QoS control expedites the transmission of voice packets in a combined voice/data network. Call admission control acts similar to a "fast busy" by not allowing additional calls to be set up in a congested network.

Packet loss concealment (PLC) algorithms are built into the G.729, G.729A, and G.723.1 codecs and may be externally added to other codecs. These types of algorithms are effective at masking 40 to 60 ms of missing speech, but gaps longer than 80 ms are typically muted. Quality drops slowly with increasing amounts of packet loss.

Adaptive jitter buffer control can be used to adjust jitter buffer lengths to match effects of network congestion. This minimizes the number of lost packets when the network is congested but avoids adding unnecessary delays when congestion eases.

## TABLE 2

### Typical Characteristics of Common VoP Codecs

| Codec | Type | Bit rate (kbps) | Frame size (ms) | Total delay (ms) | Comments |
|---|---|---|---|---|---|
| G.711 | Pulse code modulation (PCM) | 64 | 10 (varies) | 10 | High-quality, PSTN–type voice |
| G.726 | Adaptive differential pulse code modulation (ADPCM) | 32 | 10 (varies) | 10 | Used for multiplexing on 64 kbps channels |
| G.729 | Conjugate structure algebraic code excited linear prediction (CS–ACELP) | 8 | 10 | 25 | Good delay characteristics; acceptable voice quality |
| G.729A/B | Conjugate structure algebraic code excited linear prediction (CS–ACELP) | 8 | 10 | 25 | Reduced complexity version of G.729 with same quality; includes VAD/CNG |
| G.723.1 | Multi-pulse maximum likelihood quantization (MP–MLQ) | 6.3/5.3 | 30 | 67.5 | Total delay compromises VoP delay budget; used in legacy toll-bypass clients |
| GSM_EFR | Algebraic code excited linear prediction (ACELP) | 12.2 | 20 | 40 | Wireless codec |

## VoP Benchmarking

Implementations of VoP systems may include multiple processors or processor types. The computationally intensive voice processing algorithms represent a significant layer of the benchmarking strategy for VoP, and these algorithms are often implemented on a DSP for greater efficiency. Although general-purpose processors can handle these algorithms, the restriction will be on the number of channels that can be simultaneously supported. Jitter buffer processing and session initiation protocol (SIP), for call setup/teardown, can be implemented on a DSP–, RISC–, or CISC–type processor. Packet processing is typically assigned to a RISC or CISC processor because of the large data memory spaces needed.

The first layer of VoP benchmarks should be at the algorithm level of the voice processor. The popular G.729A/B voice codec and the G.168 line echo canceller are part of the *EDN* Embedded Microprocessor Benchmark Consortium's (EEMBC's) strategy for VoP, since these algorithms account for more than 50 percent of the cycles in a typical voice processor. Optimized versions of these algorithms require specialized expertise and are often developed by third-party vendors. Benchmarking at the algorithm level would allow comparison of stand-alone voice processors and comparison of third-party applications on a single voice processor. Bit exact performance is expected from G.729A/B, but an echo simulation will be needed to verify G.168 line echo canceller compliance.

The next layer of VoP benchmarking should be isolated to customer-premises equipment (CPE). By simulating wide-area network (WAN) effects and limiting the scope of benchmarking to CPE, a common configuration can be selected to support VoP benchmarking. The configuration should have digitized voice input/output (I/O) at one end and a WAN I/O at the other end. Benchmarking should include quality measurements pertaining to delay, speech compression, and packet loss. The common VoP configuration should be connected to a network simulator that models WAN effects of jitter, delay, and packet loss. A SIP–based call simulator can be used to support setup and teardown of calls. A performance measurement should be made to determine the number of active voice channels that can be supported. This can be accomplished by incrementally adding channels and noticing the point where packets start to get dropped as the processor cannot keep up with the delivery rate. This common VoP configuration would allow benchmarking across various vendors and various types of VoP products.

One of the challenges is that each vendor would be required to implement the common VoP configuration with its own product solutions. But the benefits of this type of benchmark are that it resembles a real system and can be implemented with hardware and software components that vendors are actually trying to sell.

### ZSP Case Study

ZSP is a division of LSI Logic that provides a fully synthesizable family of software programmable DSP cores. ZSP cores are optimized for low power consumption, which is ideal for portable applications such as wireless VoP handsets.

LSI's HomeBase VoIP/asynchronous digital subscriber line (ADSL) gateway design is based on the ZSP family of cores. HomeBase allows manufacturers to bring integrated access devices (IADs) to market quickly and with minimal effort. The ZSP VoP software suite includes all the key algorithms required for voice and telephony applications, including the following:

- G.711, G.723.1, G.726, G.728, G.729x, and GSM
- G.168 line echo canceller
- Dual-tone multifrequency (DTMF) detector and generator
- Caller ID generator
- Voice activity detection/comfort noise generation
- Telephony functions such as call progress, fax detection, and automatic gain control (AGC)
- Jitter buffer and real-time transport protocol (RTP) encapsulation
- T.38 fax relay

The effect of WAN impairments on the HomeBase VoIP/ADSL gateway was evaluated using mean opinion scores (MOS) via the perceptual analysis measurement sys-

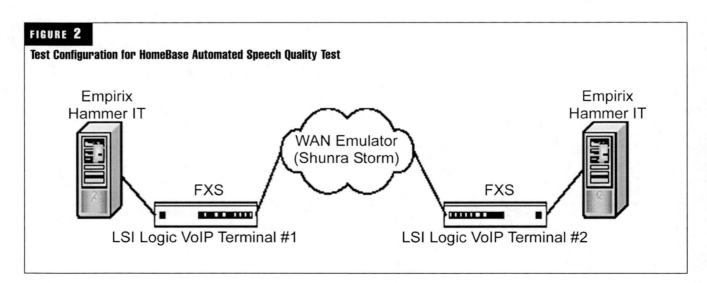

**FIGURE 2**

**Test Configuration for HomeBase Automated Speech Quality Test**

Empirix Hammer IT

WAN Emulator (Shunra Storm)

Empirix Hammer IT

FXS

FXS

LSI Logic VoIP Terminal #1

LSI Logic VoIP Terminal #2

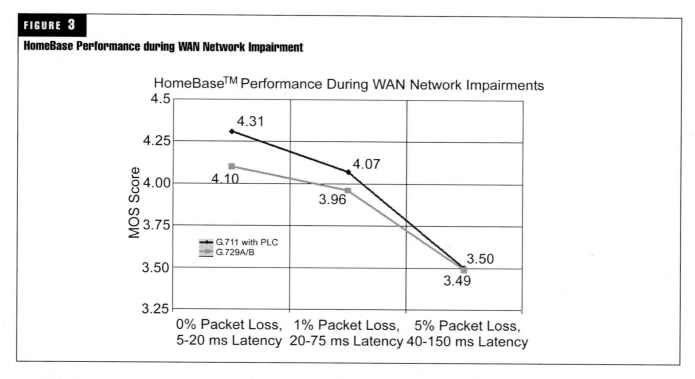

**FIGURE 3**

**HomeBase Performance during WAN Network Impairment**

HomeBase™ Performance During WAN Network Impairments

tem (PAMS). Automated speech quality tests were executed using the test setup shown in *Figure 2*, and the results of this performance testing is shown in *Figure 3*.

Equivalent MOS scores above 4.00 fall in the toll-quality range and are generally considered "business-quality voice." The other test points fall in the "communications-quality voice" range, which indicates that they are acceptable for normal conversations. This information can be used to predict voice quality at a customer location.

## Conclusion

The benchmarking methods described in this paper provide a common configuration for testing a wide variety of VoP systems, including those that use a combination of DSP,

RISC, and CISC processors to share the processing load. This layered approach to evaluating processor and system performance will allow engineers to analyze and appropriately design VoP systems with varying complexity and channel capacity.

## References

Nortel Networks, "Voice Over Packet – An assessment of voice performance on packet networks."

Chan, T.Y., Greenstreet, D. "Building Residential VoIP Gateways: A Tutorial Part One: A Systems Level Overview."

Yancey, G., Devlin-Allen, K. "Building Residential VoIP Gateways: A Tutorial Part Two: VoIP Telephony Interfaces."

Nortel Networks, "Next-generation broadband voice services: What broadband consumers want and are willing to pay for!"

# Remote CPE Management System Can Ease Complexity for Multi-Play Activation and Customer Service

## James Morehead

*Vice President of Product Management and Marketing*
SupportSoft

As global DSL service providers roll out voice over Internet protocol (VoIP) and IP television (IPTV) services to complement high-speed Internet access, residential gateways and other access devices are becoming critical points of failure for service availability. Whether the device is a router, set-top box (STB), VoIP adapter, wireless fidelity (Wi-Fi) access point, or a combination thereof, managing customer-premise equipment (CPE) can no longer be left only to the subscriber.

Service providers are responding by investing in centralized management solutions for all services and vendor equipment, as well as extensible methods for remotely managing CPE service parameters, operational performance, and firmware. Key success metrics will be customer ease of use, accelerated time to market, fewer truck rolls, problem avoidance, and simplified, less costly customer support.

CPE management platforms have traditionally been provided as tightly integrated, proprietary solutions bundled with specific CPE devices. As service providers move to adopt a multivendor CPE strategy, demand is increasing for a centralized vendor-agnostic CPE management system capable of integrating, automating, and scaling the delivery of advanced IP services.

## Market Environment

Remote CPE management plays a fundamental role in the cost-effective scaling of advanced IP services in a highly competitive broadband, triple-play, or multi-play environment. Service providers are specifically trying to address the following pain points:

- Complexity and expense of deploying and supporting advanced services
- Variability in the quality and reliability of advanced services
- Time to market for new services

## Managing the Complexity and Cost of Deployment

To differentiate service offerings and increase market share, many service providers have invested in the delivery of advanced, IP–based services that by their very nature have increased provisioning and support requirements. Profitably deploying these services becomes increasingly challenging for the following reasons:

- *Increased scale*: As consumers adopt new services, an increasing number of endpoints require management by service providers. While most traditional operations support system (OSS) management applications have dealt with managing devices in the tens of thousands, today's broadband revolution means system managers must cope with millions of endpoints. This device proliferation affects all levels of a service provider's business, from bandwidth availability to IP address space.

- *Added features*: Technology evolves, increasing the capabilities and functionality of current and future endpoints. A service provider's ability to determine and control what features are installed, enabled, or disabled is key to product differentiation, incremental service revenue, and cost control.

- *Complex provisioning*: As services become more distributed, CPE devices have increasingly complex feature configuration. Telephony service attributes such as dial tone, ring cadence, digit maps, number assignment, and emergency dial plan have migrated from a central switch to the endpoint. For example, managing a complex configuration such as an emergency dial plan may affect a service provider's timetable for launching new services.

- *Increased variety*: As more businesses move to IP–based service delivery channels, there is an increase in the types of endpoints receiving these services, including those from service providers and third-party con-

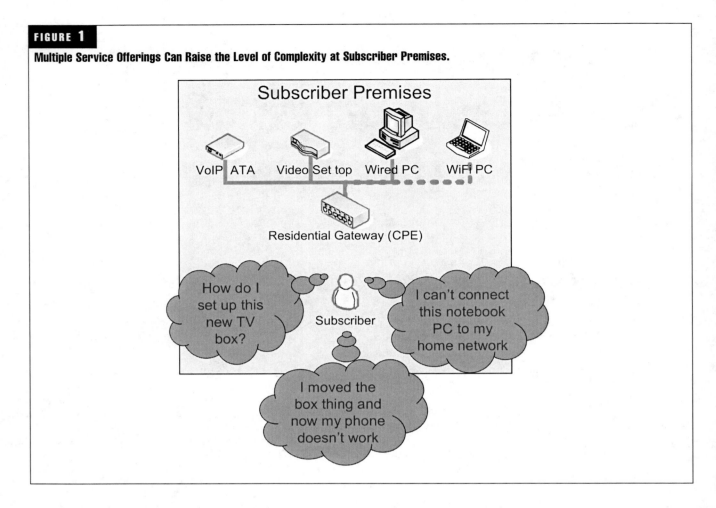

**FIGURE 1**

**Multiple Service Offerings Can Raise the Level of Complexity at Subscriber Premises.**

sumer devices. Device types may range from analog terminal adapters, STBs, and home gateways to IP–based home appliances, security, and medical monitoring systems.

- *More vendors:* As endpoints proliferate, there are an increasing number of suppliers in the equipment vendor value chain. More vendors equate to more vendor competition and reduced CPE prices. With increased competition, the choice of a low-cost CPE vendor may change from purchasing cycle to purchasing cycle.

### Customer Service: The Cost of Excellence

The delivery of increasingly complex IP services is increasing the number of customer support calls to service providers. Currently, when a customer contacts a call center, customer service representatives (CSRs) may not be able to perform even preliminary diagnostics such as isolating the problem to the subscriber's home or the service provider's network. Typically, CSRs have a limited number of tools to deal with the problem—often the only option is to forward complex problems to more experienced technical resources or dispatch a truck to the customer's household. Both options increase the cost of the service and time to resolution and may not be the right course of action. Worse, rolling a truck when the problem is subscriber confusion or network impairment is a waste of a service provider's resources.

Any service call is expensive, not only because of the actual cost involved with sending a field technician, but also because of the indirect cost of a customer waiting for the technician. If a technician is unable to fix the problem (as might be the case if the problem is in the network), then the customer is even more dissatisfied because of the time wasted waiting for the field technician.

Support is made more complex, particularly in increasingly multi-play environments, partly because installing or repairing a service can affect other services. For example, in a house with more than one service, the efforts by a broadband field technician can degrade IPTV, and vice versa.

### Managing the Quality and Reliability of Services

There is strong consumer demand for next-generation, triple-play services. Service providers are already seizing the opportunity to not only realize the financial benefits, but also to displace competition. Being first comes with opportunities and obstacles. Seizing a market opportunity often means moving quickly on product delivery, whereas rushing often means that early products and services may not be implemented with the desired level of product usability and stability. For example, VoIP and IPTV services, while well-designed in principle, are still proving themselves in the field.

Subscribers moving from traditional services to IP–based services may experience a less-than-optimal product experience. For example, there may be differences in how a subscriber uses the service (i.e., channel surfing) or how the service performs (i.e., a phone crash). The consumer's exposure to steep learning curves and instability in the first wave of advanced service deployments may likely have a negative impact on customer satisfaction and product adoption rates.

As service providers deploy new equipment and services, the ability to address a majority of these quality issues via remote CPE management provides the shortest path to increased revenue and profitability. Managing quality and reliability, however, becomes challenging for the following reasons:

- *Product ease of use*: Because of market pressure to deploy services early, there is generally an inadequate level of equipment usability across services. The fact that many equipment and software vendors contribute to an end-to-end solution means that the result is typically non-standardized. From a service provider's perspective, this can impose additional operational overhead because CSRs, installers, and technicians must be trained to deal with multiple user interfaces for a variety of CPE devices. From the subscribers' perspective, it may mean that services need to be managed using different and somewhat confusing interfaces and tools. These facts contribute to the perception of a steep product learning curve, resulting in too much complexity, customer confusion, and a poor customer support experience.

- *Device stability*: Early-to-market pressures also promote high defect counts in initial firmware releases. In addition, some defects are problematic only under certain conditions, such as when running a specific service or when connected to a particular vendor's equipment. Many devices perform adequately on a success-path basis, but when placed in exception scenarios or burdened through usage, they may exhibit problems that result in poor performance, unpredictable behavior, device crashes, or device failures. Providers must take steps to ensure that this reality does not detract from a well-earned reputation in service reliability.

- *Minimal automation*: The importance of CPE in a service provider's business model has increased dramatically in recent years. For the most part, the back-office systems required to efficiently manage a profitable CPE–driven business have yet to emerge. Without such a system, service providers do not have automation tools for capturing domain knowledge and orchestrating endpoint business processes. A significant area of concern is the customer service desk, which greatly benefits from an ability to automate problem identification, analysis, and resolution. In addition, deploying next-generation services, beyond what we can anticipate today, will depend on the service provider's ability to rapidly configure and automatically deploy that service.

## *Market Requirements for Service Excellence*

To facilitate customer support, CPE management solutions need to be accessible and manageable by subscribers (via

self-service portals), CSRs (in integrated dashboards), and field technicians (via workforce automation tools, dispatchers, and laptops). Not only does data have to be presented appropriately for these user roles, but these users also need, in a protected fashion, to have access to essential functionality that facilitates problem diagnosis and resolution.

In the future, we can expect an increasingly diverse range of broadband devices and a higher number of devices connected to home gateways. We can also expect that many of the services provided by these devices will be deployed in homes without a personal computer (PC), making a wide-area network (WAN) management system imperative.

Some lessons learned from broadband deployments of high-speed data (HSD) services indicate that they can quickly evolve beyond the original design. As a result, the business must quickly adapt to control costs and capture new revenue opportunities before the competition. The convergence of data services into multi-play bundles—i.e., voice, video, data, and wireless services—realistically means more managed devices, or endpoints, for the service provider.

Strategically, a service provider's ability to manage each of these endpoints is fundamental to increasing average revenue per user (ARPU), while avoiding lost revenue through device downtime. As services converge, consumer choice increases. With new heterogeneous environments, consumers may choose endpoint solutions and devices from a wide range of competing vendors. From this perspective, several fundamental device management requirements emerge. Service providers must have the following:

- Real-time remote diagnostics adaptable to user level
- Data-driven, automated best-practices problem identification and resolution
- Open interfaces into CPE and the business process environment
- Commitment to industry standards and interoperability

### *Real-Time Diagnostics Adaptable to User Level*

It is important to use a CPE's remote management interface to execute real-time diagnostics on the subscriber's home network. Many CPE manufacturers support utilities such as ping and local log alarms on their devices. They can be used to provide support staff with additional telemetry data that assists in reducing call times and truck rolls when subscriber equipment requires only configuration changes or customer education. A higher level of sophistication includes flexible data collection, thresholding, and decision tools for analyzing data and resolving problems. CPE diagnostic level of detail should vary based on the user's skill level. Service providers should use expert-level domain experience to help non-technical people rapidly solve high-occurrence technical problems. This may typically include the CSR, installer, or even the subscriber.

By measuring the current state of all–IP–based services in the house against a set of best-practices thresholds (called whole house verification), CSRs can view a complete picture of the customer's environment to aid in efficient diagnosis and resolution. CSRs can be empowered to perform these tests remotely, and technicians can perform them on-site or before arriving at the household. Either way, the ideal remote CPE management solution would provide the CSR or technician with an attack plan, based on best practices for each situation, guiding them through the repair or installation process. Once the repair or installation is complete, and before hanging up the phone, the remote CPE management solution assures service provider personnel and the subscriber that all services are working correctly.

By isolating customer problems and receiving the relevant information, CSRs can determine what they can repair and what actually requires a truck roll. CSRs can quickly and easily see if the problem is at the house or in the network, if there is any course of action that can be taken to resolve it, and what the proper escalation path should be. This can prevent unnecessary—and expensive—truck rolls and the associated subscriber appointments, making both the service provider and subscriber happier.

A service provider should be evaluating CSRs and asking: Are quality and reliability improving? Is the number of truck rolls after a call being reduced? Is the mean time between CSR contact and resolution decreasing? Are there fewer installation or repair-related calls? What are the comparative success rates of first- and second-time tests, and how do trends change over time? Do I need to have more CSR training? Can the workflow be optimized further?

### Data-Driven, Automated Best-Practice Resolutions

A solution-centric approach to best practices starts with measurement data, adds important configuration data such as firmware version, helps the user determine appropriate next steps to resolution, and then verifies that all services meet operational standards, all while keeping operations management informed of performance over time. It recognizes that service providers have an understanding of potential problems at customer homes and within the network.

It also recognizes what data is needed for problem analysis and the best practices that lead to a resolution. Furthermore, CSRs, technicians, and customers approach problems differently—often inefficiently and sometimes erroneously. This is mostly because of the ignorance of the correct process or answer, mistakes, hurried work, or inappropriate shortcuts. The benefits of a solution-centric approach to best practices include the following:

- Automated tests run consistently every time. Automation minimizes reliance on user input or decisions, while maximizing diagnostic data collection. It also eliminates the ability to take dangerous shortcuts. Starting from the problem avoids unneeded steps.

- Tests and resolution are ordered to maximize efficiency. The first tests run in a few seconds and use only a small amount of resources, while subsequent tests require more resources and time. This minimizes impact on the network and customer. At the same time, overcoming the need for customer input prevents inaccurate results.

- Extensible data collection capabilities built using open, standards-based technology such as Web services and documented application programming interfaces (APIs) allow for flexibility and seamless integration.

Interaction and data sharing with external resources, including network management or billing systems, are straightforward via the extensible markup language (XML). In addition, a bi-directional framework can push configuration and firmware information directly to the CPE while pushing support content to customers, CSRs, and technicians.

### CPE Remote Management by Tier-1 CSRs

A key method for improving time to resolution, as well as lower customer-care costs, is to leverage best practices and provide tier-1 CSRs with the ability to remotely manage the CPE directly. Typically, because of lack of automation and repeatable best practices, tier-1 CSRs find themselves having to escalate these types of problems to more expensive, tier-2 technical CSRs. The ideal remote CPE management solution would combine the necessary data with appropriate best practices and tools to streamline the resolution process and empower front-line CSRs to fix problems that typically were escalated without threatening network integrity.

### Operational Snapshots

On all customer-facing events, including contact with a call center, installation, or service initiation, a solution should capture an operational snapshot of the equipment's current known working state. This information facilitates diagnosis and resolution by providing a baseline for comparison during future customer interactions. It can highlight differences between current conditions and the last-known good condition. A golden record allows pre-emptive service even if a customer calls about an unrelated issue. For example, if a subscriber has a billing question, the CSR may notice that the customer also has a signal quality issue with the CPE. The CSR can ask the subscriber if new equipment such as a wireless router has been installed and solve the problem from that point. The solution should capture and store all operational snapshots, building a valuable repository of customer and network conditions over time, which can be used for trend analysis and other operational and marketing insights.

### Detailed Operational Reporting

The CPE management solution gathers and reports on valuable data from both a service performance and personnel performance perspective. As mentioned above, operational snapshots provide a way to look at problems that may be the result of the network architecture or standard operating procedures. By analyzing this information, issues can be addressed—for example, thresholds used to pass or fail a service can be further refined.

From the perspective of personnel performance, it is important to know if CSRs or field technicians are doing their jobs correctly, not only at the time of each customer interaction, but also over time. Key indicators can be based on how often they pass the first, second, or third time they test a service. That data can also be looked at periodically to make sure that job performance is heading in the right direction or if further training or process changes are needed.

### Open and Standard Northbound and Southbound Interface

Delivering advanced services increases the requirement for back-office integration points to enable dynamically derived information to flow between OSS and business support system (BSS) layers to service activation and operation.

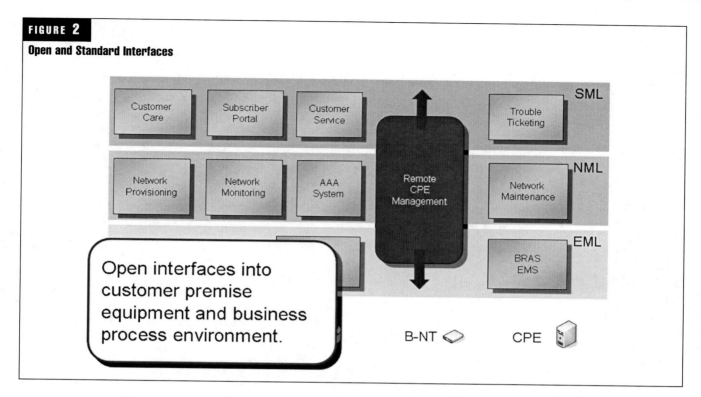

**FIGURE 2**

**Open and Standard Interfaces**

For this reason, as *Figure 2* indicates, a remote CPE management solution must provide flexible, open interfaces to business process and customer-premise environments.

Increasing system complexity is also driving the need for open APIs in the following ways:

Multiservice environments link devices in service-chain interdependencies. New capabilities need to be activated without disrupting existing services—service data must exist in easily manageable, parameterized formats.

Subscribers are expected to use a secure Web portal to configure their devices and services. This means authoritative device configuration will move from local devices to the service provider's remote management software.

As multiple audiences emerge within a service provider's network, support systems will require the ability to see and configure the devices associated with unique groups. In addition, support systems will need to facilitate automation of trouble detection and resolution flows across these unique groups.

*Commitment to Industry Standards and Interoperability*
The continual improvement and support of industry standardization can reduce operating costs and simplify the customer experience. Standards are also important for multivendor interoperability. To manage the diversity and number of CPE devices on the market, it is critical for a solution to undergo rigorous interoperability testing with CPE

vendors. It is also important to participate in industry standard testing (i.e., DSL Forum), which further ensures interoperability.

A dedicated independent software vendor (ISV) is a neutral, noncompetitive partner with CPE and traditional OSS vendors. This unique position enables them to work quickly in developing, testing, and validating standard-based and proprietary integrations.

*Conclusion*

Service providers are rapidly moving advanced services off the drawing board and into market realities. In comparison to existing HSD services, these newer services contain more components, which lead to an increase in the number of CPE types, features, and configuration options. When these variables are compounded with the number of devices that require per-endpoint management, the permutations for provisioning and support are staggering.

The ability to efficiently manage complexity has a direct impact on the adoption rate of new services and customer satisfaction. To maintain high customer satisfaction, service providers must provide a consistent level of product quality and reliability. Where possible, they should build on existing reliability strength by developing the capabilities to capture and automate remote CPE management best practices. These best practices will improve the usability, quality, and reliability of new service rollouts, thus dramatically reducing the cost of service fulfillment and assurance activities.

# The IP Revolution Is Under Way

## Jack Waters
*Chief Technology Officer*
Level 3 Communications, LLC

### A Time of Dramatic, Industry-Wide Change

The telecom industry is in the midst of breathtaking change. A fundamental technological shift from electrical to optical technology, fixed to mobile communications, and circuit-switched to voice over Internet protocol (VoIP) networks underlies these changes. For nearly 100 years, voice was telecommunications' "killer app." But today, voice is just one important application among many. Evidence of this change can be seen every day as students use Google to do research for school papers, as teenagers send text messages at parties, and as travelers watch sitcoms and respond to e-mails using handheld digital devices. People want to communicate, work, and be entertained on their terms—and that means anytime, anywhere, and in any format.

In the late 1990s, a warranted excitement about these new technologies drove massive investment into the telecom industry. Unfortunately, these investments proved to be overly optimistic. After an unprecedented surge of growth, telecom businesses began filing for bankruptcy, merging, and suffering from declining revenue. We estimate that from May 2000 to June 2003, 162 telecommunications companies in North America filed for bankruptcy or exited operations. That is an average of 52 bankruptcies per year—a huge increase over the period between 1985 and 2000, which averaged a mere 2.5 bankruptcies per year.

Contributing to this gloomy picture, U.S. long-distance revenue is expected to decline at an average of 9.2 percent per year between 2004 and 2009.[1] Forecasters estimate that revenue for consumer long distance in the United States will drop from more than $16 billion in 2004 to less than $10 billion in 2009. Even basic wireline local revenue is expected to decline. IDC predicts an average drop of 5.7 percent as revenue slips from over $157 billion in 2004 to an estimated $117 billion by 2009.[2]

The players in the telecom industry as we know it are undergoing significant transformations. Recent mergers between SBC and AT&T, Verizon and MCI, and Nextel and Sprint all speak to fundamental changes in the telecom landscape. Other consolidations reflect the change that is under way at all levels of the industry. We have acquired assets from Genuity, Allegiance, ICG, KMC, and Sprint, and recently announced a deal to acquire WilTel. Broadwing and Focal have merged. And a number of CLEC restructurings provide further illustration.

These tumultuous changes may seem to indicate an industry in trouble. But underlying all of this news are the same exciting trends that drove the optimistic investment in the first place. Those innovations and new applications are real and are driving tremendous opportunity and potential for those companies that remain.

### Soaring Demand Is Fueled by New Applications

Even as bankruptcy and consolidation continue industry-wide, overall traffic growth is exploding. Across the board, new applications are driving demand. As an example, wireless adoption in the United States grew to 177 million subscribers in 2004, up from 139 million subscribers in 2002.[3]

With the Internet, a revolutionary new set of applications is driving extraordinary growth. As an example, RSS—a family of extensible markup language (XML) file formats for Web syndication used by news Web sites and Web logs—is among those applications that are making a difference. Consider a scenario in which a major news service updates its site once per hour. If each update results in a nominal five kilobytes of RSS-generated data, and we assume that 500,000 simultaneous users poll at the top of the hour, this would require a total data transfer of more than 2.5 gigabytes per hour or 60 gigabytes of data every day. Multiply this across the thousands of Web sites with similar applications for an idea of how new and innovative applications are driving traffic growth.

Podcasting is another new technology that's contributing to the increase in demand. The technology, which involves transmitting an audio file from a podcaster to a media device, allows users to access thousands more media files than could otherwise fit on a hard drive. According to Apple's June 30, 2005 press release, in the first two days after the release of the Apple iTunes® 4.9 application program, iTunes customers subscribed to more than 1 million podcasts from the new iTunes Podcast Directory. And as media giants and advertisers begin to explore the possibilities of video-on-demand (VoD) technology, this additional large new market — with its large video file sizes — will create further demand.

Peer-to-peer (P2P) file sharing is another factor: it's the single largest consumer of data on Internet service providers' (ISPs') networks. P2P traffic significantly outweighs Web traffic and is continuing to grow. According to CacheLogic Research, P2P traffic represented 60 percent of Internet traffic at the end of 2004, outstripping every other communication protocol—and it's still growing.[4]

BitTorrent, the free, open-source file-sharing application, is a major factor in P2P traffic growth. The application is effective for distributing very large software and media files. Though Hollywood is actively trying to curb users' ability to use the application to download pirated movies, television shows, and other large files, it remains a technology of significant potential for wider use.

IP is the clear winner to support these new applications. Formerly distinct networks and applications are moving to IP for its higher quality, lower cost, and better functionality. IP supports applications as diverse as VoIP, consumer broadband, wireless fidelity (Wi-Fi), grid computing, video offline, cellular, e-commerce, IP virtual private networks (VPNs), diverse business applications, gaming, digital music, and home automation.

This leads us to ask, what is fueling the changes in telecom if demand hasn't slowed?

## IP Adoption and Technology Are Driving Change

IP adoption is one of the primary factors fueling the changes in telecom. According to Deutsche Bank, U.S. cable modem households have doubled in quantity since 2002, and cable broadband households are estimated to grow from about 11 million in 2002 to nearly 26 million in 2005. In addition, forecasters predict that U.S. DSL subscribers will grow by 4 million this year, bringing total subscribers up from about 6 million subscribers in 2002 to more than 17 million in 2005.[5]

VoIP is one of the new technologies that are starting to command consumer attention. The IP–based alternative to a traditional phone line is beginning to lure subscribers in large numbers with its price and features, which include voice mail, conferencing, and Wi-Fi. In fact, according to its September 2005 press release, Vonage has more than 1,000,000 subscribers to its VoIP offering. Skype, which offers PC-to-PC, PSTN-to-PC, and PC-to-PSTN calling, reports 215 million downloads since its inception in June 2003.[6] And the company is growing by a rate of roughly 170,000 to 190,000 downloads per day.[7]

Economics is also serving as a change factor, as lower cost is making IP more and more compelling. The cost to transmit a CD's worth of information across the United States via IP has dropped 46 percent between 1998 and 2004, according to our estimates. And as incumbents are raising prices on services while offering more bundled local services, IP is becoming even more attractive to consumers. Using recent price data, we estimate that by using VoIP, consumers can get broadband plus local and long distance for the same price as traditional local and long distance alone.

Technology is also doing its part to drive change in the telecom industry. Over the past 20 years, computing and information storage have improved exponentially, while communications technology has improved very slowly by comparison. It wasn't until recently, with the shift to IP–enabled communications, that the door opened for the telecom industry to experience rapid growth similar to that seen in computing and storage.

While IP technology has proven itself capable of driving dramatic speed and cost improvements over the past several years, improvements in optical technology, with higher integration of components, have led to improvements in cost and operability as well. High bandwidth demand allows operators to introduce superior optical technologies as needed. For instance, we recently upgraded to a photonic integrated circuit technology on its optical platform to support network growth. The industry as a whole will continue to seek out innovation as service providers work to meet explosive demand and drive costs out of the services they deliver to their customers.

Softswitch technology will also play an increasingly larger role as communications continues to shift to IP. Voice transmission has traditionally relied on circuit-switched technology, with its integrated functionality, proprietary code, proprietary service development, and traditional TDM interfaces. Softswitch technology enables distributed functionality and open platforms. Integration of IP enables IP origination and termination. In conjunction with open interfaces that enable new services, this means communication technology is open to innovation. Features such as find me/follow me, video phones, and other complex network routing capabilities create new ways of communicating and doing business. And development of these technologies takes place in a shortened life cycle that hasn't been possible with closed standards.

## Where Is the Industry within This "Massive Change," and What Is the Future Outlook for Market Growth?

IP technology is continuing to modify and replace traditional telecom models, and the industry is right in the midst of the change. In fact, Gartner forecasts that Internet telephone connections will increase by 31 million lines in the U.S. consumer market between 2004 and 2009, while traditional wired phone lines will decline by 30 million.[8]

As demand continues to increase, new applications are adopted, and technology continues to deliver better economics and functionality, IP will continue to drive change in telecommunications. Other Gartner forecasts predict that data and Internet services will generate 50 percent of telecom revenue by 2009.[9] The companies that are poised to take advantage of this IP revolution are in a position to realize a tremendous opportunity.

### Notes

1. Gartner, *Market Focus: Future of Long-Distance, Evolution vs. Dissolution*, June 24, 2005.
2. IDC: *Telecom Black Book*, November 2005.
3. Deutsche Bank, *2Q05 Preview*
4. CacheLogic Research, *Peer-to-Peer in 2005*, August 2005
5. Deutsche Bank, *U.S. Telecom Data Book*.
6. Skype, *www.skype.com/products*, December 5, 2005.
7. Computer Business Review Online, *Skype in talks with Chinese authorities*, November 17, 2005.
8. Gartner, *Forecast: Consumer Telecommunications and Internet Access, United States, 2003–2009*, July 1, 2005
9. Gartner, *Forecast: Fixed Public Network Services, United States, 2003–2009*, May 17, 2005

# Internet Protocol Television (IPTV)

# IPTV: Technology and Development Predictions

## Lars Bodenheimer
*Managing Consultant*
Detecon, Inc.

## Eckart Pech
*Chief Executive Officer*
Detecon, Inc.

## Patrick Pfeffer
*Consultant*
Detecon, Inc.

## Introduction

Many have ventured to devise the ever-illusive killer application for the broadband networks that telcos and multiple-system operators (MSOs) are deploying around the world. Most efforts have failed, and predictions have been received with yawning. So far only bandwidth has appeared to be a good application, but it is not the "killer app." While more bandwidth is better, network operators cannot be reduced to broadband pipes. A value-added application with high average revenue per user (ARPU) is necessary to justify digging up the streets of Los Angeles, Hamburg, or Tokyo. By all accounts, this new application must be based on Internet protocol (IP), must give the end user total control, and must be of high entertainment value. IP television (IPTV) fits the bill. IPTV is not an application per se; it is a host of applications centered on IP, user choice, and rich content.

Most telecom broadband network operators are planning to offer IPTV. While IPTV is at different stages of product definition, field trials, or early deployments, it is becoming apparent that it is poised to become the framework of incremental revenues for fiber-to-the-x (FTTx, where x stands for curb, node, premises, etc.) deployments. In addition to the quest for better ARPU, rapid progress in IPTV is fueled by the competition between telcos, MSOs, and new network operator entrants (municipalities, loop reseller, etc.) for the largest share, if not the totality, of the home entertainment wallet of consumers.

This paper analyzes the current technical landscape of IPTV vendors and assesses how their solutions influence the network. IPTV had several false starts during the past decade. The video server technology was immature, the bandwidth available on access network was inadequate, and the cost of CPE was too high. We consider that those barriers are now only hurdles that can be avoided if negotiated with care. We

hope that this paper will present some hope for a new beginning at telcos around the world. Building an IPTV network is within range of the telcos' technical ability, but it requires a dramatic change of culture.

The goal of this paper is to shed some light on the meanderings of the IPTV route. In the first section, we offer key findings and opinion statements on IPTV, and we propose a definition for IPTV and introduce a generic architecture. The core components of an IPTV end-to-end system are explained. We follow by two deep dives into the issues of channel change and video codec. In the proceeding sections, we cover Microsoft IPTV activity from marketing and technical standpoints. We investigate the requirements IPTV imposes on the network: additional bandwidth and multicasting support, and we describe three IPTV initiatives: Belgacom, SureWest, and SBC.

## Key Findings and Opinion Statements

- *Regional Bell operating companies' (RBOCs') last chance*: The RBOCs are experiencing an erosion of their access line and voice revenue. It has been compensated by the DSL growth and, for BellSouth, SBC and Verizon, the health of their cellular properties. However, as MSOs finally deploy VoIP, they become triple-play providers. The RBOCs must add video to their voice and data offering. It is a pivotal moment for the RBOCs; they have no choice but to succeed. The RBOCs already tried video service in the mid-'90s; they failed miserably and at great expense. This time around, anything other than complete success will spell their demise.

- *Renewed capital expenditures (CAPEX)*: The current asymmetric digital subscriber line (ADSL) network cannot support IPTV. It must be overhauled. Trenches have to be dug, fiber/copper have to be installed, and new outside electronic equipment—new routers, a new OSS/BSS, a video head end, a video server, a set-top box (STB) or set-top terminal (STT)—have to be purchased. This heavy investment should allow the telecom equipment manufacturers that have survived the nuclear telecom winter to enjoy some springlike weather. It will also lead to further concentration in the industry.

- *IPTV – Microsoft's Trojan horse for the living room?*: Microsoft has won major IPTV deals, making its MSTV middleware solution the de facto reference. After dominating the office desktop, it is clear that Microsoft is positioning itself to dominate the living room. Microsoft is offering one-stop shopping for acquisition server, delivery server, video server, and digital rights management; but at its core, it is an operating system company. It will integrate features into the operating system of the STB or STT as it sees fit and will most likely exit peripheral businesses (video server for example).

- *Small is beautiful*: There are many successful deployments of IPTV across the world. Most of them are by small or medium-size operators. Large operators are struggling.

- *Five facts about middleware*: 1. More than 50 percent of the effort to develop and integrate middleware with the rest of the system is spent on the STB. 2. Middleware scales well, and supporting hardware is not a cost factor. 3. Middleware database is small by today's database standard. 4. Scripting language in the STB is too slow. Java is better, but embedded C/C++ offers the best performance. 5. A middleware client program has a small object footprint (below 32 MB).

- *The last 100 meters*: Telcos have a good understanding of the core/metro network and the outside loop. The challenges of the last 100 meters remain. Most IPTV providers confess that it still takes an average of four hours to set up a customer (including network access termination). A solution must be found. This solution will vary from continent to continent because of different building codes.

- *The last 10 centimeters*: The STB presents major deployment/integration issues and creates CAPEX and operational expenditures (OPEX). 1. More than 50 percent of integration dollars are spent on STB integration with the rest of the system. 2. The STB is a difficult device to debug—it has no output port, lacks VT100 display, and has limited amount of memory. (3) STB cost for early deployments will be more than $100 for an entry-level model. 4. The remote management of the STB is a thorny issue. It will require, among other things, the training of telephone support staff.

- *VoD – The future of television?*: Video on demand (VoD) might be the killer app. Video is not restricted to movies; it also includes TV programs recorded automatically or by the user with a network personal video recorder (NPVR). It offers the flexibility of the Internet with the entertainment value of cable TV.

- *VoD – The end of the network?*: If VoD becomes a prevalent form of TV viewing, this means that the current network that telcos are building to support IPTV will not have enough bandwidth; most of the video streams will be unicast, thereby consuming more aggregation and core resources than are currently set aside for a mostly multicast model.

- *IPTV needs standards*: There are too many discrete elements in an IPTV system. An integrated solution such as Microsoft is an option, but clear standardized interfaces between the elements would benefit the industry in the long run.

- *Business users are high-value customers*: Telcos are focused on the residential market. It is important that the new network supports residential and business users. The business community fears that IPTV might disrupt the stringent service-level agreement (SLA) it relies on. It is paramount that telcos address those concerns. If they are not addressed, a new breed of business-centric Internet service providers (ISPs) will provide data services, along with VoIP and video over IP, to the business community, depriving telcos of a stable and high-profit customer base.

## IPTV Architecture

### IPTV Definition

IPTV is a transmission and control technique to deliver broadcast and VoD video streams to an STB. The use of IP as a video delivery mechanism is omnipotent. What is novel is the use of pure IP signaling to change channels and control other functions. This dogmatic definition of IPTV implies the use of a point-to-point networking infrastructure that supports broadcast video using multicasting techniques. The FiOS project at Verizon is, therefore, not an IPTV implementation.

The access network is still a bottleneck, and telcos have two options[1] to address the twisted-pair (TP) engorgement: either improve the copper infrastructure or abandon the TP for fiber-to-the-home. Most telcos, with the exception of Verizon and NTT, have decided to keep the copper infrastructure for the last few hundred feet, with fiber-to-the-home being only considered for new builds. Verizon, on the other hand, has engaged in an aggressive fiber-to-the-premises overlay network and plans to retire all TPs. The FiOS architecture shares many similarities with a hybrid fiber coax (HFC) system with a passive optical network (PON) instead of a coax cable.

Regardless of the access network, an IPTV network follows an N-tier architecture: national head end, regional head end, and customer premise (see *Figure 1*). The centralization of back-office functions and content with a nationwide audience reduces the number of tiers, while the need to bring one-on-one content or content with a small number of viewers increases the number of tiers.

Very large operators use two national head ends to provide redundancy and to avoid long-hauling video feeds. For example, Verizon has two national head ends—one in Illinois and the other in Florida.

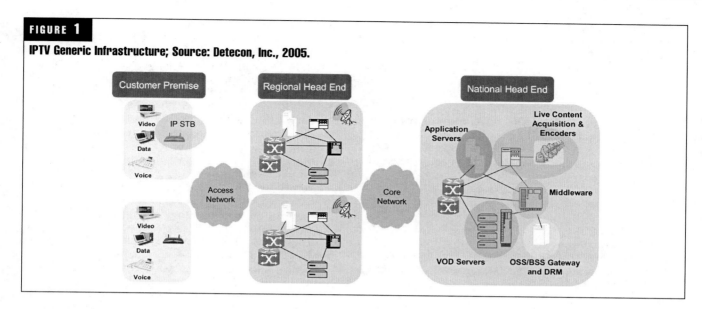

**FIGURE 1**

IPTV Generic Infrastructure; Source: Detecon, Inc., 2005.

## IPTV Components

### Video Encoders

Video encoders are responsible for transforming an input stream that can be of various formats into a digital compressed stream. The video output is either in MPEG–2, MPEG–4 AVC, or WM VC–1. Video encoders can also be responsible to encapsulate the video streams into a transport format that can be an ATM adaptation layer or IP packets. Real-time video encoders that encode live TV feeds used to be very expensive. They have decreased in price dramatically and therefore they no longer represent a large portion of the capital investment of a video head end, but the investment per digital channel is still at least $75,000. Most video encoders for live TV are in the national head end; however, they are also present in regional head ends for local programming. The key technical attributes of video encoders are quality of the encoding, compression rate, variety of encoding algorithms, and support for statistical multiplexing.

Some video encoder suppliers are Harmonic, Tandberg, Thales, and Tut Systems.

### Video Server

Video servers are computer-based devices connected to large storage systems. Video content, previously encoded, is stored either on disk or in large banks of RAM. Video servers stream video and audio content via unicast or multicast to STBs. A 1 RU$^2$ video server can support about 200 video streams at 3 Mbps. Typical storage systems range from 5 terabit (Tb) to 20 Tb. Video servers are mostly used for VoD; however, they are also used for NPVR, which allows subscribers to record shows remotely on a device at the operator site. The key technical attributes of video servers are scalability in terms of storage and number of streams, management software, and variety of interfaces.

Some video server suppliers are Bitband, Broadbus, C-COR, Entone, and Kasenna.

### Middleware

Middleware is the software and hardware infrastructure that connects the components of an IPTV solution. It is a dis-

tributed operating system that runs both on servers at the telco location and on the STBs. Among other things, it performs end-to-end configuration, provisions the video servers, links the electronic program guide (EPG) with the content, acts as a boot server for the STB and ensures that all STBs run compatible software. The key technical attributes of a middleware are reliability, scalability, and ability to interface with other systems.

Some middleware suppliers are Microsoft, Myrio/Siemens, Minerva, Orca, and Thales.

### Conditional Access System/Digital Rights Management

A conditional access system (CAS) allows for the protection of content. Historically, a switched digital video network did not require CAS, since the network would perform content entitlement. In theory, it could still be the case if the device that performs the multicasting function could also determine whether the user is entitled to view the content. In several early IPTV trials, the content was not protected; however, this content was not very "fresh." As IPTV becomes more mainstream, content providers are mandating CAS and digital rights management (DRM), which not only controls the real-time viewing, but also what happens to the content after it has been viewed once. Generically, most CAS/DRMs are a combination of scrambling and encryption. The video feed is scrambled using a control word. The control word is sent over an encrypted message to the decoding device. The CAS/DRM module on the decoding device decrypts the control word that is fed to the descrambler. The key technical attributes of CAS/DRM are: smart card versus soft client; security; server scalability; and integration with encoder, video server, and STB.

Some CAS/DRM suppliers are Irdeto, Microsoft, Verimatrix, and Widewine.

### STB/Terminal

The STB is a piece of customer-premises equipment (CPE) that is responsible for interface with the user, its television and the network. For live TV and VoD, the STB supports an EPG that allows the users to navigate through the programming. The STB transforms a scrambled digital compressed signal into a signal that is sent to the TV. The STB hosts the middleware and is poised to become the center of the communications infrastructure within the home. The first generation of STB offers minimal features (EPG, decoding and, optionally, some PVR) to keep the price down (around $100). The key technical attributes of a STB are reliability, decoder support, size of internal drive, and variety of external interface for add-ons. The cost of the STB is potentially the most important factor for any IPTV operator. MPEG–4 system on the chip (SoC), with its high level of integration, should push the cost of an STB down.

Some STB suppliers are Amino, Motorola, and Scientific Atlanta.

### Key IPTV Technologies

The decision by most telcos to leverage the TP for the last few hundred feet and to rely on IP for transmission and control has the following consequences:

- Broadcast video look and feel must be implemented using multicasting. Internet group management protocol (IGMP) is the control protocol of choice, and rapid channel change must receive special attention.

- The lack of bandwidth available on TP prohibits the use of MPEG–2. MPEG–4 AVC or WM VC–1 are the only viable options.

We analyze both channel change and encoder in the next two sections.

### IGMP/Channel Change

Most applications on IP networks are unicast, which means a data connection is set up between two hosts and only two hosts. IPTV requires one to many connections; therefore, data must be multicast from a source (called a root) to end devices (called leaves). IP multicast allows a host to send packets to a "virtual" address that is not directed at a particular host. If an application on a host is interested in that data, it can listen in by requesting the network software to find that data. The multicast address is used as a logical identifier for the content. To support this mechanism, most IPTV deployments use the Internet Engineering Task Force's (IETF's) IGMP. IGMP was introduced as RFC 1112 and then was improved in RFC 2236 and RFC 3376. IGPM allows for the creation of multicasting trees built from the leaves up. Applied to IPTV, it works as follows: When a subscriber wishes to view a specific selection of video content, an IGMP "join" message is generated from the STB at the subscriber's location to the network. The message travels upstream until it reaches a multicast node. When the "join" request is detected by the multicast node, that content is replicated and a copy is forwarded to the requesting STB. Similarly, when a user wishes to change the channel, it first sends an IGMP "leave" message to the multicasting node. Only after it has effectively left a multicasting group can a user join another group.

*Figure 2* depicts a simplification of how IGMP works.

Switching and/or routing multicast groups can take a significant amount of time (100 ms to seconds). Making this happen fast enough for channel-hopping is a major challenge in

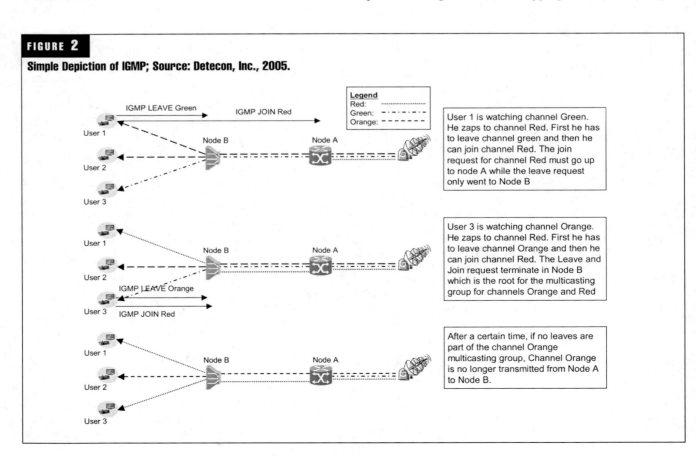

**FIGURE 2**

**Simple Depiction of IGMP; Source: Detecon, Inc., 2005.**

the design of IPTV networks, even for viewers accustomed to the delay on a MPEG–2–based digital TV network. Other factors can slow this process down, such as the time for the video decoder to buffer data. Finally, if a conditional access system is used, the time for the first control word to be generated adds time to the overall channel change.

Without special care, channel change in a switched digital video environment can take more than 10 seconds. The channel change time typically includes the following:

- Time guard against false channel change
- Channel change request
- Modification of the multicast group
- MPEG decoding process (includes demultiplexer, reorder, decode, and conditional access word)

Most IPTV middleware solutions attempt to reduce channel change time by tightening the steps while remaining within IGMP generic principles. On the other hand, Microsoft is pioneering a proprietary solution: to mitigate the time it takes to fill up the MPEG buffer in the receiver. The Microsoft software starts any multicast join with a "unicast" at a data rate above the standard rate. The goal is to fill up the decoder buffer very quickly to save several seconds during channel change. Anecdotal evidences reveal that the scheme has created havoc in some networks during trials. Microsoft channel-change implementation relies actually on several other techniques, including a low MPEG–4 I frame frequency and the concurrent transmission of low-bit-rate adjacent channels.

IPTV is more than broadcast TV. Attempting to provide similar channel change time on a hybrid fiber coax (HFC) network will be difficult and costly. IPTV should redefine the TV experience: better choice (less channel-hopping), better EPG, and more. VoD should diminish the need for constant channel changing. Advertisers and TV network executives would like that. We believe that the attempt to reduce channel change time at all cost is a misguided effort.

*Video Codec*
IPTV video encoding techniques include MPEG–2, MPEG–4 AVC, or WM VC–1. MPEG–2 video compression has been around for about 15 years. It is the encoding of choice for digital cable and digital satellite systems. It is well understood, carrier grade encoders exist, and the cost of MPEG–2 decoders is low (less than $15 for 10 K units). The compression efficiency of MPEG–2, while sufficient for HFC networks or pure fiber networks, is not sufficient for TP loops. MPEG–4 AVC or WM VC–1 is required.

MPEG–2 and MPEG–4 AVC rely on the principle that occasional pictures compressed spatially are interleaved with pictures that predict and describe the motion. The superiority of MPEG–4 AVC over MPEG–2 is largely due to substantial improvement of the motion compensated prediction. MPEG–4 also extends the adaptive field or frame encoding mechanisms. MPEG–2 uses picture-adaptive field or frame coding. MPEG–4 AVC adds the tools to allow the field or frame coding to be adapted on a macro-block basis.

Microsoft is in the process of standardizing Windows Media (now known as WM VC–1 or VC–1, through the Society of Motion Picture Television Engineers [SMPTE]). Because of the standards effort, Microsoft has revealed that the underlying technology is quite similar to MPEG–4 AVC. Some say WM VC–1 differentiates itself by adopting a toolset that is optimized toward decoding on personal computers. We believe that the major difference between MPEG–4 AVC and WM VC–1 is that there is an option for WM VC–1 streams to be decoded on a high-end (more than 1 GHz) Intel processor without the need of special hardware.

MPEG–4 AVC and WM VC–1 are the only encoding solutions for telcos that have opted to conserve TP for the last few hundred feet. Both encoding standards are still new; we expect to see dramatic compression and quality improvements in the next few years. We do not expect to see any compression improvement for MPEG–2. *Table 1* shows our compression rate predictions.

**TABLE 1**

**MPEG–2, MPEG–4 AVC and WM VC–1 Compression Rates; Source: Detecon, Inc., 2005**

| Type of streams | MPEG–2 | | MPEG–4 AVC and VC–1) 2005 | | MPEG–4 AVC and VC–1) Horizon 2008 | |
|---|---|---|---|---|---|---|
| Real Time | SD | 3 | SD | 2.5 | SD | 1.5 |
| | HD | 15 | HD | 10 | HD | 7.5 |
| Pre-Encoded | SD | 2.5 | SD | 1.5 | SD | 1 |
| | HD | 12 | HD | 6.5 | HD | 5 |

While MPEG–4 AVC is appealing, it is important to understand that it is an optimization of MPEG–2. It improves what MPEG–2 does well by doing it better. On the other hand, it is still weak for fast-move sequences or scene cuts.

## Microsoft IPTV

### Marketing Consideration
Microsoft's IPTV strategy, a.k.a. MSTV, was announced in October 2003 by Bill Gates during an International Telecommunication Union (ITU) conference. Microsoft announced that it intended to become an IPTV leader and would only pursue large deployments with millions of end devices. Microsoft has been extremely secretive about its architecture and technical options.

Most middleware systems, while having a unique implementation, offer similar features and have open interfaces to the other components of an IPTV network. Microsoft set itself apart from the rest of the industry by offering a mostly closed system. In addition, while MSTV is often referred to as a middleware, MSTV also covers acquisition server, video server, delivery server, DRM, and more.

Table 2 is a side-by-side comparison of Microsoft middleware with generic middleware offered by other companies.

Microsoft has been able to acquire many large accounts because of the following main marketing/sales motives:

- *Microsoft brand name*: Microsoft is here to stay. It has a track record of following and then dominating new technologies. Wall Street feels reassured that the telcos, which have failed their entry into television at least once, have partnered with a company that knows how to execute on its plan. In addition, Bill Gates and Steve Balmer have been personally involved in the making of some deals.

- *A partner, not just a vendor*: Microsoft has accepted to share some of the risks of the IPTV deployments with the telcos. Microsoft, in some cases, will participate in a co-marketing effort. We can imagine a campaign with the logo "Microsoft Inside." In some cases, the network operator is relying on Microsoft to source content from Hollywood, and Microsoft accepts to share some liability in case the content would be misused.

## TABLE 2

**Side-by-Side Comparison between Microsoft Middleware and Generic Middleware; Source: Detecon, Inc., 2005**

| Category | Microsoft Middleware | Generic Middleware |
|---|---|---|
| Offering | One-stop shopping—Closed System | Open interfaces |
| System Size | Large | Small to medium |
| Architecture | Distributed with specialized servers | Monolithic with single middleware server |
| QoS Assumption | Weak | Strong |
| Network Interface | Strong | Weak |
| Encoding | WM VC–1 | MPEG–2 or MPEG–4 AVC |
| Transport | RTP | UDP |
| User Plane | Involved (D-Server) | Not involved |

| Category | Microsoft Middleware | Generic Middleware |
|---|---|---|
| STB | Custom port | Open |
| Video Server | Microsoft | Open |
| DRM | Proprietary, but to the liking of content providers | Reliant on partners |
| DRM Management | Centralized | Distributed |

- *The herd mentality of large telcos*: After Microsoft signed its marquee contract with SBC for a whopping $400 million, other telcos in the United States and Europe followed.

### Technical Considerations

The marketing might of Microsoft only partially explains its success; Microsoft has invested more R&D dollars than anyone else in the industry, and as a result it offers a compelling, technically advanced solution. We consider that its four major assets are the following:

- *Instant channel change*: Microsoft, by using proprietary mechanisms, claims it can reduce the end-to-end delay for channel change well below one second. It was a major selling point with the telcos, whose initial ambition is to offer feature parity with the MSOs.

- *WM VC–1*: Microsoft has submitted Windows Media 9 (WM9) as a straw man document to the SMPTE, which has created a standard known as VC–1.

- *Weak network QoS requirements*: MSTV solution uses delivery servers (a.k.a. D-Servers), which mitigate the negative side effects of packet-based networks (specifically, delay and jitter).

- *Digital right management*: MSTV DRM inherits from the Paladium secured computing platform. It is available for video, audio, and picture or executable on STB, personal computer, or mobile devices.

We present in *Table 3* our opinion on the strengths and weaknesses of MSTV technologies.

### Track Record

In late 2003 and early 2004, Microsoft signed up Swisscom (Switzerland) and Telstra (Australia) as its first customers. In early 2005, Swisscom and Telstra announced major delays in their IPTV projects. Microsoft was swift to blame the network operators for not being ready and the operators rebutted the assertion. As it is often the case, the truth lies in the middle. We believe Swissom, Telstra, and Microsoft had to revisit their partnership for the following reasons:

- *Microsoft priorities*: While Swisscom and Telstra are national carriers, they remain small players. After the announcement by SBC that it will use Microsoft TV for its Lightspeed project, resources at Microsoft were shifted from the rather small national Swisscom and Telstra to the multi-regional giant SBC. SBC's requirements and timeline define the availability of core features and additional features besides those required by SBC are supported as time permits.

- *Microsoft platform maturity*: Microsoft offers an end-to-end solution. It appears that neither the overall system nor the components were ready for prime time. The video servers did not scale, the redundancy mechanism failed to meet the requirements, the STB that must support both Windows CE and WM VC–1 existed only in sample quantity, and the end-to-end system was very fragile.

---

**TABLE 3**

**Strengths and Weaknesses of MSTV; Source: Detecon, Inc., 2005**

| Feature | Strengths | Weaknesses |
|---|---|---|
| Instant channel change | Low latency, broadcast-like experience | Proprietary scheme<br><br>Extremely complicated and costly |
| WM9–VC–1 | Multi-platform (STB and PC)<br><br>Good compression rate and picture quality | Secondary standard<br><br>Licensing unclear |
| Weak network QoS requirements | Allow suboptimal non-video-friendly network to support IPTV | Long-term interest to build well-behaved network in order to not be limited to broadband pipe |
| Digital rights management | Multi-format<br><br>Multi-device<br>Favored by MPAA | Locks operators<br><br>Centralized architecture<br>Target for hackers |

- *Underestimation of effort by operators and Microsoft*: Both sides underestimated the system integration and project management effort. Not enough personnel were assigned to the different tasks and there was a lack of end-to-end supervision.

## Network Infrastructure

Today, telcos offer high-speed Internet access. Based on a flavor of ADSL, the guaranteed bit rate rarely surpasses 3Mbps and there is no support for QoS. In addition, as we alluded to previously, the network infrastructure supports only unicast. To provide IPTV, telcos must upgrade their network with higher and more controlled bandwidth (i.e. higher bit rate with QoS) and support for multicast.

### Improving Bandwidth
#### Legacy Networks
Figure 3 illustrates the major issues with today's ADSL networks that telcos have deployed since the late '90s. The problems exist in the following key areas and, as it has been the case since the advent of residential broadband network, the bottlenecks still reside in the access network:

- *Legacy loop topology*: The length of most TPs, whether served from a digital subscriber line access multiplexer (DSLAM) in a central office (CO) or a digital loop carrier (DLC), is superior to 8 kilo-feet (kft). It is not possible on such length to offer reliable bandwidth superior to 4Mbps with either ADSL or ADSL2+. Finally, even if many premises have several pairs, current deployments do not take advantage of it.

- *Legacy concentrator*: DSLAM and DLC were mostly designed to support a large number of first-generation ADLS subscribers with a very high level of oversubscription, since only the best effort was required. While the number of subscribers per uplink is large, those links are either DS3 or OC3.

- *Legacy Aggregation*: The aggregation network is used to combine DSLAM or DLC traffic. It connects the access network to the broadband–remote access server (B–RAS). It is often a small asynchronous transfer mode (ATM) switch with OC3 and OC12 ports. Again, a high level of oversubscription takes place. The switching fabric is commensurate with the type of line card.

It is clear that the current ADSL network can not support IPTV, and so telcos must upgrade their network.

#### New Networks
Figure 4 shows an example of architecture of the new copper-based network.

DSLAM or DLC are disappearing in the new IPTV network. They are replaced by optical network units (ONU), which are outside equipment that transform an optical signal into an electrical signal.

- *New loop topology*: The ONUs are placed very close to the end user (less than 500 feet for BellSouth and less than 3,000 feet for SBC). They support a variety of DSL flavors, but specifically the new ADSL2+ and VDSL2. Those new modulation techniques use a wider spectrum than the first DSL generation and therefore can provide bandwidth well above 15 Mbps over short loops (less than 3,000 feet). In some cases, when the loop is too long to support a high bit rate, two or more pairs can be bonded. Since most premises have several pairs, this solution may become very popular, though it is more costly than a single pair. Pair bonding will create a single aggregate connection from several pairs, even if the pairs have different characteristics. Supporters of ADSL2+ and VDSL2 claim the network will support bandwidth well over 30 Mbps over a 3 kft loop. We do not share this optimism. We consider that only 10 percent of non-conditioned pairs will support bandwidth above 20 Mbps; the rest will be in the teens at best. Pair bonding will allow almost double the bandwidth (for two pairs), especially for short loops.

- *New Concentration*: The ONUs are small enclosures powered from an external source; the closer they are

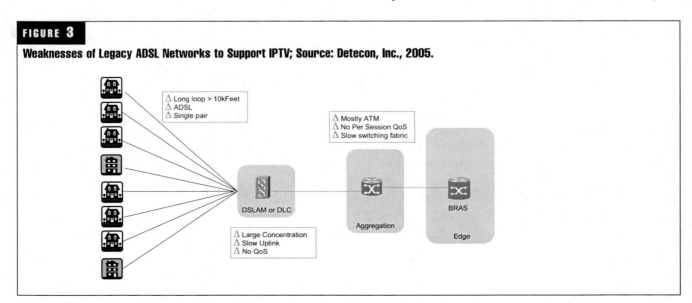

**FIGURE 3**

**Weaknesses of Legacy ADSL Networks to Support IPTV; Source: Detecon, Inc., 2005.**

Δ Long loop > 10kFeet
Δ ADSL
Δ Single pair

Δ Mostly ATM
Δ No Per Session QoS
Δ Slow switching fabric

DSLAM or DLC

Aggregation

BRAS

Edge

Δ Large Concentration
Δ Slow Uplink
Δ No QoS

---

**FIGURE 4**

**Improvements of Legacy Networks to Support IPTV; Source: Detecon, Inc., 2005**

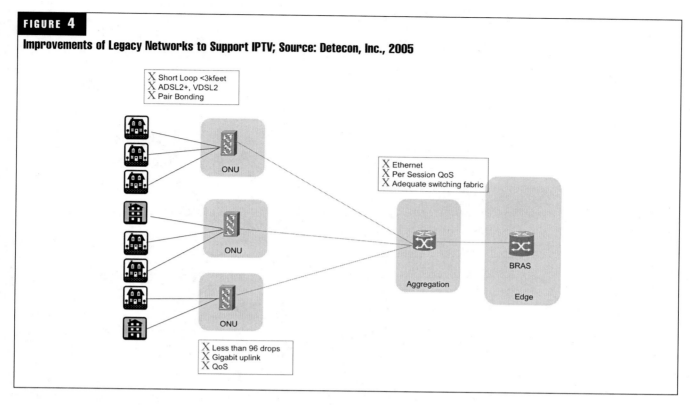

---

from the premise, the smaller they are. We expect to see ONUs ranging from 8 to 128 drops. They will usually be fed by multi-gigabit link (initially it will be 1 gigabit Ethernet, moving in the future to 10 gigabit Ethernet for large ONUs). Some ONUs will support ATM and Ethernet, with ATM slowly disappearing. Small ONUs might not need to support QoS, while large ones do.

- *New Aggregation*: The aggregation network clearly demonstrates the advent of pure Ethernet networks. Most carriers are favoring to replace their multi-service switches (support for both ATM and Ethernet) by pure Ethernet switches. This new switches are simpler, offer wider switching fabric, and support virtual LAN (VLAN) features key to the management of QoS.

## Support for Multicast

Many DSL–based networks use a point-to-point protocol (PPP) encapsulation to transport the traffic onto a Layer-2/3 tunnel between the B–RAS and the CPE. PPP over Ethernet (Ethernet is a common choice), or PPPoE, which is adapted from the dial-up Internet world, is used for centralized termination of subscriber sessions. The PPPoE model works well to have a secure session between the subscriber host and the access server. PPPoE enables a wholesale model, e.g. forwarding of client sessions via tunnel protocols to other service providers, and it leverages the existing remote authentication dial-in user service (RADIUS) platform, session-based billing and policy management, etc. The distribution of IPTV content requires multicast streams to be replicated at different network nodes. In today's network, the B–RAS, which terminates the PPPoE session, would be the closest node to the subscriber capable of performing multicasting. This would generate a significant traffic burden on the aggregation network. New methods must be implemented to allow multicast groups to be originated closer to the end users. We present in the next four sections a couple of alternatives that are being considered by telcos around the world.

*ATM–Based DSLAM with No Multicast Capabilities*
We present the architecture shown in *Figure 6* as an option for completeness. It is not realistic but could be used in a lab or for a limited field trial (low customer take-up per DSLAM). Its appeal lies in the ability to support IPTV with little to no changes in the current network.

Every user who subscribes to IPTV (and specifically Live TV) is connected to the B–RAS via a permanent virtual channel (PVC). While this PVC should be constant bit rate (CBR), one could allow some level of oversubscription even at the risk of failing connection admission control (CAC). Channel change requests would be performed out of band encapsulated in UDP packets. The video stream would be carried over ATM adaptation Layer 5 (AAL5).

The provisioning would be trivial.

Most DSLAMs support at least a primitive version of QoS. It is often implemented as a dual queue per drop: one queue for low-priority traffic (specifically high-speed Internet [HIS]) and one queue for high-priority traffic (specifically video). The queue management protocol is trivial but efficient. The high-priority queue is always served as long as there are cells available; the low-priority queue is only served when the high priority queue is empty.

The network architecture option presented in *Figure 7* reuses existing network components, but requires a SW upgrade of both the DSLAM and the B–RAS. The rationale for this solu-

## FIGURE 5

**Point to Point Nature of Legacy ADSL Networks; Source: Detecon, Inc., 2005.**

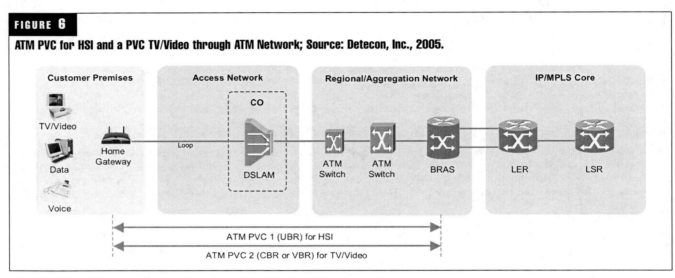

## FIGURE 6

**ATM PVC for HSI and a PVC TV/Video through ATM Network; Source: Detecon, Inc., 2005.**

## FIGURE 7

**ATM–Based DSLAM Supporting Layer-2 Multicast; Source: Detecon, Inc., 2005.**

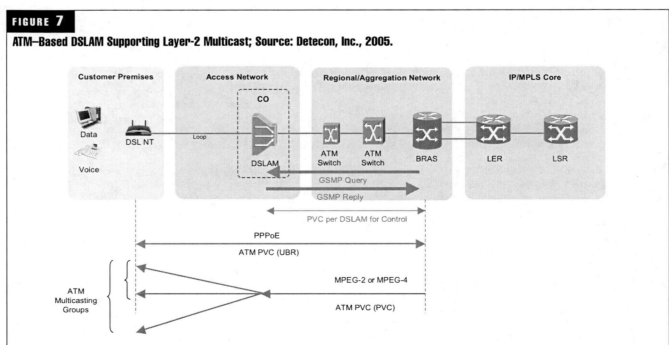

tion is to reuse the current provisioning system and concentrate the intelligence onto the B–RAS.

IGMP joins are originated by the STB and sent unaltered through the DSLAM to the B–RAS. The B–RAS analyzes its multicasting routing table and sends a general switch management protocol (GSMP as defined in RFC 3292) connection request to the DSLAM. The DSLAM adds a branch to a new or existing multicast PVC. The channel identification for the DLSAM is performed through a mapping between channel and virtual channel identifier (VCI). IGMP leaves are handled similarly. GSMP is a simple master-slave protocol. It is adequate to be used when there is a clear separation between the control and the user planes.

From a control-plane standpoint, the architecture presented in *Figure 8* is very similar to the architecture in *Figure 7*. They mostly differ by the use of Ethernet in the aggregation network and the subsequent use of virtual LAN (VLAN). We depict the use of single PVC in the local loop, but configurations with multiple PVCs are possible as well. They simply require additional provisioning.

Based on IGMP joins and leaves, the B–RAS communicates to the DSLAM which channels are to be replicated for a particular subscriber based on their multicast MAC address. In the alternative case (Option 2 in *Figure 8*), the DSLAM processes IGMP joins and leaves, builds multicast replication tables and performs the packet or cell replication function. Subscriber multicast entitlement information might be sent from the B–RAS to the DSLAM at subscriber session establishment or when a change is made to the subscriber entitlement.

The architecture depicted in *Figure 9* shows a pure Ethernet/IP end-to-end network. The architecture, by removing all the ATM aspects, simplifies the layering and achieves higher throughput. In addition, the all Ethernet approach allows for the breaking of the demarcation line between access and aggregation networks.

*Figure 9* shows in the CO a service edge router. The ONU can be in the CO or in the outside plant. It is mostly an issue of loop length. We show the use of two VLANs as an example, but many other options are possible. The network is a pure IP/Ethernet network and therefore RSVP, differentiated services, and 802.3ah, ad, q, or z can be used.

Generically, when an STB requests to join a multicast group, the ONU intercepts the join request and checks whether it is already receiving the multicast session. If it is, the ONU simply forwards that multicast channel to the STB. If it is not, it forwards the join request upstream where it is processed by the next service edge router (SER), or to the next … or to the source at the video head end.
It is clear that having the multicast taking place very close to the STB speeds the channel change time. Bandwidth in the aggregation network is conserved because individual video channels are only forwarded to the nodes that have a need to send them downstream.

## IPTV Initiatives

There are IPTV initiatives both in Europe and in the USA, but Europe is more advanced. The figure to the right (Source: Datamonitor, 2004) shows the growth of IPTV as reported by the operators. We consider the numbers quite inflated, but the trend remains certain. The services offered are live broadcast TV, time-shift TV, VoD, and NPRV. Other services, such as gaming, karaoke, or commerce, are embryonic. TV functions and IP/Internet functions are dichotomized: the STB allows to watch TV or to "surf the net"; there is no blending between the two paradigms.

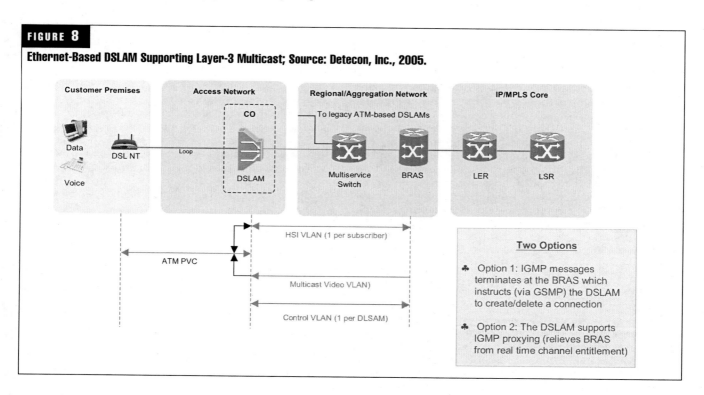

**FIGURE 8**

**Ethernet-Based DSLAM Supporting Layer-3 Multicast; Source: Detecon, Inc., 2005.**

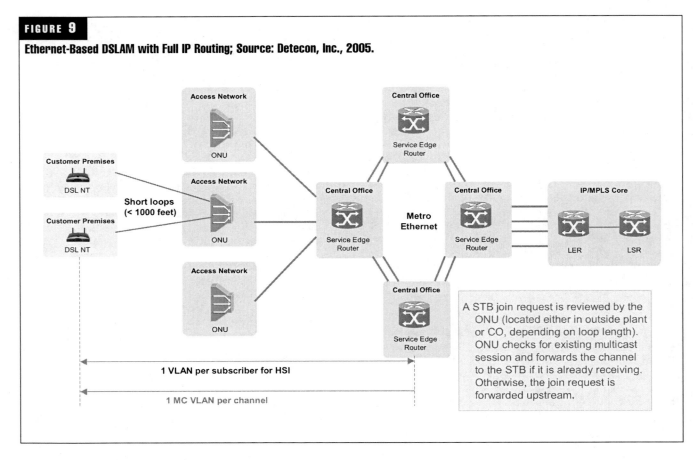

**FIGURE 9**

**Ethernet-Based DSLAM with Full IP Routing; Source: Detecon, Inc., 2005.**

A STB join request is reviewed by the ONU (located either in outside plant or CO, depending on loop length). ONU checks for existing multicast session and forwards the channel to the STB if it is already receiving. Otherwise, the join request is forwarded upstream.

Whether called commercial deployments or trials, today IPTV initiatives are still small and parceled. We propose a snapshot by looking at three initiatives: Belgacom, SureWest, and SBC Lightspeed. We have covered and built profiles for most IPTV initiatives in the world. We offer some thoughts based on early results in the last part of this section.

*Belgacom*
More competition, a clearer regulatory environment, and short access loops are the key reasons for Europe's more advanced IPTV market. A typical example is Belgium's national operator, Belgacom. In recent years, Belgacom has deployed more than 500 optical nodes, between the local nodes and the street cabinets, allowing the company to optimize its investment in the copper-feeding network and to reduce the length of the copper lines. By the end of 2004, more than 1,400 VDSL remote units were active, yielding a footprint of more than 250,000 lines. The aim of Belgacom is to achieve VDSL coverage of 46 percent by the end of 2006, whereas ADSL covers already approximately 98.5 percent of the country.

Equipped with a very modern network, Belgacom commercially launched its IPTV offering, known as BelgacomTV, in June 2005. Alcatel is providing the access network equipment. Siemens has been awarded the end-to-end system integration. The Linux STBs are being supplied by i3 Micro Technology and are fitted with a customized user interface based on Myrio middleware (recently acquired by Siemens) and a browser from Espial. The head-end equipment is provided by TANDBERG and the video server by C-

COR/nCube. Finally, Verimatrix offers content protection. BelgacomTV offers all major TV stations, a pay-per-view system, VoD, PVR, and interactive services. Belgacom has secured rights for soccer retransmission. We believe such rights are key in the European market. BelgacomTV has more than 5,000 subscribers. Belgium (like the Netherlands, where KNP is also deploying IPTV) has two infrastructures (TP and HFC) that can offer broadband to more than 90 percent of the population. Such a competitive environment is a driving force behind BelgacomTV.

*SureWest*
Several small independent (often rural) phone companies have deployed IPTV with some engineering success. At this time, the successes can only be qualified as engineering since it is not yet clear how those IPTV deployments will generate enough cash flow to obtain a decent return on investment (ROI) and fuel growth. We present below an example of such small carrier: SureWest.

SureWest is a small carrier in the Sacramento, Calif., region. It shares territory with Comcast and SBC. SureWest has about 67,000 subscribers, and 16,000 of them are broadband. SureWest entered IPTV in June 2002 when it acquired the video head-end assets of WINfirst. Valued at more than $400 million, SureWest bought a network operations center, a full-featured call center facility, and a satellite farm for gathering video transmissions from multiple sources for a little above $12 million. SureWest has a very advanced access network. It uses a combination of Calix and Occam Networks equipment. Calix's gear is mostly used for POTS and data DSL, while Occam's gear is used as a DLSAM for the triple-play offering. SureWest also uses Cisco equipment to provide fiber-to-the-home with active Ethernet. IPTV is offered both on twisted pair with ADSL2+ and on fiber.

The main IPTV components chosen by SureWest are MV50 video encoders from Harmonic, IP grooming and encapsulation broadband multiservice router from BigBand network, iTV Manager middleware from Minerva, PLsys CAS from Irdeto, video servers from Kasenna, and STB from Amino. SureWest recognized that it took more time and effort than it originally planned, but today SureWest offers a bouquet of 260 Live TV channels, unlimited access to the last 24 hours of Live TV programming, and an ever-increasing library of movies, concerts, and how-to videos (more than 400 hours).

### SBC

In June 2004, SBC outlined plans to invest up to $6 billion over five years to make SBC the second-largest video provider within its fiber footprint. With this new infrastructure, SBC believes it can fully compete with cable network operators. The name for this large-scale deployment is Project Lightspeed. SBC plans to reach 90 percent of its "high-value" residential and 70 percent of its "medium-value" customers by 2007. Twenty-five percent of SBC's customers are considered "high-value," spending $160 to $200 a month on telephone, broadband, cellular, and cable TV services. It is not clear what the percentage of SBC's "medium-value customers" is, which consists of those who spend $110 to $160 a month. The combination of high- and medium-value customers makes up 75 percent of the total revenue received by SBC from its residential customers.

In June 2005, SBC announced that U-Verse will be the brand for its suite of IP–based products—IPTV, very-high-speed Internet access, and VoIP. SBC has awarded most of the network contract to Alcatel and the IPTV middleware, video server, DRM to Microsoft. Scientific-Atlanta will provide encoders, satellite dishes, and video routers. The STB will be provided mostly by Motorola and secondarily by Scientific-Atlanta. Amdocs will provide billing, customer-relationship management (CRM), and ordering and payment mediation products. SBC's success is tied to the ability of Alcatel and Microsoft to deliver. It is still very early for SBC to announce the bouquet of programs in any details. SBC had plans to conduct a controlled launch in 2Q06 followed with a scaled launch in 4Q06. However, delays have pushed back commercial deployment of SBC IPTV into 2007. In such large and complex endeavors, such delays are very common and should be inconsequential in the long run if SBC stays focused.

### Small Is Beautiful

As we surveyed the operators around the world, we realized that the most successful at IPTV were not the traditional incumbents—either former PT&T or RBOCs—but smaller (such as SureWest) or newcomers (such as Free in France or FastWeb in Italy). It is interesting to ask the fol-

lowing question: Why are small operators successful in their deployment of IPTV while large operators seem to lag behind and struggle? We do not know the answer. One could debate that small operators are not successful or that large operators do not struggle more with IPTV than they do with other complex technologies. However, it is fair to draw the following inferences on IPTV current attempts and assess how they relate to the size of the operator:

- *IPTV works*: Some telecom operators are operating IPTV systems, that shows it can be done.

- *Network infrastructure for IPTV*: Small regional or local operators have been able to deploy IPTV either over passive optical network (PON) or xDSL. They favor simple network architectures and employ equipment from "second-tier" vendors: access network (Calix, Occam, Tut, Zhone), middleware (Minerva, Myrio), and video server (Kasenna, Bitband).

- *A best-of-breed solution works*: Small operators have not relied on a single provider, but have selected the best of breed for their specific needs. The integration was painful but successful.

- *Legacy systems are a major problem*: Integrating IPTV with an existing network infrastructure that supports only high-speed data access is difficult. Starting from scratch appears to bring faster results.

## Conclusion

IPTV is ready for prime time. Telcos are at the confluence of technical and market forces. The technology of most IPTV components is mature and the cost of equipment is far more reasonable than it was 10 years ago. The content providers understand that TV is changing and something called the Internet has created new ways to distribute video content. They can no longer dictate the conditions of the delivery of their assets. Finally, the MSOs have built a formidable network that can support voice, video, and data. While a complex endeavor, the deployment of IPTV is within reach of the telcos, and we are convinced that, with continued focus and additional CAPEX (we do not believe that current investments are sufficient), telcos will be successful. The telcos will deploy IPTV and transform themselves to offer both communication and entertainment to their subscriber. If they fail; they will disappear at the favor of other facility-based operators (MSOs or new entrants) or ISPs such as Google or Yahoo.

## Notes

1. See "Fiber Diet for Triple Play Course: A Healthy Habit," Patrick Pfeffer, Detecon, 2005.
2. An RU is a rack unit. It mostly describes the height (1 RU = 1.75 inches) of a device that can be installed in a telco rack (the width of a rack is 19 inches in the United States and 23 inches in European Telecommunications Standards Institute [ETSI] countries).

# IPTV Distribution

## A White Paper Discussing IPTV in Broadband Networks

## Peter Gustafsson

*Director of Engineering*
PacketFront, Inc.

The widespread adoption of Internet protocol (IP) technology in recent years has irreversibly changed the networking landscape. Once a technology purely intended to host data networks, IP now influences the design of other communication networks.

This document describes the technical concepts behind the distribution of television content in a common, IP–centric infrastructure. The document specifically describes the use of an environment denoted as "triple play," a network in which three baseline services—data, voice, and video—share the same underlying infrastructure.

This document also provides insight into the market logic and commercial consequences associated with different architectural approaches to TV distribution over IP networks. For many network operators, it has been a major dilemma to decide the extent to which an existing physical cable or network structure should be kept and enhanced. Some have already decided to implement a modern, IP–based network architecture based on twisted-pair cabling or, preferably, an all-fiber network. This document recommends this solution due to the easy addition of services and, in this way, increasing network revenue.

The key success factors for distribution of TV services discussed in this document are as follows:

- A guarantee that only the end users who subscribe to and pay for a particular channel receive that channel
- The assignment of high priority to the TV services to ensure high quality for end users
- A cost-efficient approach to the configuration of access nodes and the distribution of TV services
- A clear ownership of the customer premises equipment (CPE). This is important in order to decide where service functionality should reside, whether it should be embedded in the CPE or not, and who is responsible for this functionality

Discussions concerning IPTV distribution often labor under confusion surrounding the Internet and its underlying technology platform (the transmission control protocol [TCP]/IP family of networking protocols). It is important to notice the distinction between Internet connectivity and IP–based networking, Internet connectivity, IP telephony,

and IPTV distribution, which are all separate services. The only thing they have in common is that they all use the same highly efficient underlying IP network.

Properly designed, TV distribution using IP technology can be easily separated from an end user's Internet access. Hence, licensing issues related to Internet distribution and measures to restrict uncontrolled redistribution of content over the Internet do not need to become a major concern as they are in situations in which all traffic is handled as a single service. This is the case when offering Internet access as a single service, supplemented by Webcasting or Internet streaming.

### Background

Networks for the distribution of television have been based on analog technology, merely relaying TV transmissions over a wireline distribution medium. These networks usually constitute separate network islands, each with its own terrestrial and satellite receivers, and a shared coaxial cable. TV channel management was wide-meshed and often inadequate in such networks, since the individual mix of channel subscriptions was often defined in a set-top box (STB) that was either hardwired or configured by means of the periodic distribution of decryption keys to end users. The outcome of this is common knowledge: it quickly attracted creators of counterfeit encryption engines, fake smart cards, and similar fraudulent equipment. A major obstacle has been the inability to create a truly tamper-proof mechanism of restricting access to premium subscription channels such as movie channels. Creating this mechanism is difficult in a network environment in which many TV channels are distributed to large groups of households.

Star topologies with individual cables connected to a local hub or switch were introduced in the early years of cable television networks, enhancing the possibilities of offering user-specific services. In combination with smart STBs, sometimes provided with additional functionality to allow Internet connectivity, these devices often did their job fairly well.

Several recent factors have created a need to reassess the situation. These factors include the Internet revolution, a rapid increase in the performance of fiber infrastructure and a corresponding fall in its price, and the advent of IP telephony.

Network operators today are becoming increasingly aware of how crucial multiservice offerings will soon become. Other related issues have arisen, including the need to address the flexible configuration of end-user devices on a mass scale and the possibility of hosting a number of service providers in the networks. The amount of headroom available for expansion is another important topic in terms of the number of services, the capacity of each service (which determines, for example, the maximum number of television channels that can be included), priority issues, and the total number of users that can be handled by one network operations center (NOC).

The answer to these requirements in the context of the current technology has been rather unclear. Networks based on TCP/IP have traditionally lagged behind dedicated distribution networks for television in terms of real-time performance. However, the situation has now changed. With correct network design, containing priority mechanisms and multicast functionality (intelligent handling of bandwidth-consuming real-time bit streams), an IP networking infrastructure is now the most attractive alternative for realizing the vision of true triple-play networking. The most demanding challenge has undoubtedly been to host multichannel television distribution. The discussion below will explain why this is the case and how the challenge can be met.

## The Technical Challenge

Television is a highly demanding network service. TCP/IP was not designed to cater to such time-critical and band-

width-hungry bit flows. TCP, the most common transport mechanism on the Internet, was primarily designed for reliability and for a situation in which 10 ms or even 100 ms of extra delay did not affect the overall impression of service performance (such as the transmission of an e-mail message or a large data file). The prime concern was to correctly transfer each and every bit of information.

Extra delay is, of course, unacceptable when dealing with real-time traffic such as telephony and television. Real-time traffic protocols for the Internet have therefore been developed, with less demanding transport mechanisms, where a continuous bit flow has been a more important concern than receiving every bit correctly. Consequently, Internet streaming has been designed to accept a fairly large level of packet loss (dropped packets).

### Priority

Premium voice and video services cannot operate acceptably when subject to the level of dropped packets that is usually experienced when streaming traffic on the public Internet. It is therefore necessary to employ other solutions. One important step has been the introduction of priority mechanisms. In modern IP–based triple-play networks, the dominating technology that determines priority is differentiated services (DiffServ). Time-critical traffic is given a priority labeling when it enters the network, and the traffic is then handled accordingly by routers and other equipment along the path to the end user.

Telephony (voice) traffic is commonly labeled with the highest priority in the network, followed by video and audio services. Video frames are generally not as time-critical as voice synchronization, which is a factor worth mentioning here.

## FIGURE 1

### The Priority of the Types of Services Offered to the End User

| Service | Bandwidth | Priority |
|---|---|---|
| Voice over IP | 128 Kbps | Priority 1 |
| Digital TV | 8 Mbps | Priority 2 |
| VPN | 2048 Kbps | Priority 2 |
| Home surveillance | 1024 Kbps | Priority 2 |
| Environmental controls | 256 Kbps | Priority 3 |
| Internet service 1 | 512 Kbps | Priority 4 |
| Internet service 2 | 10 Mbps | Priority 4 |

### The Importance of Headroom

Other factors such as latency (the total, accumulated network delay) and jitter (a variable packet reception rate) also affect quality. These quality issues are generally solved by adding large-capacity headroom to the network (additional data bandwidth). This is fairly easy to achieve today with respect to IP telephony, since telephony traffic only requires data capacity in the region of 10 to 100 kbps. The accumulated bandwidth requirement for TV distribution, however, rapidly becomes unmanageable when the number of channels increases. Each digital video bit stream requires 1.5 to 15 Mbps, depending on the content and the predetermined quality level. It is simply impossible to implement more than a few additional TV channels in the network solely by adding more extra headroom. We may add here that this is not strictly a question of technology, rather one of operational economy.

### Multicast Distribution

The solution to the accumulated capacity problem is a procedure known as "multicasting," in which bandwidth-hungry traffic such as TV channels are sent once and directed only to the receiving party specifically requesting them, thereby minimizing unnecessary distribution. This can be compared with traditional traffic in an IP network, which is of unicast type, where every user requests his or her own bit stream from the source.

Initially, multicast was intended to revolutionize the Internet, but it did not achieve widespread acceptance in the interwoven structure of independent networks on the Internet, mainly because multicast traffic did not fit the volume-based business model of most IP network operators. (The setting of priority was another "failed revolution," by the way. It proved to be difficult to set priority when "high-

priority" traffic crossed administrative borders in the public Internet.)

In a private network, such as a triple-play access infrastructure, the situation is different. Priority labeling and traffic optimization by means of multicasting have become key factors and are crucial to offering real-time services and keeping capacity development within reach.

Another important feature of multicast distribution is the increased security that it makes possible. Multicasting also implicitly brings forward the need to control traffic at the network layer (also referred to as Layer 3 [L3]) in the access segment of the network, but it will not lead to the same complexity in the network terminals (STBs, etc.) installed at each end user as a Layer-2 (L2) approach would do. *Figure 2* and *Figure 3* illustrate the differences between an L2 and an L3 approach.

### The Importance of Easy Administration

L3 is the recommended architecture when designing a triple-play network that should offer flexibility and allow several service providers to share the same network. A major advantage of an L3 approach over a traditional link layer (L2) approach is the increased level of control that it offers, while avoiding much of the headache traditionally associated with the configuration of end-user equipment.

In the L2 architecture, the end user receives services based on the definition of the VLANs. The end-user terminal will usually have one VLAN for voice (telephony) and another VLAN for Internet access. More VLANs must be assigned when video services such as TV channels are added. This

---

**FIGURE 2**

**The Layer-2 Approach. In This Approach, Advanced Intelligence Is Required in the Customer Premises Equipment. Each Service Offered to the End User Requires Configuration of the CPE to Establish a New VLAN. Complexity at the End-User Premises Will Increase as More Services Are Added.**

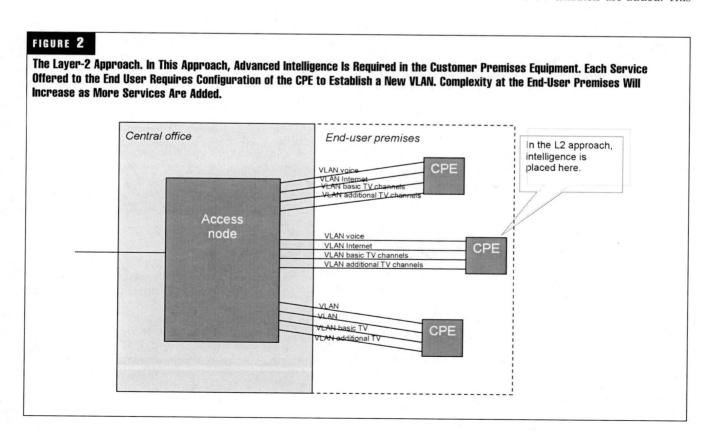

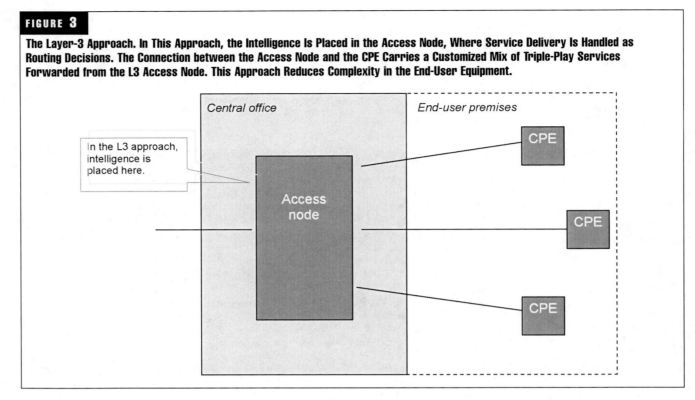

**FIGURE 3**

The Layer-3 Approach. In This Approach, the Intelligence Is Placed in the Access Node, Where Service Delivery Is Handled as Routing Decisions. The Connection between the Access Node and the CPE Carries a Customized Mix of Triple-Play Services Forwarded from the L3 Access Node. This Approach Reduces Complexity in the End-User Equipment.

approach increases the complexity of the end-user equipment, and the task of simultaneously upgrading services (reconfiguring) by 10,000 or more end-user boxes is highly demanding.

In contrast, the task of configuring new services becomes much easier when using a priority labeling mechanism, using the multicast technology, and controlling the traffic at the network level all the way out to the access node. This is true for both individual configuration for each end user and for configuration on a mass scale, such as that necessary during major network upgrades or permanent changes in the content offerings.

In fact, the difference is so significant that it will result in a network in which each service can be controlled in a second-by-second manner. No TV channel will be distributed to an end user unless it has been subscribed to, and a number of new business cases can be foreseen, given that a channel can both be configured and deactivated in a matter of seconds.

Another advantage of the L3 approach is that it can easily cope with many service providers. In an L2 environment, the end-user equipment must be reconfigured for every change in subscription status. Administrative problems may therefore arise when several service providers use such a network if every provider is to be allowed access to end-user devices.

## The Critical Choice of CPE

The most critical question for any access network operator today is undoubtedly whether service configuration should reside in the end-user device (customer premises equipment [CPE]) or not. This will affect not only which individual

services will have the potential to become profitable, but also the associated price-floor for new services (defined by the cost for setting up and maintaining the service). Furthermore, the decision will influence the level of control over services and security and reliability issues.

It is not wise to leave these critical devices in the hands of the end user. CPE units can be dropped on the floor or disassembled and may be costly to replace. Another important question is how much should be hardwired in the CPE. One of the worst cases for IP telephony may prove to be voice over IP (VoIP) hardwired or awkwardly implemented directly into the CPE. This is a good example of the unfortunate situation in which the network owner's equipment will dictate how service providers can implement their services.

The best solution is to establish a clear boundary between the responsibility of the network owner and the responsibility of each service provider at the outgoing ports of the CPE. The complexity of the hardware is reduced and the network investment will be secured for any future changes in the service offering.

## The Technology

Three factors are critical to making access networks designed for triple play and TV distribution truly efficient—quality of service (QoS), the possibility of multicasting, and a reliable authentication mechanism. QoS offers a reasonable way to ensure that there is enough capacity available for every bit stream in the network at any given moment. Multicasting is a highly efficient method of avoiding unnecessary traffic. Authentication is crucial to control and restrict access to network resources such as premium television channels for individual end users.

**FIGURE 4**

Any Functionality Embedded in the CPE Will Decrease the Flexibility and Make Administration of This Box More Complex. It Becomes Even More Complex in an Open-Access Environment in which Several Service Providers Need Access to the Configuration. The Issue of Responsibility is also Cumbersome, and Such Questions May Arise as to whether the CPE Is the Responsibility of the Network Owner or the Service Provider.

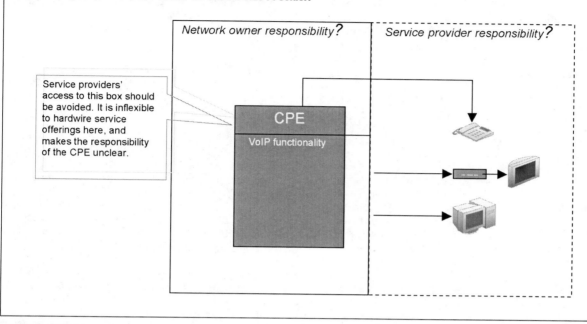

**FIGURE 5**

The CPE Should Not Have Embedded Functionality for any Service, Due to the Unclear Issue of Responsibility. Here, the CPE Is the Responsibility of the Network Owner, and a Clear Boundary Is Drawn between the Network Owner and the Service Provider.

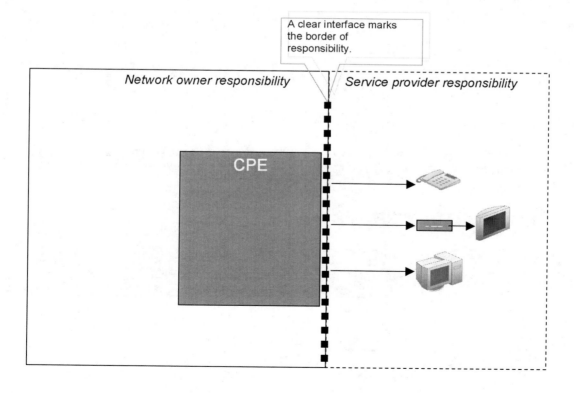

## Priority (QoS)

The Internet Engineering Task Force (IETF), the technical standardization body behind TCP/IP and the Internet, has been discussing efficient mechanisms for setting priority for more than 10 years. These discussions have led to the specification known as DiffServ, which allows a network administrator to manage network traffic in several categories. Unlike earlier solutions to the problem of priority, which allowed only a basic traffic-class management, DiffServ allows the administrator to add a certain kind of metric value to every packet in the network in a way somewhat similar to modern routing metrics.

The value added to the packet is known as the DiffServ code point (DSCP), and it is added to the type of service (ToS) field in the IP packet header when a packet enters the network, either from the boundary of another network or from an access connection (an individual end user).

A well-known problem with priority is that of administrative control. This is also why QoS mechanisms have not been very successful in network environments in which traffic must pass between many administrative domains, such as networks run by different network operators. The Internet is a good example, formed by a large number of independent domains called autonomous systems (ASs).

It is important to verify all DSCP values to maintain a reasonable level of control. A "web of trust" is created in this way between the routers in the network. Any packet not originating from a trusted source will be examined and, if necessary, relabeled. This applies, for example, to traffic from individual end users. This trust relationship ensures that an end user cannot gain a higher priority for his or her own traffic in the network simply by labeling it with higher priority. This avoids the sorry situation that has arisen with spam e-mails, in which every message is marked with highest priority.

It should be noted that it is much harder to reach this level of control in a network in which control is placed in the link layer (an L2 network).

## Multicasting

In a traditional cable TV (CATV) network, every television channel is distributed over the complete physical cable infrastructure, and the reception in the individual end user's home is controlled by a simple filter or encryption device. This design is simple but inefficient. It is also limited in that it is sometimes necessary to group several TV channels to match a certain filter, and access is commonly based on a shared encryption key rather than individual encryption keys.

The optimal method of distributing a large number of television channels over a network is, obviously, to distribute a video channel only to the end users who explicitly request it. This is not possible in a traditional TCP/IP network.

Multicast is a method in which the source sends the packets once, and the packets are then distributed in a tree-like fashion only to parties that explicitly subscribe to a particular service. Multicast was specially designed to optimize traffic distribution, and it thus solves the problem with congested networks that is caused by inefficient distribution of TV content. Network routers along the route must be multicast-enabled for multicast to function, and a number of rendezvous points (RPs) for the traffic must be established.

To receive a certain television channel, an end user sends a request to the nearest RP in the network to "join" the multicast stream that contains this channel, and the distribution tree is then instantly extended to include that end user. Similarly, a user may request to "leave" the channel. The association is then immediately removed and the tree no longer contains this end user. These requests are sent with a routing protocol called the Internet group management protocol (IGMP), and the multicast enabled routers use a router-to-router protocol called protocol-independent multicast (PIM) to exchange information about changes in the multicast tree structure. See *Figure 6* for an illustration of multicast.

The use of multicast has been somewhat limited until recently, due to its technical (routing) complexity and a lack

---

### FIGURE 6

Multicast Uses a Logic-Tree Structure. The Data Is Sent Only Once, and Then Branched off at Certain Router Nodes—Called Rendezvous Points—in the Network. Individual End Users Can Quickly Join or Leave a Multicast Channel by Sending a Request to such a Rendezvous Point.

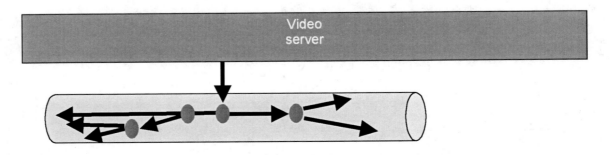

---

of interest among public IP network operators for the function. The problem of routing complexity, however, can be greatly reduced within a well-structured administrative domain such as a triple-play network run by an access network operator. This means that multicast has become an attractive choice to keep traffic volumes low and reach a high level of cost-efficiency in the network.

The ability to keep a tight control over who may join the different multicast channels in the network is, of course, critical. For this reason, a reliable authentication mechanism that is hard to bypass must be added.

### Authentication and Security

Access to TV channels must be restricted in almost every network that distributes such channels. Early solutions included simple filters that scrambled certain frequency ranges or brought unsubscribed premium channels out of sync for the receiving party. These are, however, rather basic solutions to the access problem and have been replaced to a certain extent by various encryption devices and regularly distributed shared encryption keys. This has, to some extent, increased the level of security.

The ultimate way of controlling television content would be to use a method in which only the exact numbers of subscribed channels are distributed from the access node to the end user—and only after a robust means of authentication.

A multicast-enabled L3 network that passes routing information all the way out to the access nodes provides exactly this capability. Here, the network operator is offered full control over which packets are allowed over the cable to each individual end user. Moreover, the control is carried out at the access node rather than at the user device, thus simplifying the CPE.

A very efficient method of authenticating end-user equipment—and ensuring that the connected device is an authorized one—is to use its MAC number as identification. The device issues a dynamic host configuration protocol (DHCP) request to the network to obtain an IP address, and the access node adds some information about the physical connection (cable) over which the request was made. This combined information can then be checked against centrally stored information about the end user and the related individual mixture of services. This information is sent back to the access node, which enables the services. In this way, the user is not involved at all in the authentication process. No manually entered encryption keys are used, and it is almost impossible to manipulate the network to deliver unsubscribed services.

### Summary

Distribution of television channels is an activity that consumes large quantities of bandwidth, despite modern video compression formats. Therefore, it is essential to optimize the way television signals are handled in an IP–based triple-play network.

This paper outlines an architectural approach to triple-play networks based on L3 (network level) switching from the core of the network out to each access node connecting individual end users. This design makes administration easier than it is in L2 (link level) networks and reduces the need to invest in complex end-user devices. It also paves the way for the use of multicast technology to minimize network load while giving the network operator unprecedented control of service delivery to each end user.

# Toward Less Homogenous TV

David Howard
*Principal*
Up Periscope

On August 16, 2005, Veronis Suhler Stevenson started shipping its 2005 Communications Industry Forecast report. Within days, media types nationwide were engrossed in more than 500 pages of text, charts, tables, and spreadsheets detailing trends and forecasts on spending in industry segments such as yellow pages advertising; consumer coupons; public relations; and broadcast, cable, and television advertising. Copyright law prevents verbatim reproduction of Veronis Suhler Stevenson's material in this space, but it is fair to say that the report resonates with stories in the popular press and anecdotal reporting on viewing and spending trends that are of interest to participants in the pitched battle between carriers, broadcasters, and cable operators for television subscribers.

One trend that comes to light, and which has also been reported on in the *Wall Street Journal*, is the increasing sameness from cable network to cable network, channel to channel. Fox Networks has been taken into litigation by the traditional broadcasters on accusations of out-and-out copying of certain reality TV program concepts. Putting aside the allegations of copying, certainly the average viewer recognizes similarities between two shows on different networks each concerning up-and-coming pugilists, or between *CSI* and *Law & Order*. Further, cable operators have been suing networks for breach of contract over format changes that blur distinctions between channels and foul the operators' lineup. Presumably such format changes and alleged copying are motivated by the pursuit of higher ratings and, subsequently, higher advertising rates. But as the multiple-system operators (MSOs) and satellite broadcasters increase the number of available channels, should viewers not expect more varied content?

This situation presents an opportunity for traditional telephone carriers entering the Internet protocol television (IPTV) space to differentiate their services from cable. And not just services for subscribers, but for advertisers too. It is worthwhile to consider the advertising case first—advertising spending in the United States across broadcast television networks, stations, and cable and satellite networks is forecast by Veronis Suhler Stevenson to exceed $65 billion in 2005, surpassing cable and satellite subscription fees, which are forecast to reach just over $60 billion. And growth in cable television advertising spending is forecast to significantly exceed gross domestic product (GDP) growth for the next several years.

## The 18-to-34 Demographic

Much of the homogeneity of content across channels is due to advertising dollars chasing the free-spending 18- to 34-year-old audience demographic. The problem facing advertisers in recent years has been the development of advertising capabilities for new vehicles that naturally attract this demographic, such as Internet advertising (almost $10 billion in 2004) and video game advertising (roughly $100 million in 2004). How should the national multibillion-dollar-per-year advertiser allocate its dollars across television, Internet, and video game venues? The broadcasters and networks try to "help" the advertiser decide by developing more content that appeals to, or adjusting the format to reach, this key demographic. The result is the sameness from network to network than can turn viewers off.

Carriers deploying IPTV can use this state of affairs to their advantage by combining their experience in Internet services with powerful new IPTV technology that implements what has been learned about Internet advertising techniques. The technology of IPTV is grounded in the same technology as the Internet, it is inherently bi-directional between viewer and source, and there is every reason to believe that advertising tools will migrate from Web-based content delivery to IPTV multimedia delivery. The advertising targeting and tracking capabilities that were developed for the Internet—paid search advertising, contextual ad placement, pay per click, conversion tracking, referral reporting, and so forth—have been developed on Web-based platforms because, until now, it was possible there and nowhere else. Who would have refused these features for conventional linear television as development for the Web began 10 years ago if it could have been done? In addition, the media experts continue to develop new ways to hold viewers' interest throughout the spot, such as shorter, smarter spots and in-context advertising, which attempts to closely match advertising content with the context of the program it is interrupting. Again, IPTV is the ideal platform for the content and ad spot libraries and the associated metadata to make it work. And carriers do not need to start from scratch to secure trial participants for these new IPTV advertising models—they have an existing roster of sponsors for their Internet-based content that can offer special programs for migration to IPTV models.

Nielsen Media Research—the granddaddy of television audience research information, which is used to price advertising and make programming decisions—has been slow, at least publicly, to jump on IPTV viewer measurement, tracking, and reporting. It seems there was nobody at the Nielsen booth at the National Cable Television Association (NCTA) in San Francisco in spring 2005 capable of "speaking IPTV." They may be forgiven for this as they have been battered recently by many camps for, among other things, underrepresenting certain ethnic markets in their measurements. So development of these tools will likely have to come from the IPTV vendors themselves. And, interestingly, the scalability of such systems, derived as they are from Web server technology, may provide the means of overcoming the obstacles faced by Nielsen by continually tracking viewer behavior based on a virtually 100 percent sample size and storing massive amounts of data for post processing, analysis, and reporting to advertisers. IPTV carriers have an opportunity to get a leg up with advertisers over their competition with behavior tracking and metric reporting that exceeds the capability of conventional television and goes far beyond simple legacy demographic models.

## Programs as the Center of Attention

If you have not already heard it, here it is again: The advent of IPTV technology virtually enables all content to be made available on demand, rendering the channel as we know it today obsolete. The Internet introduced random access and on-demand content to the masses of the world, and TiVo pioneered the digital video recorder to apply the same concepts to linear broadcast television. The initial success of TiVo has prompted the cable and satellite operators to embrace digital video recorders (DVRs) as well, as they slowly come to grips with the "evils" of ad skipping. It certainly seems that consumers will adopt video on demand (VoD) en masse, whether through IPTV or DVRs attached to traditional broadcast systems.

Telco TV providers can then use IPTV VoD as the basis for their television services, offering distinct content to niche viewer bases and thereby avoiding the whole "different channel, similar content" problem altogether. The "sameness" problem that cable operators are facing is rooted in the current organization of content around linear and real-time channels, along with the presumed demographic groups those channels appeal to. Remember that advertising dollars are used to fund, or rather, recoup costs for, program development. Better advertising reporting through IPTV, as discussed earlier, should provide for at least a couple of improvements, namely better matching of advertising dollars to target demographics and the content they consume and identification of viewer groups with fine granularity and matching them with content that can be funded by advertising dollars directed at them. There is a subtle difference here. Advertisers in general are crying out for more

effective targeting and reporting, demanding more bang for their buck, as seen in the first case. As for the second case, it was the decline in advertising rates that drove the broadcasters to examine ways to wring cost out of production, which resulted in the fad of reality TV. Further, as production technology (cameras, editing tools, etc.) continues to decline in cost, there is more and more programming of sufficient quality on the market to attract advertising spending for narrow interest groups.

More effective advertising, combined with on-demand content, should translate to a richer programming package for IPTV viewers distinct from the blandness of the multichannel universe. Telephone carriers should pay careful attention here as they develop their IPTV business models. To the extent that telephone carriers offer the same channel lineup as the MSOs, they will suffer the same lack of distinction among channels as their chief competition. While carriers need to negotiate content license agreements with the branded networks to secure content for delivery today, they need to keep one eye looking forward to the migration toward content—not channels—as the center of viewers' focus. Consider the growth in recent years of the market for television programs available for rent on DVD from bricks-and-mortar video stores. This is evidence of consumers actively seeking out television programs, not channels or networks, for on-demand viewing on their own schedule. The rental of television programs on DVD is the "sneaker net" version of what IPTV VoD services afford. IPTV providers that can accelerate the migration from a channel lineup to an all on-demand library of diverse content will be well positioned.

## In Pursuit of Advertisers and Subscribers

If carriers simply negotiate for the same channel lineups as cable operators do and stop there, they will look no different from their competition to subscribers or advertisers. However, if they build in to their networks today the capabilities for highly targeted advertising on the basis of massive viewer sample sizes, in-context advertising, individual viewer profiling, and techniques gleaned from Internet advertising, carriers will be positioned on the higher ground for coming battles with the MSOs. The top three national cable advertisers spent over $1 billion on ads last year. The top 10 spent almost $2.5 billion and that, in addition to subscriber dollars, should be the telephone carriers' target.

Fully exploiting the scalability and capabilities of the IPTV platform is the best way to keep the viewer engaged with fresh, distinct programming that does not cause their eyes to glaze over. The combination of VoD program-centric consumption with Internet-caliber targeting and tracking presents new opportunities for highly differentiated subscriber and advertiser services to give carriers an edge over cable and satellite.

# Live Video Feeds for an IPTV System

## David Lui

*Engineering Manager*
Broadband Network Systems, Ltd.

## Abstract

Video encoding and Internet protocol (IP) broadband technology for digital subscriber line (DSL) and metro Ethernet networks has advanced significantly in the past few years. This has not only made the distribution of live video and TV content to broadband subscribers feasible, but it also has transformed IP television (IPTV) into a formidable competitive threat to existing cable and satellite operators.

For an IPTV service to be successful in an increasingly competitive pay-TV environment, operators have to understand the functionalities and constraints of the video feed options—live streaming and VOD—to be able to choose the equipment and feature enhancements to fit their requirements.

This paper will cover the two key live video feeds to an IPTV system. It provides a basic and practical introduction to deployment considerations of this part of the system from an operational point of view and highlights the importance of adopting a codec-agnostic approach to video encoding.

## Overview

When considering deploying a telco IPTV service, understanding the technical implications of delivering the solution should be of paramount importance. In the overall technical parameters of an IPTV service rollout, there are four key areas, or modules, that need to be addressed to ensure a robust and scalable service delivery: content distribution, middleware, transport infrastructure, and customer-premises equipment (CPE) (*Figure 1*).

The content distribution module contains live encoding platforms, a video file repository, and IPTV video servers, which are the key elements enabling video feeds for an IPTV service.

In general, content distribution (in the head end), processing, and adaptation are all part of the functions of a TV head end. However, this paper focuses on content adaptation where the IPTV system consolidates multiple live video sources—both live events and video sources that are fed to the IPTV system live—into IPTV video feeds.

While the live video encoder and IPTV video server are conceptually considered to be part of the TV head end, they do not necessarily need to be placed at the same physical location. Multicast video sources are usually at the top level of the core network for better bandwidth efficiency, whereas unicast IPTV video sources are commonly installed at the local point of presence (PoP) level to minimize core bandwidth usage.[1]

## The Live Video Encoder: Video Encoding, Transport Stream (TS), and Delivery Techniques

### Encoding Techniques
As uncompressed digital video and audio data is far too large to be carried by most transport networks, encoding both digitized audio and video data is necessary to facilitate smooth data transport. Encoding takes away the part of the audio visual information that is redundant and not quite noticeable to humans. The compression of audio and video is handled separately in the encoder.

In the broadcast industry, audio encoding is performed according to audio encoding standards such as Moving Pictures Experts Group (MPEG) Layer II, MPEG–2 advance audio coding (AAC) (International Organization for Standardization/International Electrotechnical Commission [ISO/IEC] 13818-7), MPEG–4 AAC (ISO/IEC 14496-3), Dolby audio codec (AC)–3 2.0, and Dolby AC–3 5.1.

For video, MPEG–2 (ISO/IEC 13818-2:2000[2]) has emerged as the most popular encoding standard. However, the continuing demand for video encoding at ever lower bit rates has made more advanced video codecs such as the MPEG–4 standard, Part 2 (ISO/IEC 14496-2) and Part 10 (ISO/IEC 14496-10) increasingly popular.

### Advanced Video Coding
MPEG–2 video encoding (and other legacy video codecs) has limitations as to how far bit rates can be reduced. To overcome these limitations, advanced video coding (AVC) was introduced. AVC is a new generation of video codecs that offers high-quality video at low bit rates, usually at or below 2 Mbps. Two examples of the AVC standard are MPEG–4 Part 10 (ISO/IEC 14496-10), also known as H.264, which was proposed by the Joint Video Team (JVT[3]), and virtual concatenation (VC)–1, the Society of Motion Picture

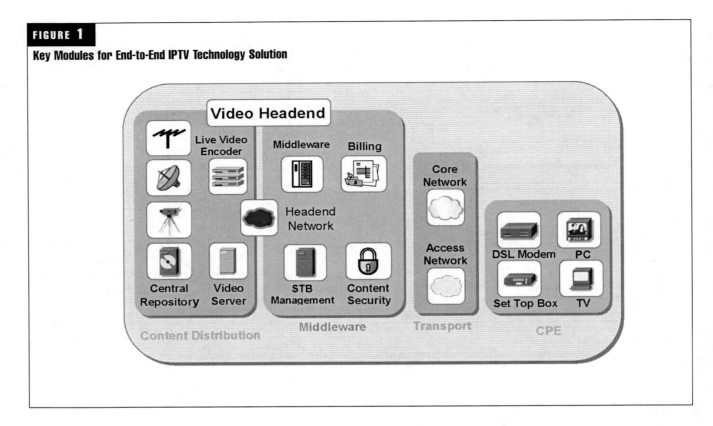

**FIGURE 1**

**Key Modules for End-to-End IPTV Technology Solution**

and Television Engineers (SMPTE) 421M[4] standard submitted by Microsoft.

AVCs achieve greater compression ratios than MPEG–2 by performing more precise and flexible inter-frame referencing structures and higher-complexity video compression analysis. Hence the processing of AVC requires greater computation load then MPEG–2. This is also why AVC encoders and decoders possess higher central processing unit (CPU) power than less sophisticated codecs such as MPEG–1 and –2.

### IPTV MPEG–2 Transport Stream Delivery
In the first phase of preparing encoded audio-visual contents, the independently encoded audio and video streams are called elementary streams (ESs). The audio and video ESs are packetized independently into packetized ESs (PESs) before they are combined (or multiplexed) to form either a program stream (PS) or transport stream (TS). Both PS and TS multiplexing techniques are defined in the ISO MPEG–2 Part 1 Standard: Systems description standard (ISO/IEC 13818-1:2000)[5]

MPEG–2 PS packets are large and variable in size, while MPEG–2 TS packets are fixed at 188 bytes. The fixed-size packet principle for MPEG–2 TS has the advantage of making the decoding process more predictable at the decoding point. Another advantage is that the network commonly used to transmit the data is inherently susceptible to bit-level errors and packet losses. Fixing the size of the packet increases data immunity against these errors.
Since one of the key value propositions of digital broadcast is reliability of service delivery, streaming MPEG–2 videos are encoded in TS format (i.e. MPEG–2 TS or simply TS) as

opposed to MPEG–2 video used for other purposes that are usually based on the PS format.

In broadcast terminology, a TS carrying a single video program can be further defined as a single-program TS (SPTS). Hierarchically, multiple SPTS programs are multiplexed to form a multi-program TS (MPTS) before the video programs are streamed out to broadcast networks. This multiplexing process is especially popular for general content contribution and digital broadcasting (i.e., content distribution). A MPTS carries different kinds of TS–packetized information for the multiple channels inside it. This includes video, audio, subtitles, and even electronic program guide (EPG) information.

Inside a MPTS, each piece of packetized information is assigned a program identification (PID) to identify the purpose of that specific TS packet. To give meaning to PIDs, tables of information are sent along the MPTS to inform the receiving entities which PIDs belong to which single TV program. These tables are called program-specific information (PSI) tables and are packetized with well-known PIDs.[6] In summary, multiplexing of encoded video and audio PESs is done at the MPEG–2 system level (ISO/IEC 13818-1:2000), whereas multiplexing SPTSs to form a single MPTS is a program-level procedure.

TS multiplexing and encapsulation is the standard for packetizing compressed audio-visual streams and metadata in digital broadcast networks. To leverage the existing MPEG–2 TS infrastructure, new generation video codecs streamed on the same network, such as H.264, will also have to be encapsulated in TS. This reasoning is behind the often-used phrase "H.264 over MPEG–2 TS," or simply "H.264 over TS." In the case of IPTV, it is "H.264 over TS over IP."

Streaming of MPTS is not necessary or practical in most broadband network–based IPTV services. Since an IP network is not a broadcast-type network, the IP transmission mechanism of video does not require a scheme such as MPTS to synchronize the flow of all streams. Also, the last mile of most broadband networks is limited in bandwidth and cannot accommodate multiple video streams embedded in the MPTS to be transported over the last mile simultaneously.

Finally, SPTS streaming gives IPTV service providers greater flexibility in controlling the flow and admission of SPTS video streams at per–SPTS levels. For this reason, the MPTS video feeds to such service are transformed into individual SPTSs, where each SPTS is mapped to a single multicast group before launching to the broadband network. The flow of multicast SPTS channel is controlled by the multicast-enabled core and broadband infrastructure, running multicast routing protocol and Internet group multicast protocol (IGMP) respectively.

### Video Encoder Input and Output

Among all video transmission interfaces, serial digital interface (SDI)[7] is most commonly used for transmitting uncompressed digital video streams in broadcast head ends. Digital audio can either be embedded in an SDI stream or carried independently on a separate interface such as the Audio Engineering Society/European Broadcasting Union (AES/EBU)[8]. Some video encoders support insertion of pre-encoded audio onto the output TS. In an IPTV system, there are a number of ways for an SDI video source to be fed to the IPTV video encoders, including the following:

- *Digital video cameras with SDI output*: Digital video cameras are deployed where local programming production is practiced by the IPTV service operator. However, the potentially substantial investment in

enabling a production capability at the head end usually prevents new IPTV operators from executing local programming production in the launch phase of a service.

- *SDI video playout servers*: To streamline the process of content ingest, the use of SDI video playout servers is a logical option. (Note: The video playout server is different from an IPTV video server, as its output is in SDI format.) These servers provide functions such as scheduled program playout automation, program insertion, and live broadcast delay playout in case of censorship requirements. Edited TV programs can be preloaded onto the SDI video playout server from video tape recorders (VTRs) or pre-production workstations. The SDI output of the playout server can then be connected to the input of the video encoders.

### Encoding and Decoding Quality

Video and audio encoding standards are designed to provide guidelines for encoder and decoder manufacturers to conform to certain procedures and specifications. However, the implementation of overall encoding and decoding processes are discretionary, enabling vendors to offer encoders with varying features and at different prices. For instance, as an enhancement feature, an encoder may implement a pre-filtering process to filter out unnecessary information in the video signal (e.g., background noise) before the encoding commences. This lowers the encoding bit rate and improves picture quality. Two-pass encoding is another technique practiced in variable bit rate (VBR) video encoding. It enables the encoder to first measure the complexity of the video scene to determine the optimal encoding settings and encoding bit rate to achieve bit-rate savings.

In contrast, some video quality enhancement features, including MacroBlock-Adaptive Frame/Field (MBAFF[9]) coding and de-blocking filtering (in H.264), require both the encoding and decoding sides to support the same feature. Encoding process delays and the freedom to adjust certain

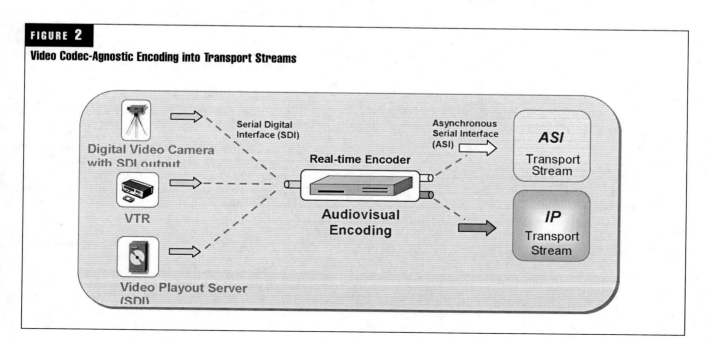

**FIGURE 2**

**Video Codec-Agnostic Encoding into Transport Streams**

Digital Video Camera with SDI output

VTR

Video Playout Server (SDI)

Serial Digital Interface (SDI)

**Real-time Encoder**

**Audiovisual Encoding**

Asynchronous Serial Interface (ASI)

*ASI* Transport Stream

*IP* Transport Stream

video encoding and network parameters are also factors that contribute to the quality of an encoder. To fulfill the network quality of service (QoS) requirements, some encoders have the capability to assign priority bits in the type of service (ToS) field[10] of the IP streaming video packet headers.

### Video Encoding Parameters
The properties of video codecs are defined by two independent parameters: profile and level. These labels were introduced to identify the different visual applications and purposes of a codec.

#### Profiles
Profile is related to the purpose or application of the encoded video. Taking MPEG–2 as an example, MPEG–2 profile defines chroma sampling, the types of frames used in inter-frame coding, and the adoption of certain enhancement features. For example, MPEG–2 video for editing and primary contribution purposes often adopts the MPEG–2 422 profile (or 422p) for its capability to be edited with comparatively low degradation.

On the other hand, for video distribution such as an IPTV service, the MPEG–2 main profile (with 4:2:0 chroma sampling) is used since the decoding and processing of 4:2:0 video is less complicated than that of the 422p. Due to the lower processing requirements in 4:2:0, it is more efficient to deploy MPEG–2 IPTV set-top boxes (STBs) to support 4:2:0 only because the investment in STBs usually makes up a substantial part of any TV broadcast service. An EBU research article has reported that the subjective video quality difference between video encoded in MPEG–2 422 profile and main profile, at bit rates lower than 5 Mbps, is small[11].

The three numbers in the 422 profile refer to the sampling scheme of the YUV components (the luminance, chrominance blue, and chrominance red components) of the original video[12]. Profile also defines the subset of the frame types used (e.g., a H.264 baseline profile does not support the use of bidirectional predictive picture elements).

H.264 handles inter-frame correlations at the slice level.[13] There are three common H.264 profiles: baseline, main, and high. For H.264 to replace MPEG–2 as the principal broadcast video codec standard, it needs to support studio and editing applications. The JVT[14] has defined a range of H.264 profiles. This new group of profiles is known as the fidelity range extensions (FRExt[15]) and comprises four profiles: high profile (HP), high 10 profile (Hi10P), high 4:2:2 profile (Hi422P) and high 4:4:4 profile (Hi444P).

#### Levels
Each profile has its own set of relevant levels that describes the resolution and frame rate of the encoded video. The most common form of MPEG–2 video is the main profile @ main level (or MP@ML), where main level represents a resolution of 720 by 576 at 25 frames per second for PAL and 720 by 480 at 30 frames per second for the National Television System Committee (NTSC).

#### Observing Compatibilities
Since the processing requirements for decoding video at different profiles and levels can vary, it is important that the targeted decoder is compatible with the chosen profile and level. The higher the profile and level of a codec, the higher the encoding and decoding loads need to be. For example, a mobile phone performing H.264 encoding for videoconferencing is expected to encode video at a lower resolution, less CPU power, and possibly lower encoding delay than an H.264 encoder running in a head end.

The mobile phone's decoding characteristics are expected to be under a similar comparison to an H.264 STB. This means the mobile phone will likely be able to support simpler processing decoding profiles and levels than the STB.

Profiles and levels differ from codec to codec. For instance, the common profiles of H.264 are the baseline, main and extended profiles, whereas the main and 422 profiles are widely used among many MPEG–2 codec applications. There are 19 visual profiles in the MPEG–4 Part 2 standard (ISO/IEC 14496-2) in which the advanced simple profile is the most popular choice for general video viewing.

The encoding bit rate required for a TS is dependent on the video/audio codecs in use, the quality of the encoder, the profile and level of the encoded video, and the scene complexity of the video content. As a general rule, a standard MPEG–2–encoded full-D1 resolution (720 by 576) PAL video with MPEG Layer II audio compression forming a TS of reasonably high quality can be encoded at around 3.0 Mbps to 5.0 Mbps in real time. For H.264 TS, the encoded bit rate is around 1.5 Mbps to 2.0 Mbps at reasonable quality in real time.

To achieve broadcast quality, sports programs usually have to be encoded at higher bit rates than news report programs. This is due to the larger amount of motion contained in complex sport programs, which reduces the inter-frame correlation between consecutive video frames.

### Integrated Receiving Decoders (IRD)

Satellite transmission is a popular method for distributing video programs from content providers to TV service providers and on to subscribers (i.e., digital video broadcast – satellite [DVB–S] subscribers). In satellite TV broadcasting, the two best-known portions of the electromagnetic (EM) spectrum used are the C-band (downstream, 3.7–4.2 GHz[16]) and Ku-band (downstream, 5.9–6.4 GHz).

After receiving C-band or Ku-band signals by a satellite dish, the signal is first converted to a lower-frequency signal (L-band) using a low-noise block (LNB), also known as a low-noise converter (LNC). The L-band frequency is also regarded as intermediate frequency (IF) since it is down-converted for the application. As the IF is in the lower-frequency spectrum, the frequency down-conversion allows video signals to travel on cables at 950–2150 MHz and at much lower attenuation than C and Ku signals.

Satellite-transmitted video streams are commonly multiplexed as MPTS. To tackle this, IRDs have the capability to de-multiplex an incoming MPTS into single/multiple SPTS, depending on the physical configuration of the IRD. *Figure 3* outlines the input and output of an MPEG–2 IRD. In this example, the IRD is capable of decoding an incoming MPEG–2 video stream and output it as an uncompressed digital video stream (SDI). Due to the popularity of IPTV,

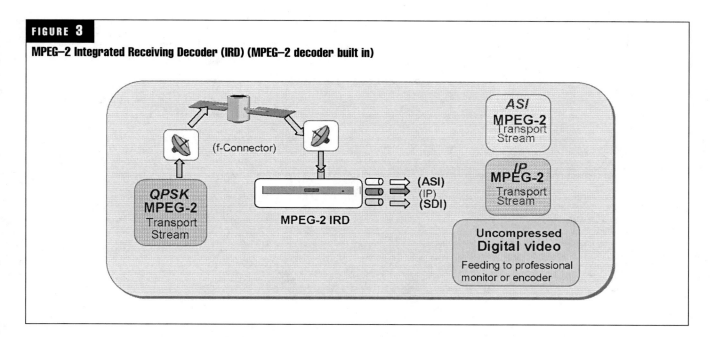

**FIGURE 3**

**MPEG–2 Integrated Receiving Decoder (IRD) (MPEG–2 decoder built in)**

most IRDs today are equipped with an Ethernet interface for MPEG–2 TS over IP output by default.

It is possible for an MPEG–2 TS feed from the satellite to be encoded at a higher bit rate than what is actually supported by the receiving MPEG–2 IPTV service provider's broadband network. More precisely, broadband networks predominantly offer IPTV services at around 4.0 Mbps or lower per channel; whereas a satellite transmitted MPEG–2 TS can be encoded at 8.0 Mbps due to the nature of contribution. In that case, a bit rate adaptation process is needed to lower the TS bit rate. The bit rate adaptation is not commonly executed inside the IRD but in a transrating device. Standalone and modular type transrating devices provide greater cost benefits than decoding and re-encoding the TS at lower bit rates. Transraters operate in the MPEG–2 domain which makes the complex process of decoding and re-encoding the content back into MPEG–2 unnecessary[17]. However, the performance of MPEG–2 TS transrating devices vary with some transrating equipment vendors claiming to be able to reduce the bit rate by up to 30 percent without significantly degrading the video quality at the service end.

### Decryption
Most incoming MPEG–2 TSs received from a satellite are already encrypted by a conditional access (CA) scheme from a particular vendor. If an IPTV service provider is using CA from one vendor that is different from the one used by the satellite, the encrypted MPEG–2 TS must first be decrypted before being encrypted again by the service provider's preferred CA. To perform the decryption, IRD commonly provides at least one slot to host the CA modules that fit the CA decryption smart card to decrypt the incoming video stream. The clear MPEG–2 video TS can then be further processed by the IPTV head end.

### De-Multiplexing
If the incoming MPEG–2 TS is in an MPTS format, the IRD has to be able to de-multiplex the MPTS and select one program for output as an SPTS. However, it is not uncommon

for the IRD to also pass through the MPTS to the IPTV head end to be de-multiplexed at a later stage.

MPEG–2 IRDs are capable of decoding MPEG–2 TS into uncompressed SDI video format so the video can be re-encoded (i.e., transcoded) to the required codec before being fed to the IP network. An example for this is converting the MPEG–2 TS video into H.264 over TS over IP for an IPTV service.

## Summary

With the advances in video encoding and broadband networking, the number of concurrent video channels that can be carried to each subscriber loop for simultaneous streaming of video channels is set to increase continuously.

To be prepared for these developments, it is important for an IPTV system to be video codec-agnostic, as this allows operators maximum flexibility for future upgrades. In a codec-agnostic system, a change to a new video codec only requires the encoder and IPTV STB to be updated via software updates or replacements without major expense and integration efforts.

## Notes

1.  For more information on non-live/off-line content by IPTV video servers and network architecture and IP video servers, please refer to "Distributed vs. Centralized Architecture for IP VoD," which can be downloaded at www.bnsltd.com.
2.  Information technology – Generic coding of moving pictures and associated audio information: Systems www.iso.ch/iso/en/CatalogueDetailPage.CatalogueDetail?CSNUMBER=35006
3.  JVT: (a joint project between the ISO/IEC and the International Telecommunication Union (ITU www.itu.int/ITU-T/studygroups/com16/jvt/JVTToR.pdf
4.  Society of Motion Picture and Television Engineers, ww.smpte.org/news/press_releases/003_06.cfm
5.  Information technology – Generic coding of moving pictures and associated audio information: Systems www.iso.org/iso/en/CatalogueDetailPage.CatalogueDetail?CSNUMBER=31537. For a brief

description of the "Part" definitions under ISO/IEC 13818: www.chiariglione.org/mpeg/standards/MPEG–2/MPEG– 2.htm

6. To manage the multiplexing of different types of information, the Digital Video Broadcasting Project (DVB) and Advanced Television Systems Committee (ATSC) have used MPTS in their broadcast transmission hierarchical model; the DVB–TR 101 154 and ATSC A/65, both extensions of the ISO/IEC 13818-1 standard for multiplexing at the MPTS level.

7. ITU–R BT.656(formally CCIR-656), www-inst.eecs.berkeley.edu/~cs150/Documents/ITU656.doc
SMPTE–259M, www.smpte.org/smpte_store/standards/index.cfm?scope=0&CurrentPage=12&stdtype=smpte

8. For simplicity, this document will not specifically discuss digital audio processing in detail as it is often an embedded entity in SDI stream or TS.

9. MBAFF, www.dspr.com/www/technology/csvt_overview.pdf

10. RFC 1349: Type of service in the Internet Protocol Suite, www.ietf.org/rfc/rfc1349.txt

11. EBU Technical Review – Autumn 1999, "MPEG–2 4:2:2 and 4:2:0 – comparative subjective tests at low bit rates," Anders Nyberg (Sveriges Television, SVT, Sweden), www.ebu.ch/en/technical/trev/trev_281-nyberg.pdf

12. The interpretation of the three numbers goes beyond the scope of this report and will therefore not be discussed in this report.
For comparisons between the quality of 422 and 420 MPEG–2 videos at different encoding bit rates and resolutions, visit www.ebu.ch/trev_279-cheveau.pdf (For a full description of the different MPEG–2 profiles and levels, please refer to an MPEG–2 profile-level table).

13. A slice is a macroblock or a set of macroblocks that make up the picture of the screen; this is analogous to the frame level in MPEG–2.

14. JVT is a joint project between the ISO/IEC and the International Telecommunication Union (ITU), www.itu.int/ITU-T/studygroups/com16/jvt/JVTToR.pdf

15. www.imtc.org/docs/LiaisonFRExtAmendment.doc

16. www.ntia.doc.gov/osmhome/allochrt.pdf, allocated by the Federal Communications Commission and National Telecommunications and Information Administration, United States

17. This is only valid for an MPEG–2 IPTV service. For an H.264 IPTV service, the incoming MPEG–2 TS is required to be decoded into an uncompressed SDI format and re-encoded into H.264 TS.

# Requirements of the New IPTV Network: Beyond Optimization

## Daniel Marcus

*Broadband Product Marketing Manager*
UT Starcom, Inc.

Wireline carriers realize they can no longer incrementally enhance or "optimize" existing network elements to support Internet protocol TV (IPTV) services. Unlike high-speed Internet access, IPTV must compete, in both quality and services breadth, with a long-entrenched incumbent standard bearer: cable TV. To compete successfully, carriers will have to overhaul their existing network architectures to enable greater capacity and flexibility. Unlike their cable competitors, whose network equipment investments suggest gradual optimization, carriers have a unique opportunity to leapfrog current offerings by delivering a radically improved TV viewing experience to subscribers. Features such as time-shift TV and massive video on demand (VoD) libraries can have a significant impact on network design and are all the more reason why carriers must build their networks wisely.

## Bringing Home the Bandwidth

IPTV is a critical element of the digital home, an evolving term that describes the trend toward home subscriber environments, including inter-networked elements that enable seamless user experience regardless of content origin for TV, music, phone, and high-speed Internet. Enabling these new services requires the ability to deliver sufficient network capacity.

For carriers to predict bandwidth requirements and their consequences on equipment selection and deployment, it is useful to have a starting point for simultaneous service delivery to the digital home. A safe initial assumption would include three TV streams—two standard-definition (SD) and one high-definition (HD)—three voice over IP (VoIP) phones, and streaming digital audio and music.

With advanced compression standards (H.264), this digital home scenario suggests a minimum bandwidth requirement of 15 megabytes[1]. This is a conservative estimate based on technology available today, but this will surely increase as HD content becomes ubiquitous and HD–capable displays become increasingly commoditized. This is critical for carriers to bear in mind because three simultaneous HD streams alone require 24 megabytes, without even considering the implications of upcoming applications such as video

telephony and personal broadcast. This could drive the bandwidth requirements for the digital home to 50 megabytes and above, which in turn will have a number of implications for carriers on selecting appropriate access technology solutions.

## Endless Channels, Multiple Delivery Options

Carriers have options for delivering this high bandwidth to the digital home. Digital subscriber line (DSL) technologies have grown at a tremendous rate[2] in recent years, and analyst projections suggest this trend will continue. DSL access multiplexer (DSLAM) platforms must accommodate current bandwidth requirements while simultaneously providing a path for future growth. This means providing a backplane sufficient to support today's and tomorrow's highest-speed DSL variants. These include asymmetric DSL (ADSL) 2+ (here today, with up to 26 megabytes downstream and 3 to 6 megabytes upstream), and very-high-data-rate DSL (VDSL) 2 (already in trials, with up to 100 megabytes downstream and 50 to 100 megabytes upstream).

Next-generation DSL standards are an important consideration for carriers in their selection of appropriate DSLAM platforms. Equally worthy of consideration is DSL's inherent distance limitations and the potential for new technologies to provide remedies. Environmentally hardened DSLAM platforms designed for outdoor deployment are now available. These DSLAMs incorporate the latest DSL technologies in hermetically sealed, line-powered enclosures that enable carriers to effectively deploy DSL to subscribers and communities via 3-kilo-feet and shorter copper loops, enabling data rates in excess of 50 Mbps.

Passive optical network (PON) technology provides a compelling complement to DSL for fiber-to-the-node (FTTN) applications, and a compelling alternative in the case of fiber-to-the-home/business (FTTH/FTTB) applications. An ideal solution where fiber is available in the access network, PON technology's symmetrical nature anticipates new types of interactive, bandwidth-intensive subscriber applications such as video telephony. PONs combine the high capacity of fiber with the scalability of point-to-multipoint network topologies. Two PON variants stand to play an important role in access networks serving digital homes: Gigabit Ethernet PON (GEPON) is available today, and ATM–based Gigabit PON (GPON) will be available in the future. The currently available GEPON variant enables a single gigabit Ethernet uplink to be split between 32 subscribers, offering 30 megabytes of symmetrical bandwidth to each connected digital home.

## IP Technology Frees Subscribers from Schedule-Driven Programming

Bandwidth capacity is not the only predictor of an operator's success in accommodating the requirements of the evolving digital home. The operator's access network platforms must enable flexibility and bandwidth management to an extent previously reserved only for equipment residing in the network core. Time-shift TV is an excellent example of a service that necessitates increasingly intelligent access solutions. Time-shift TV, also known as network personal video recorder (nPVR), coupled with network-wide VoD, expands the concept of "watching TV" to include higher levels of interaction and control. Essentially, time-shift TV enables subscribers to pause and rewind live TV programs, adding fast-forward functionality up until the point at which a subscriber reaches parity with the system-wide live broadcast. By coupling time-shift TV with network-wide VoD, carriers are poised to offer subscribers full control over their viewing experience—any program, at any time, with full VCR-like control. Applications such as time-shift TV create a whole new set of demands. Carriers must effectively meet these demands by deploying intelligent access solutions that enable seamless transition between multicast and unicast streaming—two methodologies for streaming media across a network.

As the newest market entrants in the TV services space, carriers are uniquely positioned to deploy high-bandwidth access networks capable of rapid transition between multicast and unicast, depending on the behavior of each subscriber. Multicast uses a shared access topology to relay a single instance of content to multiple verified subscribers on the network at the same time. Clearly, for multicast networks, the number of programs available at any one time must be limited by the capacity of the network employed to deliver them. By contrast, unicast delivers a single unique media stream to each individual user on the network. Consequently, the potential arises to offer vastly more viewing options in addition to interactive services such as time-shift TV. The downside of the unicast model is that the network must accommodate all of this new content and can become bogged down if it does not support multicast and unicast transition throughout (for both core and access).

The challenge for carriers is how to provide the greatest possible flexibility in their network so that whenever a subscriber requests a program that is already being delivered to other subscriber(s), that request will result in shared distribution of that program (the subscriber joins the appropriate multicast group). This will conserve bandwidth by eliminating the need for identical instances of a program to traverse the network. Internet group management protocol (IGMP) can solve this problem.

IGMP is an Internet protocol that enables DSLAMs, PON optical line terminals (OLTs), and routers to passively "snoop" subscriber traffic to identify and properly assign multicast group membership. An access platform with this functionality checks IGMP packets passing through it, picks out the group registration information, and configures multicasting accordingly. Without IGMP snooping, multicast traffic is treated in the same manner as broadcast traffic, that is, it is forwarded to all ports. Via IGMP snooping, multicast group traffic is only forwarded to ports servicing members identified as belonging to that particular multicast group. IGMP snooping generates no additional network traffic, allowing carriers to significantly reduce network congestion.

Bandwidth management is another critical component of successful IPTV deployment—the most bandwidth-intensive deliverable to the digital home. To facilitate consistency with their network management and load-monitoring practices for other less bandwidth-intensive applications such as telephony and high-speed Internet access, carriers must select IPTV systems with tools that enable them to carefully monitor and predict the results of network oversubscription rates. These tools must alert them ahead of time to adjust the oversubscription ratio/capacity to meet their service commitments. Average subscriber viewing behavior, relative to peak viewing behavior (major sporting events), along with unpredictable peaks and surges (disaster coverage), must be anticipated and bandwidth allocation made flexible. These monitoring and modeling tools must communicate with a comprehensive network-management system (NMS) that enables rapid alerts and response to potential network problems.

## The Distributed IPTV Model

The carrier's goal is to minimize the amount of traffic that must traverse the core. According to this logic, the worst possible scenario from a bandwidth management or utilization perspective for a large-scale deployment is a centralized model. Ironically, this model is often depicted in network topology diagrams illustrating IPTV deployments. In this model, a centralized "super-headend" combines encoders, back office servers, and VoD servers. Regardless of whether the network is delivering live TV, VoD, or time-shift TV (essentially the same as VoD the instant it switches from multicast to unicast), all content and network traffic resulting from subscriber requests must traverse the entire network from the super-headend all the way to each subscriber's set-top box (STB).

By changing their network topologies to encompass a flexible, distributed model, carriers will realize huge advantages, particularly for VoD and time-shift TV. Unlike the centralized model described above, wherein single-source encoding takes place at the super-headend and is then multicast throughout the network, a distributed model uses regional headends so local content (community interest and news) is only distributed in-region. This unburdens the core and the access networks. While an IPTV network does have the advantage of enabling local content to be viewable outside a region, there is substantially less demand for this programming outside the respective region and would be more suitably delivered via unicast stream. Most of the actual traffic that traverses an IPTV network is going to be from live TV and subscriber requests (authentication/billing information/electronic programming guide [EPG]).

A final component of successful IPTV delivery concerns the content storage and distribution mechanism itself. Since maximum distribution is the key to creating flexibility, it stands to reason that a segmentation scheme, in which sequential content segments are distributed piecemeal, would radically reduce network congestion. Rather than sending multiple iterations of content in their entirety, from one storage and streaming server to another, in a segmented content solution, adjustable segments replace large files. Content segmentation at the edge, coupled with innovative protocols such as broadband media distribution protocol (BMDP), which enables the IPTV system to adjust storage and distribution heuristically according to trends in subscriber behavior, provide a buffer against network jitter and greater tolerance for peak bursts in traffic.

IPTV means new revenue for carriers, and its differentiated services suggest a huge potential benefit for carriers and the subscribers they serve. With the coming of virtually infinite VoD and time-shift TV, the industry is poised to witness the first radical advances in television since digital cable. However, while subscribers are accustomed to occasional service quality variance due to traffic latency for streaming video applications on their computers, they will have limited tolerance for similar issues that affect their long-familiar TV watching experience. After all, people do not subscribe to services based on technology, they subscribe based on quality of service, value, differentiation, and convenience. IPTV requires careful selection of new technology solutions that will ensure successful initial implementation and scalability, enabling carriers who plan correctly to move on to their own cycle of optimization.

## Notes

1. Even with the advanced compression rates promised by H.264, conservative projected requirements for the digital home begin at 15 megabytes and rise quickly:
   1 HD stream with encoded audio requires 8 megabytes
   2 SD streams require 2 megabytes each
   3 VoIP lines require 64 kilobytes each
   2 channels of digital audio music require 128 kilobytes each
   High-speed Internet access requires 3 to 20 megabytes, depending on access technology

2. The number of broadband households has doubled in the past two years and is expected to double again by the end of 2006 (Merrill Lynch).

# IPTV Success: Lessons from the Front Line

Arjang Zadeh

*Managing Partner, Global Network Practice*
Accenture

The delivery of television programming and interactive services over a converged IP network—called video over Internet protocol (IP) or now, more frequently, IPTV—will be a critical part of keeping high performance flowing in the communications industry over the next decade.

As broadband penetration continues to expand—both in connectivity and in products and devices—a "land grab" is coming as telcos, satellite companies, and cable providers vie to attract and retain customers.

To respond to this competitive environment, one where the price and bandwidth ratio keep dropping, telcos are putting a new IP infrastructure in place so they can offer additional bandwidth at as low a cost as possible. They also want to be able to introduce new bundled products such as IPTV.

The business case for IPTV adoption is strong. Not only do companies get an increase in average revenue per user, but telcos can also put together a bundle that helps them compete more successfully with cable offerings. Telcos can also reduce churn when customers begin to seek VoIP options.

In spite of the benefits telcos can get from IPTV, the fact is that no operator has yet launched IPTV services that are scalable, stable, and high-quality, and that deliver an acceptable return on investment. Obviously the challenges are substantial. However, based on our experiences and lessons from the early adopters, here are the most important factors to be addressed with an IPTV strategy.

## Quality Is King

Some of the excitement about IPTV services focuses on the interactive and on-demand capabilities that can be delivered over the high-bandwidth IP network. Companies do need to be adopting those capabilities in pursuit of differentiation and increased customer loyalty. However, even in countries such as the United Kingdom—which has the most advanced interactive services in cable, satellite, and terrestrial technologies—interactivity is not yet generating significant revenue uplift. This situation is likely to continue, at least until a "killer app" emerges that leads to a dramatic uptake in demand for interactive, on-demand services.

What is the killer app for IPTV? The TV programming itself. Everything must begin from that foundation. The experiences from providers in the United Kingdom and the United States have made clear that the TV service itself must work as well as the consumers' existing level of service. No "me too" TV service will succeed—regardless of its interactivity—unless it functions as well as or better than existing TV services.

## The Double-Edged Sword of the Bundle

There is no turning back today from the imperative to bundle products and services to customers. A customer using multiple services from a single provider becomes much more entrenched and is less likely to churn if the value and quality of the entire bundle remains high. And there is the rub.

Looking at the extensive experiences of operators in Europe that have employed a bundle or solution-selling strategy, bundling is, in fact, a double-edged sword. If the bundle is good, with value differentiators built into the bundle as a whole, customers are less likely to be lured away from competitors for fear of losing the value of the bundle. However, studies have shown that operators that have trouble with the quality or value of one of the components or services in the bundle find that their customers are actually as much as 60 percent more likely to churn than a customer of a single service.

So service providers that aggressively launch a low-quality video service bundled with voice and/or data may not only lose that customer as a video buyer, but also as a buyer for voice and data services as well. And that can be a very expensive blow to the customer base. What to do? First and foremost, operators must monitor and rigorously measure both customer satisfaction and the stability of services being offered. This may require ensuring that customer-relationship management (CRM) capabilities are fully integrated throughout the entire solution that is being sold.

## Platform

Service stability is the most important ingredient to IPTV success; and that means stability of the platform and architecture itself. If the IPTV service is unstable, high customer churn will result, and operators may end up with a customer base where churn negates their customer acquisition efforts.

A truly comprehensive IPTV solution encompasses the systems, video infrastructure, and network elements required for an end-to-end solution, as well as definition of the processes to operationalize the video services being offered. The most important success factor, in Accenture's experience, is to create a stable and scalable IPTV service over a broadband multiservice platform. This has proved to be challenging to almost every operator. All the network and service control issues required for quality of service (QoS) and bandwidth control between the different services over the same broadband access requires significant engineering know-how and an understanding of multiple broadband services and their characteristics. The architecture that enables the video service must also be stable over the access network and home access gateway using IP. This is not a trivial matter; design of such an architecture requires unique understanding of video as well as broadband.

The other important element of IPTV success is the set-top box (STB) and its integration and stabilization. In simpler broadcast environments, stabilization of the STB will take significant effort, time, and costs. The cost of the STB is often the determining factor in establishing the overall business case. Optimizing this element of the overall IPTV strategy requires an experienced and deep understanding of issues related to the deployment of STBs and applications, and also IPTV specifics related to such a platform.

Finally, the complexity of delivering an end-to-end IPTV platform is not just in the technical challenges in integrating all the video network components and systems involved, but also in the integration with operations support system (OSS)/business support system (BSS) platforms to be able to offer video services in tune with the operator's customer experience goals.

## Meeting the Challenges

IPTV represents important competitive opportunities for communications companies, and operators should act now to address their most important operational challenges. For example, service provisioning and customer service are extremely important, as are storage and management of content and developing true, end-to-end quality of service in a complex network.

A robust and stable platform is the most important ingredient to success. Without such a platform, and the architecture that underpins it, operators cannot develop and retain the customer base necessary to succeed.

# Internet Protocol Multimedia Subsystem (IMS)

# Session Border Control in IMS

## Jonathan Cumming

*Director, VoIP Product Management, Network Protocols*
Data Connection Limited (DCL)

## Executive Summary

There is a lot of controversy and press coverage over the role of session border controllers (SBCs) and the design of the Internet protocol multimedia subsystem (IMS). In this environment, it is difficult to determine what the real issues are for each technology, let alone how they need to work together.

This white paper is aimed at equipment manufacturers looking at building SBC functionality into their product range to target the IMS market, and carriers and consultants looking to understand how an SBC fits into an IMS network.

It explains why IMS networks need session border control and what alternatives are available. It also looks at how these requirements are likely to evolve as services and access methods change and discusses the function that products targeting this market require.

SBCs are described as both a cure-all for next-generation telecommunications networks and an unnecessary attempt by carriers to stop their business from becoming a simple bit-carrying commodity. This paper seeks to explain how these different views arise and the varied roles that SBCs play in IMS.

## Introduction

IMS defines the functional architecture for a managed IP–based network. It aims to provide a means for carriers to create an open, standards-based network that delivers integrated multimedia services to increase revenue, while also reducing network capital expenditures (CAPEX) and operating expenditures (OPEX).

IMS was originally designed for third-generation mobile phones, but it has already been extended to handle access from wireless fidelity (Wi-Fi) networks, and is continuing to be extended into an access-independent platform for service delivery, including broadband fixed-line access. It promises to provide seamless roaming between mobile, public Wi-Fi, and private networks for a wide range of services and devices.

Moving from a centrally managed network with control over the core and access networks to an open network with

soft clients represents a change in the applicability and deployment of IMS. Previously, it was aimed at centrally managed networks with significant control over the core and access networks and the clients. Now it is moving to a much more open network model, where previous assumptions about the sorts of connecting networks and clients break down. This introduces the need for session border control at the network boundary to provide security, interoperability, and monitoring.

This article examines these evolving requirements for IMS and where session border control fits in the IMS functional architecture. It also assesses the evolution of existing equipment to handle these new requirements and the likely future evolution as the market and technology mature.

The first section provides an overview of session border control and IMS. Sections 2 and 3 cover the requirements for session border control in IMS and how this function fits into the IMS architecture. Sections 4, 5, and 6 discuss what products need to address these requirements, how this market is likely to change in the future, and what conclusions can be drawn from this. Section 7 provides a list of references to additional information. Section 8 contains information on data connection and its products.

## Overview of Session Border Control

Session border control is not a standardized set of functions. Instead, SBCs have evolved to address the wide range of issues that arise when voice and multimedia services are overlaid on IP infrastructure, including the following:

- Security and prevention of service abuse to ensure quality of service (QoS)
- Monitoring for regulatory and billing purposes
- Maintaining privacy of carrier and user information
- Resolution of VoIP protocol problems arising from the widespread use of firewalls and network address translation (NAT), and the vast array of differing protocols and dialects used in VoIP networks

These issues are relevant for access to both carrier and enterprise networks, and on both user network interfaces (UNIs) to end users and access networks, and network-to-network interfaces (NNIs) to peer networks. *Figure 1* shows where SBC function is typically required.

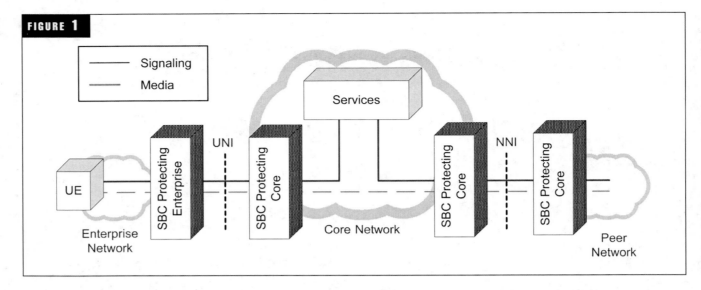

**FIGURE 1**

Signaling
Media

Services

UE

SBC Protecting Enterprise

UNI

SBC Protecting Core

SBC Protecting Core

NNI

SBC Protecting Core

Enterprise Network

Core Network

Peer Network

*Figure 1* depicts a single device at the edge of each network (a traditional SBC), but there is actually great flexibility in how this function is distributed. Examples include the following:

- A device in the access network might perform initial user authentication
- An edge device might enforce access policy to limit denial of service (DoS) attacks and prevent bandwidth theft
- Core devices might limit the total usage for a particular group of users and detect distributed DoS attacks

The location of each function will depend on the overall system design, including the availability of processing resources and the level of trust between the devices.

The following sections describe each of the SBC functions.

### Security

An insecure network cannot charge for its use or provide a guaranteed QoS service because unauthorized users cannot be prevented from overusing limited network resources.

SBCs can provide security and protection against the following:

- Unauthorized access into the trusted network
- Invalid or malicious calls, including DoS attacks
- Bandwidth theft by authorized users
- Unusual network conditions such as a major emergency

Typical resources that require protection are bandwidth on access links and processing capacity on network servers. In general, core network links can, at low cost, be over-provisioned to help prevent bottlenecks.

To provide this security, the SBC identifies and authenticates each user and determines the priority of each call, limits call rates and resource usage to prevent overloads, authorizes each media flow and classifies and routes the data to ensure suitable QoS, and prevents unauthorized access for both signaling and media traffic.

QoS across the core of the network is normally handled by an aggregated classification mechanism such as differentiated services (DiffServ), as this removes the overhead of reserving bandwidth for each individual flow.

The SBC may also be used to enforce QoS in the access network by signaling to the access routers or instructing the endpoint to reserve necessary resources across the access network. Alternatively, an intelligent access network may independently determine appropriate QoS for the media streams by analyzing the call signaling messages.

### Monitoring

Network usage may need to be monitored for regulatory reasons (such as wiretapping and QoS monitoring) as well as commercial reasons (such as billing and theft detection).

The monitoring devices need sufficient intelligence to understand the signaling and media protocols. They must also be at a point through which all media and signaling flows.

SBCs fulfill both requirements, as all traffic passes through an SBC to enter the network. They provide a scalable, distributed solution to this processing-intensive function.

### Maintaining Privacy

Information about the core network, which might provide commercially sensitive information to a competitor or details that could aid an attack, and user information that the user does not wish to be made public are two examples of types of information that need to be protected.

An SBC can be used to remove confidential information from messages before they leave the core network, including details of internal network topology and routing of signaling through the core network. It can also hide the real address of the user by acting as a relay for the media and signaling.

### Resolution of VoIP Protocol Problems

SBCs can also act as gateways to heterogeneous networks by hiding any differences between the protocols used in the core and access networks. This can include the following:

- Hiding access network topology, including the complexity of routing through NATs and firewall and to overlapping address spaces of virtual private networks (VPNs) or private IP address spaces
- Interworking between devices and networks of different capabilities (such as conversion between the session information protocol [SIP] and H.323 signaling, or between IPv4 and IPv6, or even different versions of H.323)
- Transcoding media flows between incompatible codecs

Putting this function in the SBC, which is close to the access device, simplifies the core network devices by limiting the range of protocol variations they must support.

## Overview of IMS

IMS is the control plane of the third-generation partnership project (3GPP) architecture for its next-generation telecommunications network. This architecture has been designed to enable operators to provide a wide range of real-time, packet-based services and to track their use in a way that allows both traditional time-based charging as well as packet- and service-based charging.

IMS provides a framework for the deployment of both basic calling services and enhanced services, including multimedia messaging, Web integration, presence-based services, and push-to-talk. At the same time, it draws on the traditional telecommunications experience of guaranteed QoS, flexible charging mechanisms (time-based, call-collect, premium rates), and lawful intercept legislation compliance.

Network operators also hope that IMS will cut their CAPEX and OPEX through the use of a converged IP backbone and the open IMS architecture. The IMS architecture defines many common components (for example, call control and configuration storage) so less development work is required to create a new service as this existing infrastructure can be reused. The use of standardized interfaces should increase competition between suppliers, preventing operators from being locked into a single supplier's proprietary interfaces. As a result, IMS should enable new services to be rolled out more quickly and cheaply, compared with the traditional monolithic design of telephony services.

### History and Evolution
IMS was initially developed as a call control framework for packet-based services over 3G mobile networks as part of 3GPP Release 5 (2003). It was then extended to include Wi-Fi roaming and additional services such as presence and instant messaging in Release 6 (2004/5).

Although originally designed for mobile networks, both European Telecommunications Standards Institute (ETSI) Telecoms and Internet-Converged Services and Protocols for Advanced Networks (TISPAN) and the Multiservice Switching Forum (MSF) have now also adopted the IMS architecture for their visions of fixed telecommunications networks. Discussions within these groups are driving the IMS extensions to cover fixed networks in 3GPP Release 7 and many of the session border control requirements that fixed network access introduces.

At this point, it should also be noted that the design of IMS Release 7 is not yet complete and there is ongoing disagreement over the scope and location of specific functions. However, although the names and details of the specification are likely to change, the principles and issues described in this document are unlikely to be significantly affected.

### Drivers
Although originally developed for mobile operators, the main interest in IMS is from fixed-line operators, as the existing fixed-line network is older and is due for replacement, whereas much of the mobile infrastructure has only recently been deployed.

In particular, the current generation of fixed telephone networks is limited to narrowband voice services and is at great risk of being displaced by mobile and Internet telephony services. An IMS–based network would enable fixed-line operators to offer a much wider range of services to help protect their market.

Despite the widespread industry support for IMS, many uncertainties remain over its value. The cost of a providing such a QoS–enabled managed network is high compared with the Internet's stateless model. Also, as the success of Vonage, Skype, and many other VoIP providers testifies, telephony services are easily provided over the Internet and the quality is sufficient for many situations.

To justify the investment in IMS, the resulting service must be significantly better than that available over the Internet and people must be prepared to pay for it. Whether IMS is a commercial success will be determined over the coming years, but competition from Internet-based providers will make this a competitive market.

### Architecture
IMS decomposes the networking infrastructure into separate functions with standardized interfaces between them. Each interface is specified as a "reference point," which defines both the protocol over the interface and the functions between which it operates. The standards do not mandate which functions should be co-located, as this depends on the scale of the application, and a single device may contain several functions.

The 3GPP architecture is split into three main planes or layers, each of which is described by a number of equivalent names: service or application plane, control or signaling plane, and user or transport plane.

### Application Plane
The application plane provides an infrastructure for the provision and management of services and defines standard interfaces to common functionality, including the following:

- Configuration storage, identity management, and user status (such as presence and location), which is held by the home subscriber server (HSS)
- Billing services, provided by a charging gateway function (CGF) (not shown)
- Control of voice and video calls and messaging, provided by the control plane

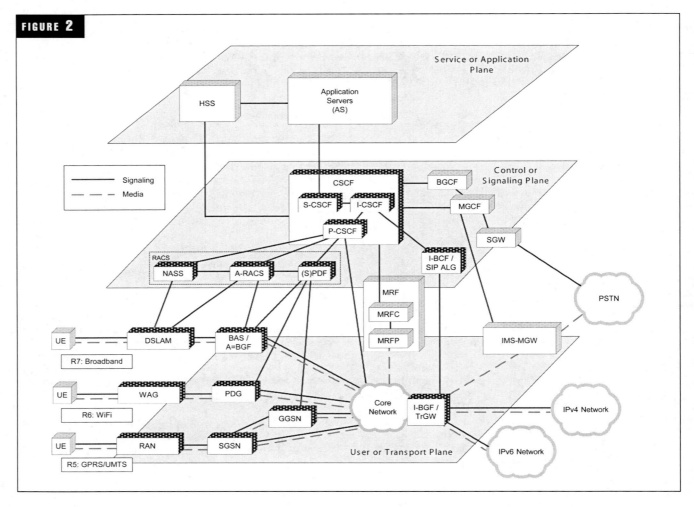

**FIGURE 2**

## Control Plane

The control plane sits between the application and transport planes. It routes the call signaling, tells the transport plane what traffic to allow, and generates billing information for the use of the network.

At the core of this plane is the call session control function (CSCF), made up of the following functions:

- The Proxy–CSCF (P–CSCF) is the first point of contact for users with the IMS. The P–CSCF is responsible for security of the messages between the network and the user and allocating resources for the media flows.

- The Interrogating–CSCF (I–CSCF) is the first point of contact from peered networks. The I–CSCF is responsible for querying the HSS to determine the SCSCF for a user and may also hide the operator's topology from peer networks (topology hiding inter-network gateway [THIG]).

- The Serving–CSCF (S–CSCF) is the central brain. The S–CSCF is responsible for processing registrations to record the location of each user, user authentication, and call processing (including routing of calls to applications). The operation of the SCSCF is controlled by policy stored in the HSS.

- This distributed architecture provides an extremely flexible and scalable solution. For example, any of the CSCF functions can generate billing information for each operation.

*Figure 3* shows the routing of a typical call in an IMS environment and the two distinct uses of the NNI.

- *Roaming* (the left-hand NNI): This is required to access services provided by your own service provider (home network) when connected to another carrier's network (visited network).

- *Interworking* (the right-hand NNI): This is required when placing a call to a customer of a different carrier network.

The call signaling flows from the caller pass through the P–CSCF in the visited network to his home S–CSCF. The signaling then passes onto the called party via his SCSCF.

The control plane controls user plane traffic through the resource and admission control subsystem (RACS). This consists of the policy decision function (PDF), which implements local policy on resource usage, for example to prevent overload of particular access links, and Access-RAC Function (A–RACF), which controls QoS within the access network.

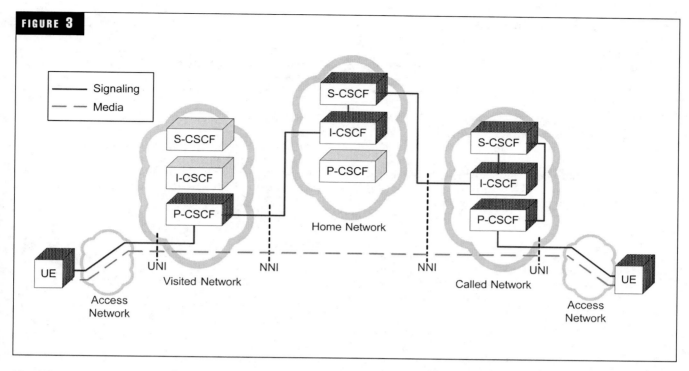

**FIGURE 3**

*User Plane*

The user plane provides a core QoS–enabled IPv6 network with access from user equipment (UE) over mobile, Wi-Fi and broadband networks. This infrastructure is designed to provide a wide range of IP multimedia server–based and P2P services.

Although IPv6 is defined for this transport plane, many initial deployments are built upon existing IPv4 infrastructure and use private IPv4 addresses. This introduces NATs at the boundary of each address domain and the associated difficulties routing VoIP calls across the boundary.

Access into the core network is through border gateways (gateway GPRS support note [GGSN]/packet data gateway [PDG]/broadband access server [BAS]). These enforce policy provided by the IMS core by controlling traffic flows between the access and core networks as follows:

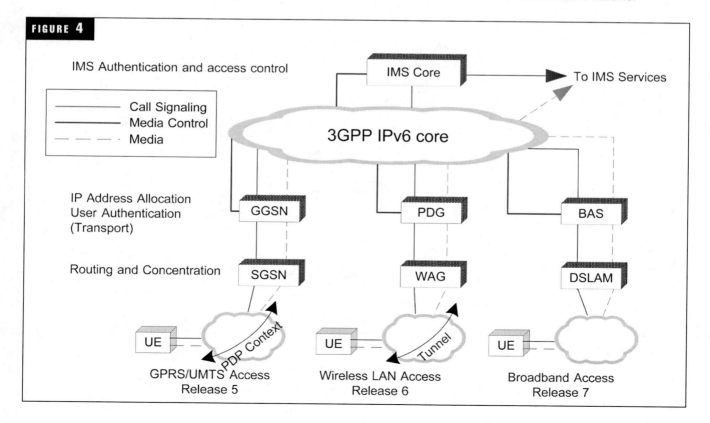

**FIGURE 4**

- With general packet radio service (GPRS)/universal mobile telecommunications system (UMTS) access, the GGSN authenticates the user equipment (UE) and controls the establishment of media channels using authenticated packet data protocol (PDP) contexts. This enforces QoS and access control through the access network to the UE.

- With wireless local area network (WLAN) access, the PDG controls the establishment of tunnels through the access network to the UE. These tunnels provide security of the message flows to the UE, but not QoS. Separately, the access network may apply QoS policy to data flowing to/from the carrier core and have a billing arrangement with the carrier to charge for use of its network.

- Release 7 adds support for IP connectivity over a range of access technologies. There is ongoing discussion over how much of this access will be covered by the core IMS specifications. For example, the ETSI TISPAN architecture envisages the IMS core connected to external networks through border gateways that are not part of the IMS specifications.

It is this change from a very controlled network with limited-access methods in Releases 5 and 6, to a much wider range of access devices in Release 7, which introduces the need for SBCs.

## IMS Requirements for SBCs

Some of the functions provided by an SBC have always been important and inherent in IMS, given its role in providing a QoS–enabled service with detailed usage monitoring to enable charging for its use. Others are only now becoming important as Release 7 expands the range of supported access methods. This chapter looks at the requirements for this functionality and how they are changing.

The IMS architecture consists of interconnected core networks belonging to different carriers, with endpoints connected through attached access networks and gateways to non–IMS networks. Border gateways control access into and out of each core network, monitoring and regulating the data flows on each interface.

This architecture is shown in *Figure 5*.

The core network needs to be protected against all the threats described at the beginning of this paper, but each interface imposes a different set of border control requirements due to differences in the attached devices and access networks.

The following sections describe the common requirements and the specific issues that each interface introduces.

### Security
Security in IMS Releases 5 and 6 is designed around an open IPv6 core with well-protected access, including the following:

- Access to the network core is protected using transport layer security on the UNI in the form of authenticated PDP contexts and tunnels.

- The NNI is an internal trusted interface in this secure core, so it requires very little security.

However, IMS Release 7 and the reality of early IMS deployments have changed this model, and have expanded the range of security needed on each interface. Nevertheless, over some interfaces, a subset of this functionality may be provided by the access network.

### User Network Interface (UNI)
The expanding range of access devices and reduced control over the access network with each new release, demonstrated as follows, has increased the responsibility on the border controller at the edge of the core network.

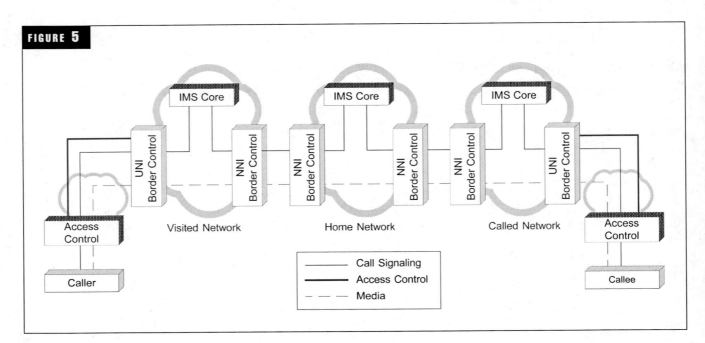

**FIGURE 5**

- In 3GPP IMS Release 5, only GPRS/UMTS access networks are supported. In this environment, the P–CSCF uses the GGSN to control access and bandwidth use through the entire access network all the way to the handset. Additional DoS controls on the signaling can be applied by the P–CSCF, but due to the controlled design and certification of handsets, there is limited scope for such attacks.

- The addition of WLAN access in 3GPP Release 6 does not greatly expand the range of protection required at the network border, as traffic into the core is controlled through tunnels managed by the PDG. In addition, Release 6 is primarily aimed at data roaming between 3G and WLAN, not the handling of voice calls over WLAN access, so there is limited function to protect.

- Release 7 expands both the range of supported access methods and the function, and as a result greatly expands the scope of attack. In addition, the core network now exerts little control over the access network, so the border gateway becomes its first line of defense.

### Network-to-Network Interface (NNI)
Early IMS deployments have identified that security on the NNI is also required to protect the core network from malicious or unexpected behavior by a peer and prevent a problem in one network core from affecting another.

### Monitoring
Government regulations and commercial reasons both require monitoring of network use.

IMS Release 5 did not include monitoring on the NNI. However, reconciliation of intercarrier charges, monitoring of service-level agreements (SLAs), and lawful intercept of calls traversing the network have all increased the need for monitoring on this interface.

### Privacy
Privacy of both network topology and user information is required on all interfaces.
Again, network topology hiding was not considered in the design of IMS Release 5 but is considered a requirement for real deployments to protect this commercially sensitive information from peers.

The requirements of telco networks impose the following aspects to privacy of user information:

- A caller may request for his identity to be hidden from the callee.
- The caller's identity must be available for emergency calls and lawful intercept regulations.

These requirements mean that user policy may modify the visible identification of the user to the callee, but that the signaling within the trusted core must continue to contain the true identity of the caller and, at the border of the trusted network, the true identity of the caller must be removed.

The border of the trusted network may be the edge of one carrier's core or contain the core networks of several carriers, depending on the regulations under which each operates.

### VoIP Protocol Problems
VoIP protocol problems were not seen as an issue in IMS Releases 5 and 6. However, the importance of this area has been raised by the use of IPv4 and other interoperability issues in early IMS trials and the inclusion of NATs and a wider range of devices and network topologies in Release 7.

The scope of VoIP protocol problems seen depends on the interface and access method, as each has very different characteristics.

### User Network Interface (UNI)
The UNI border typically has to handle a large number of connections from individual users and a wide range of equipment, so it has to deal with a wide variety of protocol variants and network topologies.

It is not decided whether responsibility for NAT and firewall traversal issues is part of the IMS architecture, but functionality is required to enable interworking across such devices.

### Network-to-Network Interface (NNI)
The NNI handles the signaling and media traffic between IMS carriers and through gateways to non–IMS carriers. A single interface typically handles a small number of high-volume connections with peer carriers.

The original IMS architecture envisaged a pure IPv6 IMS network core with minimal protection at the NNI boundaries. However, this model has changed due to the need to interoperate with non–IMS and pre-standard networks, such as IPv4 networks, and the requirement from carriers to protect their network core. The effect of this on the IMS architecture is to add an SBC on the NNI.

The NNI border controller may therefore provide interworking between signaling protocols, protocol variants, and media codecs and NAPT function.

Firewall and NAT traversal mechanisms are not required, as the peer carrier is expected to manage its own NAT/firewall.

### IPX Proxy
The GSM Association (GSMA) has identified the need for centralized interconnection of multiple carriers through an inter-carrier carrier that provides both IP connectivity and a clearinghouse for inter-carrier charges. This mimics the existing inter–GSM carrier (GRX) networks and removes the need for bi-lateral agreements between all interconnected carriers. This inter-carrier IP network is known as an IPX network.

In addition to simple connectivity, the IPX network provider can also provide a wide range of session border control functionality to its customers by providing an SBC within the IPX network. This SBC is known as an IPX proxy.

### Summary
Table 1 summarizes the original SBC function defined in IMS Releases 5 and 6, and the new function introduced with Release 7 and early IMS deployments.

**TABLE 1**

|  | **Releases 5 and 6** | **Additional requirements in Release 7 and Early IMS** |
|---|---|---|
| **Security** | Access network enforces access controls<br>Peer networks are trusted | Border gateway enforces policy at all network boundaries |
| **Privacy** | Topology hiding not considered<br>User privacy is handled by the P–CSCF | Topology hiding required.<br>Media relayed to hide end-user location<br>User privacy may be required at gateways to non–IMS networks |
| **Monitoring** | Monitoring of UNI for billing and lawful intercept | Monitoring of NNI to enforce intercarrier agreements |
| **VoIP protocol problems** |  | NAT to support IPv4 core with private addresses.<br>NAT/firewall traversal on UNI<br>Interoperability with devices with limited function |

## IMS Architecture for SBC

As discussed earlier, session border control is inherent in the design of IMS. However, unlike architectures defined by other standards bodies—for example, the Multiservice Switching Forum (MSF)—the IMS architecture does not include a device labeled "SBC."

This chapter describes how SBC function fits onto the IMS–defined functional architecture and how this architecture is evolving to handle the increasing requirements.

The IMS architecture defines separate sets of functions for each access type. However, many of these functions perform

**FIGURE 6**

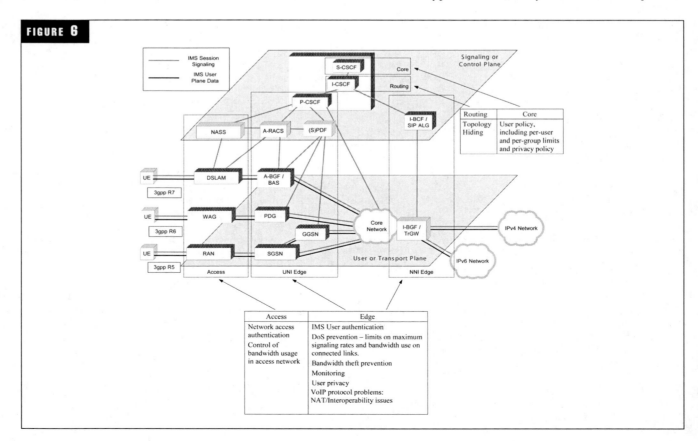

similar roles in the network and each role provides the same subset of session border control function.

The following sections describe the differences between the function required on the UNI and NNI and the set of IMS functions that may be combined into an IMS–targeted SBC.

### UNI

On the UNI, the set of IMS functions providing session border control depends on the access method. *Figure 7* shows how the IMS functions could be combined to build a single-box SBC for Release 6 and Release 7 network access.

The I–CSCF could also be part of the SBC in some situations.

The SBC function is split between the following functiso listed in Table 2:

### NNI

The SBC–related functions within the NNI have a similar

---

FIGURE **7**

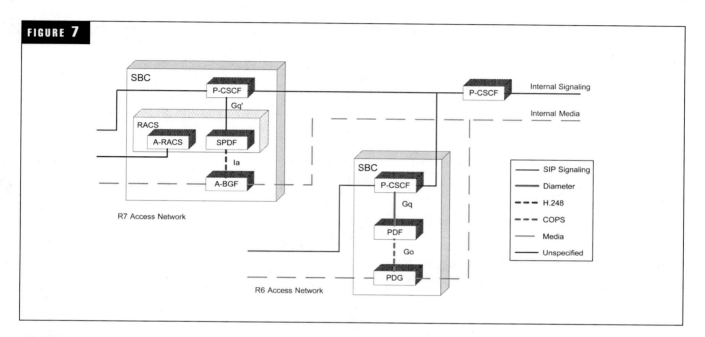

---

TABLE **2**

| | R6 – GPRS/UMTS/Wi-Fi | Additional SBC features in R7 |
|---|---|---|
| **P–CSCF** | Controls security over the access network<br>Tells the PDF what resources are required for the call | SIP ALG for IPv4 address translation and NAT firewall traversal |
| **PDF/SPDF** | Implements media resource allocation policy<br>Authorizes media resource requests from BGF | Programs the A–BGF to accept media flows |
| **A–RACS** | Function incorporated into GGSN/PDG | Controls resources within the access network<br>In IMS Release 7, the management of the access network is split out of the expanded PDF into the Access-Resource and Access Control Subsystem (A–RACS). The combination of the A–RACS, SPDF is known as the Resource and Access Control Subsystem (RACS). |
| **A–BGF** | Provides media relay for hiding endpoint address with managed pinholes to prevent bandwidth theft | Implements NAPT and NAT/firewall traversal for media flows |

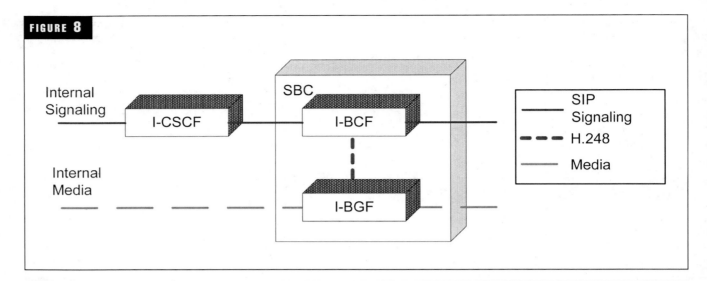

**FIGURE 8**

**TABLE 3**

| | | SBC features provided |
|---|---|---|
| **I–BCF** | | Transport-level security |
| | | Tells the RACS what resources are required for the call |
| | | NAPT function and control of NAPT in BGF |
| **I–BGF** | | Media relay for hiding endpoint address |
| | | Pinholes to prevent bandwidth theft |
| | | NAPT and NAT/firewall traversal for media flows |

architecture to the UNI. This is shown in *Figure 8*. Again, the I–CSCF could also be part of the SBC in some situations.

The I–BCF and BGF are new functions in IMS Release 7.

There is ongoing discussion whether the I–BCF and I–BGF function should be standardized within the IMS architecture. The ETSI TISPAN recommendation is that they remain outside the core specifications, providing a flexible gateway to other networks.

There is also discussion over the inclusion of a standardized RACS function between the IBCF and I–BGF, as on the UNI, to mediate requests for media resources and manage local policy.

### Reference Points
A reference point is a standardized interface between two IMS functions. It defines both the functions that it links and the protocol across the interface. Each reference point is denoted by the combination of one uppercase and one lowercase letter, e.g. Gq.

The main reference points involved in session border control are described in *Figure 9*.

### Call Signaling (Gm and Mw)
All call signaling in IMS uses the session initiation protocol (SIP).

### CSCF to PDF/RACS (Gq/Gq')
This reference point controls requests for network bandwidth from the IMS core. The original Gq interface in Release 6 enabled access control policy to be centralized in a separate policy decision function (PDF). The Gq reference point is based on the Diameter protocol and enables the P–CSCF to request an authorization token from the PDF for access for a specified bandwidth.

This reference point is being extensively expanded in Release 7 to include the direct control of access. The Gq' reference point enables the P–CSCF to program the BGF to perform specific NAPT and NAT traversal function, as well as control the access network. This revised interface may be based on Diameter or H.248.

### PDF/GGSN (Go)
In Releases 5 and 6, the Go reference point enables to GGSN/PDG to authenticate tokens received on the establishment of new media channels. This interface uses the COPS protocol.

### PDF/BGF (Ia)
Release 7 expands the role of the Go reference point to enable direct control of the BGF to program NAPT and NAT traversal function and open pinholes in the gateway. The new Ia reference point is being defined for this purpose, probably based on H.248.

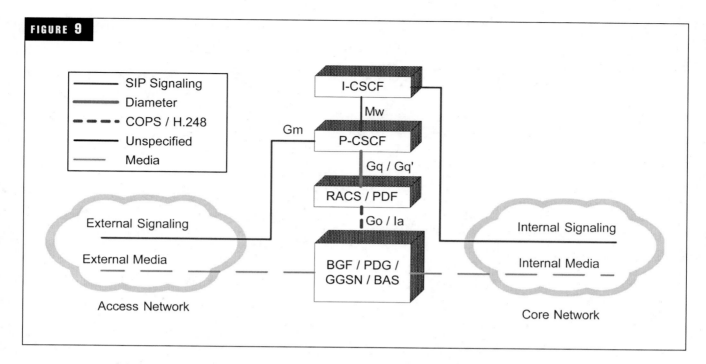

**FIGURE 9**

## IMS SBC Products

The earlier chapters describe how session border control fits onto the functions defined by IMS. This chapter considers how this function is likely to be packaged into products and how these products will evolve from existing devices.

There are three types of interfaces requiring control: UNI, NNI to peer IMS carriers, and NNI to non–IMS carriers. Although a single device could be used for all applications, it is likely that separate products will be developed to target the functionality and scale of each application.

### Scope of Function

Most current SBC products are implemented as a single stand-alone device that is placed in front of existing equipment in the path of all the signaling and media traffic on an interface. This one box includes the media (BGF) and signaling processing (BCF/PCSCF) as well as the media resource control (PDF/RACS).

Although in some small-scale applications, it makes sense to include all these functions in a single device, this solution does not scale well to the requirements of service providers. For these applications, the media and signaling processing will often be split into separate devices, with the signaling

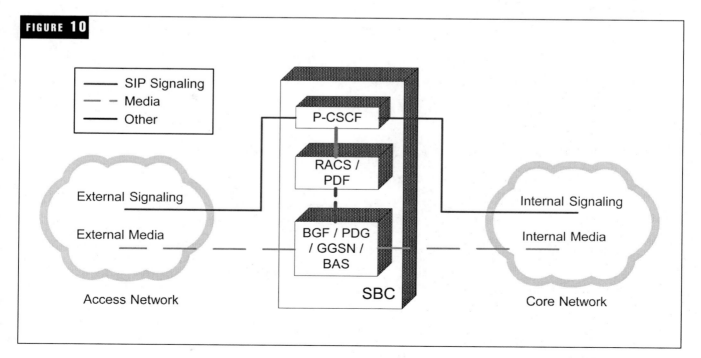

**FIGURE 10**

processing centralized into regional server farms and the media processing distributed closer to the user. This provides economies of scale on the signaling processing while maintaining direct media routing to minimize network transit delays.

The PDF/RACS is likely to remain co-located with the BCF/PCSCF in most situations. However, the function might be split along this interface (Gq) to enable specialist RACS devices to be deployed to handle each access network with a common P–CSCF, or for the RACS to be used by non–IMS applications.

In addition, the PDF requires less processing than the P–CSCF, so a small number of PDF may be able to handle multiple P–CSCFs in highly scaled installations.

### Management and Control
One of the most challenging aspects of the number of functions defined by the IMS architecture is managing the resulting proliferation of devices. User configuration is centralized in the HSS, but operator policy is implemented across many separate devices (P–CSCF, I–CSCF, S–CSCF, PDF, and others). There is no standardized way to control these devices, but an SNMP–based solution is likely to be the most effective way to centrally configure and monitor the system.

### Project Evolution
The requirements of IMS Release 7 are bringing together the following three sets of products, each lacking the complete set of function:

- *Products targeted at earlier IMS releases*: These generally lack NAPT and NAT/firewall traversal function and do not have sufficient call access control to prevent the wide range of DoS attacks that broadband devices can generate.

- *SBCs targeted at the broadband access market*: These may not be designed to fit the IMS architecture and generally do not support the IMS reference points and 3GPP–specific protocol extensions.

- *IP and multiservice routers that have traditionally targeted the IP carrier interconnect and carrier edge markets*: These products excel at high-performance routing, but need to add both the SBC function and IMS interfaces.

The short-term effect is that very different products are being promoted as fulfilling this same requirement, and the still-evolving standards are being pushed by each manufacturer to conform as closely as possible to its existing product range.

At the same time, there is huge pressure on the manufacturers to enhance their products to address the rapidly growing IMS SBC market. Depending on their situation, most are taking one of the following routes:

- Developing the function themselves, often incorporating much of the technology from independent suppliers such as Data Connection to reduce cost and improve time to market
- Partnering with complimentary suppliers

- Purchasing companies with the relevant expertise and trying to merge the product lines (for example, Juniper/Kagoor, Tekelec/IPtel, NetCentrex/NeoTIP)

Whichever method they choose, a rapidly expanding range of IMS–targeted devices incorporating SBC functionality will evolve to create an extremely competitive market. However, given the amount of investment by operators in their next-generation networks, this will be extremely lucrative for those vendors that get it right.

## The Future

Looking forward, there are two areas that we need to consider to predict the evolution and interaction of IMS and SBC products—the future of IMS and its requirements for session border control and the evolution of session border control function itself as VoIP technology matures.

The following sections look at each of these areas in more detail.

### The Future of IMS
There is huge pressure on fixed-line operators to deploy a new architecture, and IMS offers the most attractive model available. However, IMS has not yet been deployed outside trials, and the IMS standards continue to evolve to overcome flaws in the original design and provide new functionality. As a result, it is likely that IMS in some form will be deployed, but that it will not be in the form envisioned in any of the defined releases. Instead, future networks will combine elements from each of the releases with new functionality to address new market opportunities.

The following factors will have a significant influence on the direction of IMS and will change nature and location of session border control within the IMS network and its derivatives.

### Legislation
Government action to apply lawful intercept (wiretapping) and mandatory quality levels to telephony services may force all telephony service providers (including pure VoIP services) to deploy managed networks with border controls.

However, unless governments make it illegal to communicate over P2P VoIP services (as they have in China), the effect of this sort of legislation will be to increase the cost of providing a traditional telephony service and increase the use of less regulated P2P solutions.

### Consumer Pressure
Consumers will judge the value of IMS on whether the service that it provides surpasses that of alternatives. If it does, then the operator will be able to charge a premium for IMS services and reap the benefit from its investment in the IMS infrastructure. However, IMS–based operators will be at a cost disadvantage compared with operators that offer a pure IP connection without the expense and complexity of an IMS infrastructure, so they will not be able to compete on price on basic services.

Many factors influence consumers' choices, but important areas include reliability; trust, including solutions to SPAM telephony (SPIT) and SPAM messaging (SPIM); convenience and simplicity; and cost

The IMS architecture as it is designed is focused on an "operator knows best" model. The success IMS and its evolution will depend on whether consumers agree with operators' choices or require a different set of features and restrictions.

*Competitive Pressure*
The competitive landscape will be the primary driver for the introduction and success of IMS. If all the major carriers pursue the IMS model to prevent their service becoming a commodity, then there will be limited competition and pressure to encourage them to try more radical business model.

However, this scenario is unlikely for number of reasons, including the following:

- One or more carriers will choose to offer an open IP network link at low cost. There are already examples of unlimited IP price plans from fixed and mobile operators.

- If an uncompetitive market develops, antitrust legislation will force operators to open up their networks to competitive carriers, who can provide a pure–IP service at marginal cost.

- The spread of alternative network providers (such as Wi-Fi hot spots and urban Wi-Fi networks) is increasing competition in the access network.

- Internet telephony is being promoted by well-known companies with deep pockets, including eBay/Skype, Google, and Microsoft.

### SBC Function Evolution
The power of the Internet is the transparency of the network and the ability for services to evolve without the need to upgrade the network core. Many session border control functions break this transparency by requiring the SBC in the core to understand the media signaling. This both increases the cost of running the network and reduces the speed at which services can evolve, but provides additional security for the users.

Hopefully, some of the more intrusive SBC roles will diminish over time with the spread of IPv6 and VoIP–friendly NATs, but others will remain to control and monitor access to the operator's network. The following sections discuss the likely evolution of each area.

*Security*
SBCs cannot support end-to-end security, as they need to be able to understand and modify the signaling messages. If users require higher security, then they should use an encrypted P2P service across an open IP network instead. The primary driver for such end-to-end security is likely to be illegal because it could be used to avoid surveillance by the intelligence services, but it may also be used to prevent unauthorized surveillance, for example, by an intermediate carrier or competitor.

The security model provided by IMS will not change—it will remain point-to-point. Users who require end-to-end security should use an alternative service or encrypt the media to provide sufficient security for their requirements.

*NAT and Firewall Traversal*
In the small office/home office (SOHO) environment, the use of symmetric NATs[1] is likely to decrease, so the support of STUN and other NAT traversal techniques by endpoints will enable the NAT traversal technology in SBCs to be retired for many users. The use of STUN enables SIP to be used through all types of NAT except symmetric NATs.

However, in enterprise environments, SBCs will increasingly be installed at the edge of the corporate network to protect it from attack.

Corporate rules will enable NAT and firewall traversal according to corporate policy—this may limit the rollout and availability of IMS services in a corporate LAN. This is identical to the situation today with access to other Internet services, e.g. e-mail and Web browsing, from within a corporate LAN, and should not be subverted by the carriers.

*IPv4 and NATs*
The IMS network, particularly when IPv4 and VPN issues are included, is not an open transparent network, so SBC function is required to enable multimedia services to work. Endpoint NAT traversal technologies such as STUN remove the need for traversal devices at the UNI but do not provide an end-to-end solution for media traffic that needs to traverse multiple private address spaces.

The use of IPv6 or a single global IPv4 address space throughout the network core would enable media to be routed directly between endpoints without the need for SBC function on the NNI. In addition to reducing the processing required, this would also ensure that the media takes the most direct route to its destination.

However, the current generation of core networks is based on IPv4, and many early IMS deployment will run over these networks. It will be a number of years before NATs on the NNI can be removed.

*QoS*
QoS across core networks is already extremely good, and the Internet has been shown to be capable of handling serious disruptions to its infrastructure without significant effect on its performance.

However, QoS in access networks remains a challenge. Bandwidth availability in access networks will continue to increase, but new services will evolve to use any extra capacity, so QoS mechanisms in access networks will continue to be required. The design of these mechanisms will depend on the specific access medium. In some cases an SBC will be required to enforce the rules at the network boundary, but in others it will be possible to negotiate access from end to end.

It is certain that differential handling of different classes of traffic over the access network will increase, but it is not clear whether this will require session border control at the core network end of the link to enforce policy. A more flexible solution would be to allow the endpoint to determine the class of service to be applied to each stream using an out-of-band mechanism.

## Conclusions

Session border control is fundamental to the IMS proposition for both operators and consumers. The exact function required will evolve as the underlying infrastructure and customers' demands change, but SBCs will always be part of any IMS solution.

There is short-term pressure for manufacturers to enhance their existing SBC and IMS products to enable them to be used as part of a Release 7 solution. This will be a competitive area, with products being developed by many manufacturers, but is also an area that requires an unusually wide breadth of expertise, so it will challenge the expertise of many contenders.

In the longer term, the evolution of session border control in IMS networks will depend on the ability of IMS–based networks to compete with lower-cost solutions available over the Internet, and of operators to charge enough for the QoS and security that they offer.

With or without IMS, SBCs will continue to provide protection at the boundaries between managed networks. The evolution of the next generation of telco networks will just determine where and how transparent those boundaries are.

## Notes

1. Symmetric NATs set up separate mappings between the private IP address and port, and the public IP address and port, for each remote address. As a result, an endpoint cannot use STUN to determine a public address that a third party can use to route media to it through a symmetric NAT. Instead, a media relay must be used.

# The Evolution of IMS

## Jin Huang

*Director*
Huawei Technologies Co., Ltd.

## Overview

Communication and information technology (IT) are maturing at an unprecedented rate thanks to rapid technological progress in both fields. The emergence of Internet protocol (IP) technology has acted as a precursor to the sudden prevalence of Internet applications, which in turn has created a greater reliance on IP–based fixed and mobile network solutions. Consumers are now able to enjoy a variety of communication methods coupled with an increased range of services; the industry has evolved past previous limits that confined services to voice-related ones. Within the context of this new era of service provision, network operation and content operation have become separate constituents.

Telecom operators find it extremely difficult to offer new and profitable services that are cost-effective. Bearers of time division multiplex (TDM)–based networks tend to be inflexible; interoperability between modes is too complicated; and the technologies used to develop value-added services are exclusive. Next-generation network (NGN) and third generation (3G) R4m are gradually emerging at the telecommunications forefront and are attracting the attention of both fixed and mobile network operators. The softswitch-centered NGN products and solutions hailing from different manufacturers have already enjoyed large-scale commercialization and application. Milestones have been reached in IP–based public switched telephone network (PSTN) switching services, mobile circuit-domain services, and the separation between control and bearer.

To meet the rising demands relative to IP multimedia applications, the third-generation partnership project (3GPP) promotes the IP multimedia subsystem (IMS). 3GPP defines the specifications for radio access by both wideband code division multiple access (WCDMA) and Global System for Mobile Communication (GSM). It acts as a facilitator for Release 99 (R99) and R4, inclusive of antenna interface specifications, voice service specifications in circuit-switched (CS) domains, and basic data service specifications in packet-switched (PS) domains. With respect to R5 and R6 research in relation to IP multimedia applications, R5 defines the core network architecture, public components, and basic service flows of IMS. Based on the extension of some R5 components, R6 defines the key service capability of IMS, quality of service (QoS), network interoperability,

and also IMS/CS integration. The IMS architecture derived from 3GPP is broadly recognized as a reasonably comprehensive solution to the IP multimedia domain. 3GPP2 and Telecommunications and Internet Converged Services and Protocols for Advanced Networking (TISPAN) have adjusted their IP multimedia network architectures and service systems according to the 3GPP IMS model. In terms of their responsibilities with regard to IMS, 3GPP2 is handling access for CDMA 2000, and fixed networks are under the remit of TISPAN.

## Network Evolution

### IMS: The Future of the Network
The consulting company Ovum predicts that IMS technology will commence with a series of new applications before gradually evolving into fixed-mobile converged architecture.

This evolution will be subject to the steps displayed in the following figure:

### Network Architecture Evolution
To meet operational needs and provide customized, diversified, and cost-efficient services, traditional switching networks will progressively develop from a traditional voice services network provider to a network that offers additional, value-added services. The network architecture will adjust correspondingly and control will be separated from the three constituent aspects: access, bearer, and service. User data management will also change from a distributed to a centralized mode.

Traditional circuit-based switches exist in the context of an integrated and closed architecture. In broad terms, this architecture is incapable of fulfilling the diversified requirements, which renders rapid service deployment throughout the whole network impossible. Making the essential connections in this type of architecture is both complicated and expensive, and doing so therefore falls outside the operational-expenditure parameters of constructing a cheap shared network.

The expedient step of separating control from bearer is the catalyst creating a distributed NGN system. This type of architecture reduces overall network construction costs, allows easier upgrades and backward compatibility, and

**FIGURE 1**

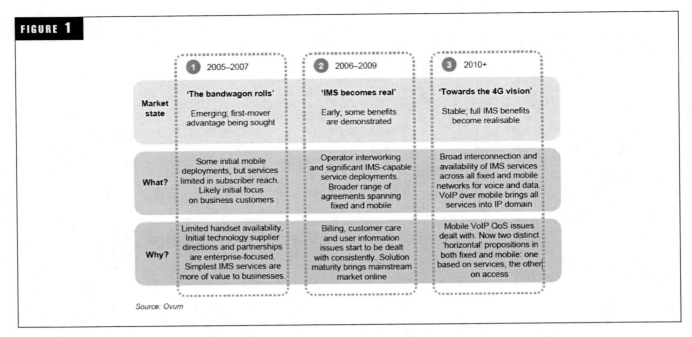

| | **1** 2005–2007 | **2** 2006–2009 | **3** 2010+ |
|---|---|---|---|
| **Market state** | **'The bandwagon rolls'**<br><br>Emerging; first-mover advantage being sought | **'IMS becomes real'**<br><br>Early; some benefits are demonstrated | **'Towards the 4G vision'**<br><br>Stable; full IMS benefits become realisable |
| **What?** | Some initial mobile deployments, but services limited in subscriber reach. Likely initial focus on business customers | Operator interworking and significant IMS-capable service deployments. Broader range of agreements spanning fixed and mobile | Broad interconnection and availability of IMS services across all fixed and mobile networks for voice and data. VoIP over mobile brings all services into IP domain |
| **Why?** | Limited handset availability. Initial technology supplier directions and partnerships are enterprise-focused. Simplest IMS services are more of value to businesses. | Billing, customer care and user information issues start to be dealt with consistently. Solution maturity brings mainstream market online | Mobile VoIP QoS issues dealt with. Now two distinct 'horizontal' propositions in both fixed and mobile: one based on services, the other on access |

Source: Ovum

facilitates the accelerated development of new services and applications. In this way, the network structure benefits from simplification, resource allocation efficiency becomes optimal, and a new service can be implemented throughout the whole network at little expense.

IMS is a network independent of access technology. Users of the network enjoy the same quality benefits and experiences whether IMS is accessed via asymmetrical digital user loop (ADSL) from a fixed terminal or through a WCDMA mobile terminal.

As illustrated in *Figure 1*, an IMS network may be divided into access interconnection, session control and application layers.

The access interconnection layer provides the following functions:

- Session origination and termination of various session initiation protocol (SIP) terminals
- Conversion of various IP packet bearer types
- Various QoS policies based on service deployment and session layer control
- Interoperability between conventional PSTN and public land mobile network (PLMN)

The access interconnection layer incorporates equipment such as various SIP terminals, wired and wireless access, and interoperable gateways.

The session layer implements and/or maintains the following functions:

- Basic session control
- User registration

**FIGURE 2**

**IMS Network Architecture**

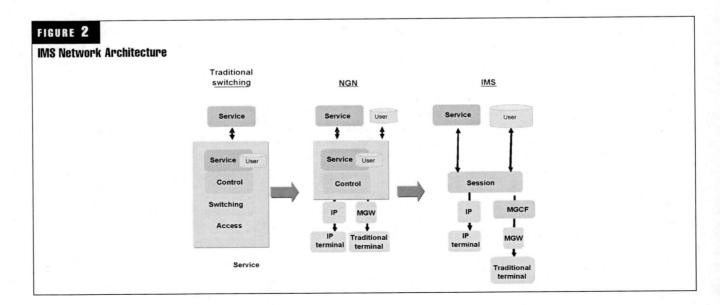

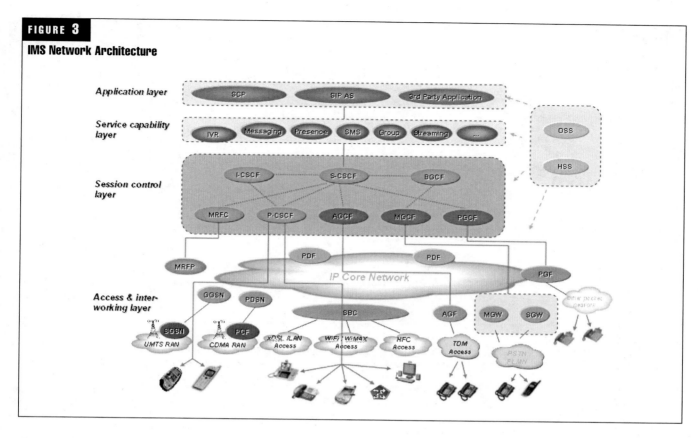

**FIGURE 3**

**IMS Network Architecture**

- SIP session route control
- Interaction with the application server to enable application service session
- User data maintenance and management
- QoS service policies' management
- Provision of a consistent service environment for all users within the application layer

The session layer includes a spectrum of functional entities such as call server control function (CSCF), multimedia resource function controller (MRFC), breakout gateway control function (BGCF) and IP multimedia service switching function (IM–SSF). CSCF features encompass proxy CSCF (P–CSCF), interrogating CSCF (I–CSCF), and serving CSCF (S–CSCF), which can be physically integrated or set. Important related factors to consider are the IMS service access mode and access point position. In terms of CSCF, capacity, capability, and user traffic requirements for service allocation and deployment during networking must be taken into account. Service allocation and deployment also relate to the operator's network topology-hiding and inter-operability requirements.

Entry for the UE to access IMS, P–CSCF implements the proxy and user agent functions in the SIP. The core positioned S–CSCF implements the following functions:

- Registration authentication and UE session control
- Basic session route function of calling and called IMS users
- Value-added services (VASs) to the application server (AS) based on the IMS triggering rule subscribed to stipulated conditions being reached by a given user

- Service control interaction

The I–CSCF acts as the gateway node in the IMS core network, and functionally allocates local domain user service nodes, routes queries and expedites topology hiding between IMS domains. I–CSCF also determines which S–CSCF will provide a given service for users via combined conditions.

Service logic for users is allowed through the application layer, which triggers conventional basic call services such as call forwarding, call waiting, and conference calling. IMS interacts with conventional intelligent network (IN) services, such as customized application of mobile enhanced logic (CAMEL) and IN application part (INAP), through the IM–SSF, thus inheriting the existing IN CS and PS services. In addition to existing CS and PS services, IMS also provides non-conventional AS channeled telecom services, including IM, PTT, and Presence. A simple API interface channeled through the open services access-gateway (OSA–GW) is also provided by IMS, which allows a third party to make use of network resources and provide secure services for feature-rich functions such as games and entertainment.

### Fixed-Mobile Convergence Trends

2005 marked the culmination of six years of research and development for Huawei's softswitch-centric NGN. The company began developing its softswitch in 1999 and, to date, large-scale applications and deployments have proven the softswitch system capable of replacing the traditional PSTN switch and the MSC in a mobile network. Although actual deployment might vary according to place and circumstance, developments in technology, interconnectivity,

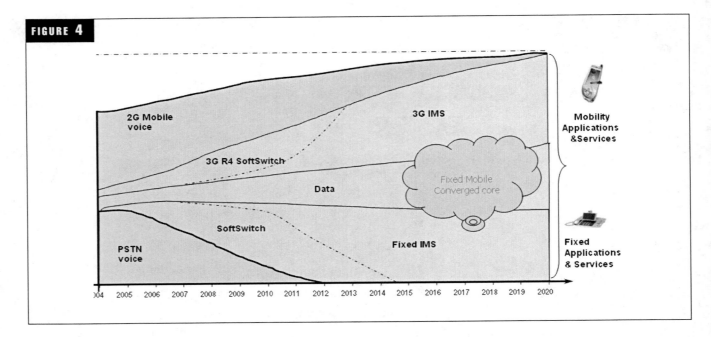

FIGURE 4

and standardization betrays the direction of network evolution. The following describes the possible phases of network evolution:

- *2005*: Large-scale NGNs will be built to replace the legacy PSTN.
- *Prior to 2010*: Softswitch is *the* solution for PSTN migration. Early IMS will focus on new service provision.
- *2010 to 2020*: Softswitch will migrate to IMS architecture. True fixed-mobile core network convergence (FMC) will occur.

The significance of TDM–based PSTN and 2G–network transformation to an IP–based NGN cannot be underestimated. As bearer networks converge, distributed networks based on standard architecture and interfaces reduce operation costs and are adaptive to the provision of new services as and when necessary. It is anticipated that, in time, the softswitch to IMS evolution process will undergo simplification, given that the process is achievable mainly by call control plane software upgrades, which in fact creates a negligible impact on the network as a whole.

## Evolution of IMS Standardization

IMS is being assessed and evaluated by a range of international standardization organizations, including 3GPP, 3GPP2, the European Telecommunication Standards Institute (ETSI), and the International Telecommunication Union Telecommunication Standardization Sector (ITU–T). The progress and focus of each vary significantly. During the early stages of their investigation, for instance, 3GPP and 3GPP2 were broadly concerned with mobile network architecture definition. ETSI and ITU–T, on the other hand, have been considering the fixed access field.

### 3GPP/3GPP2
The IMS forms part of the 3GPP standard series. 3GPP expresses the concept of the IMS in R5: the IM subsystem comprises all CN elements for provision of IP multimedia services consisting of audio, video, text, chat, etc. and a combination of them delivered over the PS domain. 3GPP TS 23.002 depicts the basic architecture of the IMS and defines CSCF functions, the key component. With respect to IMS, 3GPP proposes a series of specifications, inclusive of frame architecture (as shown in the following figure), call process, and some interface protocols.

At present, 3GPP's IMS research focuses on the following aspects:

- *Bearing IMS real-time services over the CS domain*: As part of the network architecture defined in the first stage of the IMS, signaling and user data are borne over the IP network. However, IP network QoS policy is incomplete; therefore, QoS is not guaranteed. Moreover, since many 2G–network operators already possess reliable and complete TDM networks, 3GPP is now developing specifications for bearing real-time multimedia services over the circuit domain. These specifications are suitable for traditional GSM or R99 network operators as they are able to use TDM network's real-time feature. For an R4 network operator's circuit domain and real-time service, it is irrelevant whether user data uses IP or the packet domain given that both modes are restricted by the IP network's QoS.

- *Achieving IMS services through the IPv4 technology*: Although IMS is designed to use IPv6 to avoid a shortage of IP addresses, no service requires that terminals must have public IP addresses. Therefore, the terminals can use private IP addresses to actualize different services. In short, this is not a key issue. Since IPv4 the networks are based on mature IPv4 technology, 3GPP is formulating relevant specifications.

- *IMS–based fixed-mobile convergence*: At present, fixed access methods mainly include WLAN and DSL. 3GPP is currently cooperating with ETSI TISPAN to research relevant standards, some outcomes of which appear in ETSI TISPAN–produced specifications.

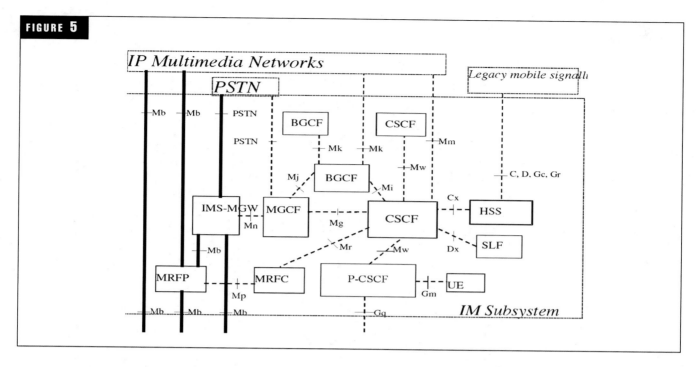

**FIGURE 5**

- Other problems relating to IMS architecture include emergency call support, IP stream-based charging, and interconnectivity between bearer independent call control (BICC) protocol and SIP.

- In general terms, 3GPP's focus is to update and perfect first- and second-stage IMS specifications. Along with 3GPP's standardization progress, the most current IMS technical draft is 6.6.0. 3GPP is planning to solve the problem contained in the fact that IMS needs to support various fixed access modes. TISPAN and 3GPP's two-day joint conference in 2004 (June 22 and 23) discussed the problems and cooperation framework that need to be considered when the IMS is reused in the TISPAN NGN. In addition, 3GPP2 introduces IMS to 3G CDMA networks, and have renamed it multimedia domain (MMD).

### ITU–T and ETSI TISPAN
ITU–T and ETSI TISPAN are focusing on the following tasks:

- IMS requirement proposals in relation to supporting fixed access modes.
- Clarifying the relationship between IMS, associated subsystems, and external networks in the context of overall NGN architecture.
- Updating specifications based on 3GPP specifications.

The following describes ITU–T Release 1 and TISPAN Release 1, which support some fixed access modes.

TISPAN Release 1 only defines the functional architecture and describes the functions of different layers, such as transport layer, service layer model, interconnectivity with other networks/domains, and user equipment (UE).

TISPAN Release 1 defines the architecture and focuses on fixed access (mainly DSL, and, to a lesser extent, WLAN).

Release 2 will be released in 2006 to optimize resource utilization.

Release 3 will be released in 2007 to facilitate further research in interoperability, roaming, and broadband access.

ITU–T defines NGN in Y.2001: "An NGN is a packet-based network able to provide telecommunication services and able to make use of multiple broadband, QoS-enabled transport technologies, and in which service-related functions are independent from underlying transport-related technologies. It offers unrestricted access to users of different service providers. It supports generalized mobility which will allow consistent and ubiquitous provision of services to users."

The architecture of NGN is defined as follows:

The first meeting of the NGN Focus Group (NGN FG) determined that to achieve their goals, the NGN core network managing sessions should be based on 3GPP IMS. The NGN FG intends to conduct research on the extensions needed to support various access modes. Organizations such as ITU–T are researching function requirements. The flows, messages, and parameters used for function implementation will be defined in follow-up specifications. The organizations considered must also modify and optimize the existing IMS specifications and formulate a series of new specifications.

### Summary on Standardization
- It is a common understanding among all the main standardization organizations that IMS will form the core architecture in future fixed-mobile convergence.

- 3GPP has conducted thorough research in the IMS field. In the mobile field, while IMS architecture has reached maturation, technical details are still subject to research.

**FIGURE 6**

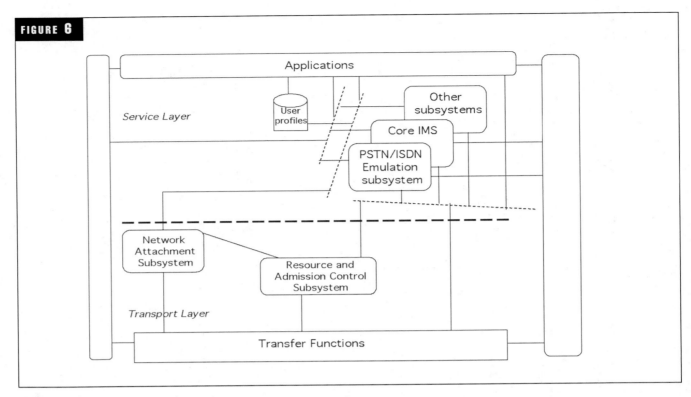

**FIGURE 7**

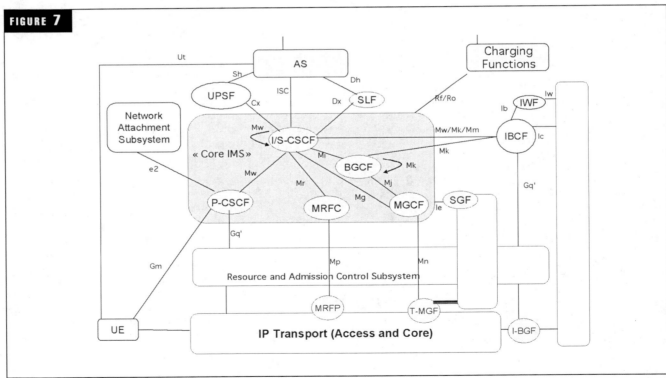

- In terms of the fixed-network IMS field, ETSI TISPAN has conducted thorough research. The IMS–based NGN architecture has been defined, and specific requirements and complementary technical research are ongoing, as the volume of yet-to-be-completed specifications is considerable.

- ITU NGN FG IMS research continues to exhibit an emphasis on requirement analysis.

Currently, network convergence IMS research is still in its infancy. With this in mind, many technical standard issues, including the following, require solutions:

**FIGURE 8**

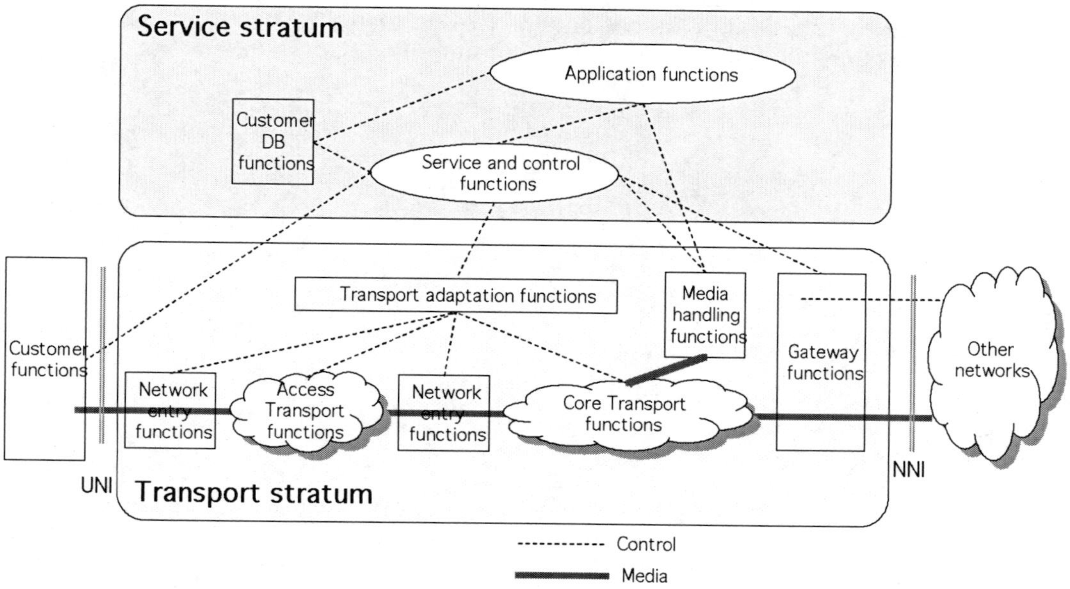

Note: Charging and billing functions and Management functions are applied to both Service and Transport strata

- Does IMS architecture need to realize all PSTN and ISDN services? If yes, how can this be achieved? How can IMS be expanded to support PSTN/ISDN? By an emulation or simulation subsystem? These questions are still far from being answered.

- SIP interoperability. Although the IMS is based on SIP, the IETF and 3GPP have each defined their own SIPs. Interoperability between these two SIPs is difficult to achieve. Furthermore, when an SIP is expanded to support PSTN services such as hooking, the problem of compatibility recurs.

- Incomplete MRFP function definition. This function was promoted by 3GPP a few years ago, and it then used the workable H.248. However, when IMS comes into being and the MRFP function is not expanded accordingly, IMS services provision is greatly hindered.

- Only basic parameters are defined for the secure SCIM function, and no detailed definition exists. The SCIM will play an important role in the complex services of IMS. While the CSCF implements the main call control function, the separation and integration of traditional services and new applications must be realized by the SCIM. Given this, numerous problems related to the SCIM details require solutions.

- The status of IP–based fixed networks raises the following question: Is an IC card needed to store a user's USIM or ISIM? Do any other methods exist? Must the fixed network support the AMR codec? Must IPv4 requirements be met? Where does NAT take place?

- The judgment required by a specific NGN charging might be different from that defined in 3G IMS.

- Is it necessary to increase security requirements for the 3GPP security system?

- How can the QoS levels defined by 3GPP be mapped onto the QoS levels defined in ITUY.1541? Will jitter occur?

## IMS Summary

The IMS–based convergence architecture describes a conceptual architecture for fixed-mobile converged network development. It is set to transform telecommunications and this explains and justifies the attention IMS is receiving. Network operators hope to use this scheme to provide identical services for fixed-line and mobile users over a unified core network. This will in turn simplify network structure, lower operational and maintenance costs, and achieve flexible service provision.

While IMS structure boasts a range of unparalleled advantages, many network-, service-, management-, operation-, and administration-related issues have yet to be resolved. Numerous functions require improvement. IMS, for instance, does not holistically consider all aspects of fixed access, and therefore the relevant specifications need to be expanded. The introduction of various fixed-access modes complicates the network, which requires further thought analysis in terms of fixed- and mobile-access network features. Otherwise, problems are likely to occur when attempting to provide services to all kinds of terminals while avoiding conflict with systems in existing networks. These factors increase the difficulties of standards' formulation. When viewing service integration, for example, the IMS is not a complete packet-based service system, so improvements are by necessity a dynamic and ongoing process. The involvement of the bearer layer also requires solutions in terms of QoS, security, insufficient address problems, user management, service management, and lawful interception. It is clear that to reach the other side of this technological cusp; much further study is required, particularly in terms of IMS application over the fixed-network domain.

# IMS–IP Multimedia Subsystem

## IMS Overview and the Unified Carrier Network

Neil Kinder

*Technical Director for Europe, the Middle East, and Africa*
Sonus Networks

## Introduction

### Background

The Internet protocol (IP) multimedia subsystem was initially defined by the 3GPP and 3GPP2 wireless working bodies. Its focus was to provide a new mobile network architecture that enables the convergence of data, speech, and mobile network technology over an IP–based infrastructure. The information management system (IMS) was designed to fill the gap between the existing traditional telecommunications technology and Internet technology that increased bandwidth alone will not provide. As a result, IMS will support operators in offering new and innovative services that will attract new subscribers and maintain their existing base.

Mobile operators have been very successful in driving penetration of their services over the past 10 to 15 years. They have used the services applications of service management system (SMS), voice mail, push-to-talk, and prepaid to support greater market segmentation, customer retention, and focus. Innovative tariffs mechanisms and the attraction of mobility have enabled mobile carriers to reach a broader customer base than the fixed-line carriers. However, fixed-line services are now changing dramatically, with the introduction of broadband and IMS a key step in helping the mobile carriers compete on the services side.

In the fixed-line world, IP has been already been deployed in many forms, from H.323 toll bypass through to softswitching for class 4 and 5 traffic types. However, the fixed-line standards bodies working on IP and voice were struggling to agree upon an architecture that maintained the important qualities of the past 100 years of public switched telephone network (PSTN) voice with the dynamic flexibility of the Internet. IMS has become attractive to both camps and, over the past four years, this architecture has developed an increased following from organizations such as the International Telecommunication Union (ITU), European Telecommunications Standards Institute (ETSI)/Telecommunications and Internet-converged Services and Protocols Networks (TISPAN), and various other important working parties. Today, operators and standards bodies see it as the umbrella architecture to both mobile and fixed multimedia communication in the 21st century.

While being focused at embracing IP, IMS was specifically created to enable and enhance real-time, multimedia mobile services. The range of services has had to be as broad as what the Internet has already exposed to subscribers on fixed-line broadband infrastructures, e.g., rich voice, video telephony, messaging, gaming, conferencing, and instant messenger (IM) services. To support the sophisticated demands of these services, many key mechanisms, including session negotiation and management, quality of service (QoS), and mobility management, have had to be developed.

However, IMS enables much more than just real-time user-to-user services. This paper focuses on the current market forces and how the IMS architecture responds to them.

## Market Factors and Trends

### A Market in Flux and under Pressure

Trends in technology, consumer/enterprise buying habits and expectations, operator cost pressures, regulatory changes, and geographical and political changes are just some of the influences being mixed in the telecommunications melting pot. Even though minutes of use (MoU) are increasing significantly, average revenue per user (ARPU) from voice services continues its downward trend, driving network operators to seek new revenue streams to build long-term growth, differentiation, and profitability. The hunt is on for new products and services that can be introduced quickly and cost-effectively, as well as services that will maintain existing subscribers, help reduce churn, and attract new users while supporting entry into previously untapped markets.

The traditional technical, geographical, and operational constraints, which segmented the roles of carriers in the fixed and mobile markets, are being removed. Today, an application service provider (ASP) or Internet service provider (ISP) is just as capable of launching a multi-country voice service across other fixed and mobile transport networks as an incumbent that owns those same networks.

### Service Silos Cannot Be Maintained

The influences of market deregulation, declining revenues/ARPU, and the variety of technological innovations in the past 20 years have led carriers to develop and deploy

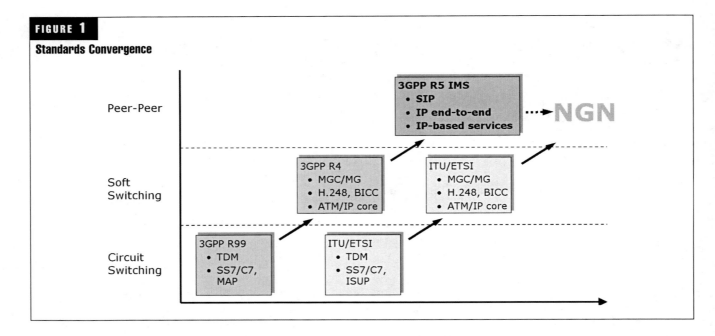

**FIGURE 1**

**Standards Convergence**

"stovepipe" or "silo" services and networks to meet the increased and broadening customer demand. This approach worked well while the increases in customer penetration and profitability continued to cover the original investment and maintain market share. However, the past five years have seen carriers finally accept that operating up to 20 transport networks (i.e., second generation [2G], frame relay, switched multimegabit data service [SMDS], asynchronous transfer mode [ATM], IP) with similar functional services mirrored across the various technologies can no longer be sustained without reducing profitability further and falling behind a market that is only increasing in momentum. Ten-year plans are being shortened to three years, and business cases need to start showing returns within 12 months to attract the attention of the board.

## Class 5 Trends

### Maturity of VoIP Technology
In comparison to all other communications technologies, IP and the Internet have continued to adapt, develop, and

grow in the number of infrastructures it can be transported across, applications it can support, and devices in which it is incorporated. Despite carriers having mixed opinions of the competencies of IP during the 1990s, it is now the case that IP is the only technology that can provide that can unify so many networks and applications.

As part of this evolution, voice over IP (VoIP) has also developed to provide the quality and reliability to rival that of the public switched telephone network (PSTN) while providing the significant cost savings of deploying a mass-market technology (at the time of writing this document, more than 11 percent of worldwide long-distance traffic [200 billion minutes] is now VoIP). Virtually, all new long-distance and tandem/transit deployments use VoIP and massive momentum is building behind VoIP in Class 5.

### Changing the Customer Landscape
In many markets, brand loyalty is diminishing and customers are becoming more discerning of their own needs. Indeed, many traditional brands and incumbents are now

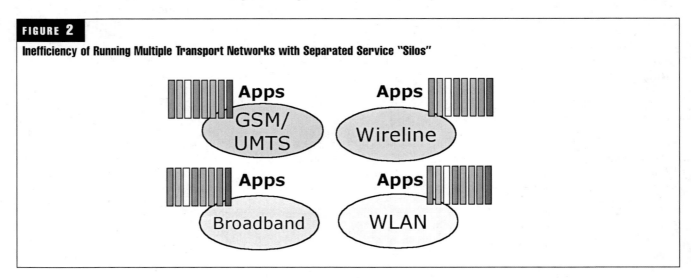

**FIGURE 2**

**Inefficiency of Running Multiple Transport Networks with Separated Service "Silos"**

developing strategies to make themselves more attractive to customers who now base their decisions on price, functionality, and ease of use.

Mobile services supplanting fixed is becoming common with both businesses and consumers. Increasingly, end users are relying purely on mobile for their voice service. In the developed parts of the world, wireline subscriber figures are stable or else declining. In the developing world, the areas experiencing voice market expansion and significant proportions of growth are in the wireless market due to the lack of last-mile infrastructure.

Mobile carriers are seeing increasing churn as the competition for the best tariffs and handsets, combined with the ability to transfer a phone number to the new service, drives customers to switch services. It is also the case that most mobile operators' services are, in the broadest sense, very similar to each other. Therefore, there is a need to support more innovative service ideas.

Subscribers increasingly see too much complexity in their interaction with utilities, including multiple accounts, multiple devices, multiple bills, and uncoordinated services. As a result, one way to gain competitive advantage is to simplify subscribers' lives through a single account, a single bill, and consistent services across all types of access.

### Voice over Broadband

Within incumbents and major fixed-line providers, major budget shifts are taking place, from expanding existing time division multiplex (TDM) voice platforms to investing in broadband, including DSL, cable, wireless, and Wi-Fi; wireless/mobile service platforms; and VoIP.

To incumbents, voice over broadband (VoB) represents the opportunity for major cost savings. This is because the investment in replacing 20-year-old concentrator equipment with intelligent, remotely configurable, and scalable access equipment can remove the costs for the "man-in-a-van" configuration changes and spiraling maintenance costs on equipment approaching obsolescence.

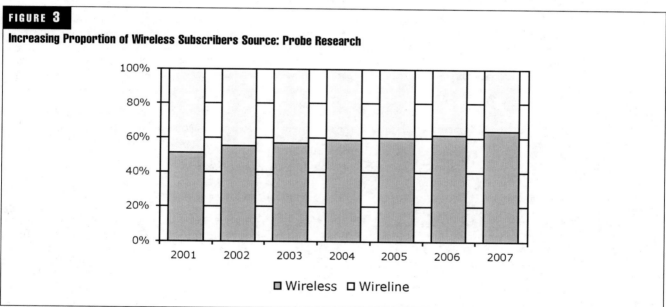

**FIGURE 3**

Increasing Proportion of Wireless Subscribers Source: Probe Research

□ Wireless  □ Wireline

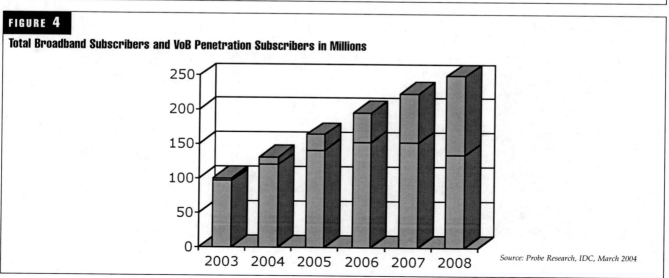

**FIGURE 4**

Total Broadband Subscribers and VoB Penetration Subscribers in Millions

*Source: Probe Research, IDC, March 2004*

To alternative operators, it represents the elimination of the key barrier to entry, that is, it removes the need to deploy an access network to reach subscribers. Indeed the service providers that are reacting quickest to this opportunity are those with significant IP expertise and investments such as ISPs and next-generation–based alternative operators. They then connect to their customers over other operators' access networks.

VoIP generates high revenues on broadband infrastructures compared to the bandwidth required. The service can also take broadband to environments less interested in the pure data solution. In some markets, the attractive factor of the marketed service is that it is VoIP.

Mobile carriers are looking toward VoIP to assist with the next phases of their services expansion. Push-to-talk, simplified services integration, and multimedia services are examples of the benefits of VoIP. Mobile carriers do not want their current services advantage to diminish when compared to VoB, e.g., IM.

Another benefit of VoIP to carriers whose existing market is mobile is the ability to take their services to fixed-line broadband networks and therefore expand their service reach. It therefore allows wireless operators to offer mobility without necessarily needing the user to always use the same handset, i.e., VoIP allows the subscriber to roam across multiple devices, both wired and wireless.

### Services Challenge
The communications industry is agreeing on the following architectural principles:

- One network, multiple access technologies
- Open and standards-based
- Internet protocols
- IP–based application framework
- Services that span the whole customer base

### Technology Revolution in Handsets and CPE
Communication and handset devices are going through a

revolution. Increasingly more complex technology is being packaged together in affordable handsets. Bluetooth, Wi-Fi, camera, office software, Internet browsing, session initiation protocol (SIP) voice client, IM client, gaming formats, etc., are enabling operators and service providers to develop services that exploit all of these user interfaces.

Likewise, fixed-line customer premises equipment (CPE) devices are benefiting from the commoditization of communication software stacks, network processors, and digital signal processors (DSPs). This is generating a reduction in unit prices and improvements in remote management and upgrades. Many of these models are also able to support voice, video, and data service feeds from the same device and will optionally include Wi-Fi, universal serial bus (USB), Bluetooth, and data storage.

### IMS Architecture

IMS broke the challenge down into three key areas, as shown in *Figure 5*:

With this model, it becomes possible to offer simplified operations, service consistency, ease of service bundling, streamlined provisioning, and common billing.

The solution is based on standards-based interfaces. Applications and services can be deployed in a much quicker fashion through simplified SIP control mechanisms. Voice, data, and multimedia can be incorporated in the services over a single IP interface and the application programmers insulated from access details. The result is the ability to quickly develop high-value applications.

### IMS Layer Overview

#### Application Server Layer
The application server layer contains the application servers, which provide the end user with service and enhanced service controls. The IMS architecture and SIP signaling is flexible enough to support a variety of telephony and non-telephony application servers. Sonus Networks has

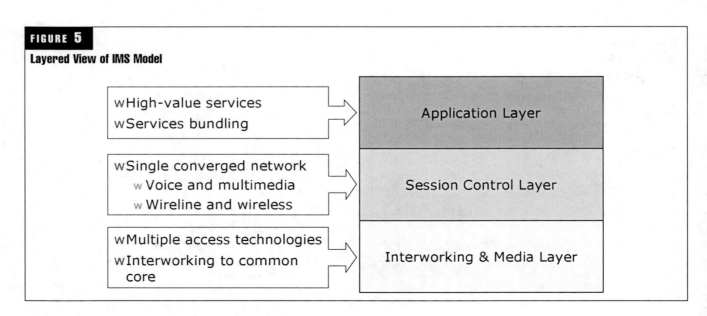

**FIGURE 5**
**Layered View of IMS Model**

been interfacing to SIP–based application servers for many years (www.sonusnet.com/contents/partners). The IMS model has defined a number of application servers. However, the basic function of them is to provide service applications via SIP.

### Session Control Layer

The session control layer contains a number of components that maintain the relationship between the applications and endpoints. These components perform the core work of call establishment.

### Interworking and Media Layer

SIP signaling is used to initiate and terminate sessions and provide bearer services such as conversion of voice from analogue or digital formats to IP packets using real-time transport protocol (RTP). All of the media processing facilities are in this layer, e.g., media gateways (MGs) are here for converting the VoIP bearer streams to the PSTN TDM format, and media servers provide many media-related services such as conferencing, playing announcements, collecting in-band signaling tones, speech recognition, and speech synthesis.

## IMS Component Overview

- *Call session control function (CSCF)*: This provides the registration of the endpoints and routing of the SIP signaling messages to the appropriate application server. The CSCF works with the interworking and transport layer to guarantee QoS across all services. There are a number of roles defined for CSCF servers, including the following:

  ○ *Serving (S–CSCF)*: This is a session control entity for endpoint devices that maintains session state.
  ○ *Proxy (P–CSCF)*: This is the entry point to IMS for devices. Whether in a home network or visited net-

work, the P–CSCF will be the first point of contact for the UE and forward SIP messages to the user's home S–CSCF.

  ○ *Interrogating (I–CSCF)*: This is the entry point to IMS from other networks.

- *Breakout gateway control function (BGCF)*: This function selects the network in which a PSTN breakout is to occur. If the breakout is to occur in the same network as the BGCF, then the BGCF selects a media gateway control function (MGCF), this will be responsible for interworking with the PSTN. The MGCF then receives the SIP signaling from the BGCF. The BGCF's role will increase in importance as networks begin to peer at an IP level for voice. As this starts to happen, the BGCF will become the peering control point in these IP–IP network boundary points.

- *Media gateway control function (MGCF)*: This interworks the SIP signaling with the signaling used by the media gateway (if required). The MGCF manages the distribution of sessions across multiple media gateways.

- *The media server function control (MSCF)*: This manages the use of resources on media servers.

- *SIP applications server (SIP–AS)*

The result is a converged architecture supporting a plethora of services over fixed and mobile access networks—a single unified network supporting all major access technologies with a single set of services that apply network-wide that are available anytime, anyplace, anywhere.

## Open and Standards-Based

For IMS to be successful, it needed to have the involvement of the standards bodies and benefit from practical deploy-

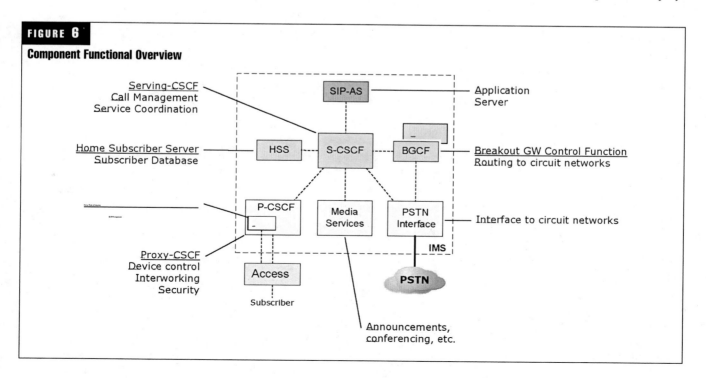

**FIGURE 6**

**Component Functional Overview**

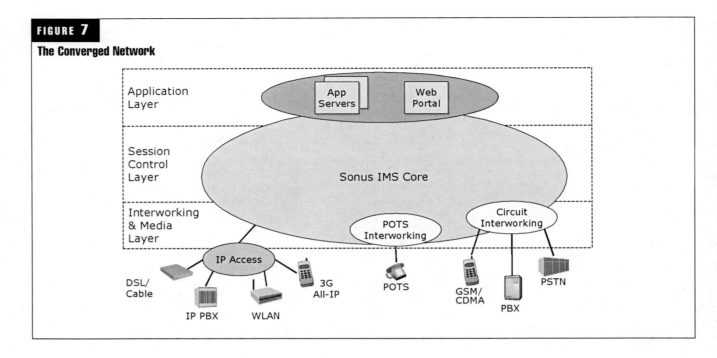

**FIGURE 7**

**The Converged Network**

ments. Fortunately, VoIP has been widely deployed over the past 20 years for many applications and, as a result, significant experience of the technology has been gleaned. The largest standards bodies have supported the development of IMS from the wireless world 3GPP/3GPP2 and from the fixed-line world, the ITU, the American National Standards Institute (ANSI), and the ETSI. The Internet Engineering Task Force (IETF) has also provided many of the base elements such as SIP. Finally, specialized organizations have added meat to the specification, e.g., Open Mobile Alliance (OMA) and the development of service applications, e.g., PoC (push-to-talk over cellular).

## IMS Application Servers

The IMS application service architecture has been defined with sufficient flexibility to incorporate a breadth of service profiles. Many intelligent network (IN) application servers relied too heavily on a tightly integrated architecture with proprietary implementation of application programming interfaces (APIs) or protocols. As a result, applications took too long to develop to meet market opportunities and were unnecessarily expensive. These service platforms were poorly integrated with the Web and Internet services.

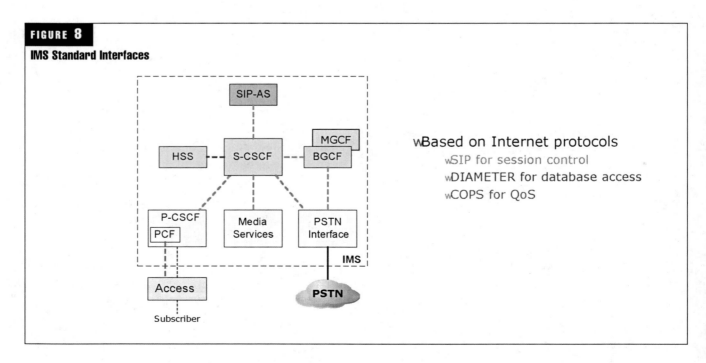

**FIGURE 8**

**IMS Standard Interfaces**

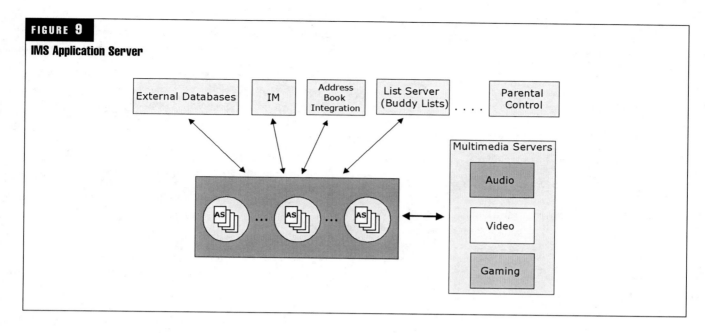

**FIGURE 9**

**IMS Application Server**

IMS application services will employ Web-enabled APIs such as call control extensible markup language (CCXML), voice XML (VXML) and Java together with standard interfaces to external databases such as simple object access protocol (SOAP)/XML, open database connectivity (ODBC), and java database connectivity (JDBC). Media servers and gateways will provide media processing when instructed by the application server. The newer implementations will cleanly separate the development environment from the run time to optimize performance.

## Benefits of IMS

### IMS Applications and Services Examples
*Push-to-Talk over Cellular (PoC)*
Push-to-talk is a mobile network–derived service for instantly communicating with the rest of a nominated workgroup. It is based on half-duplex VoIP technology. Converged PoC offers the subscriber all the benefits of two-way radio, across the country or around the world. Users can self-provision their own call groups, making changes and updates at any time. With the integration of converged presence and directory services, the user will be able to tell which group members are available for instant PoC contact at the push of a button.

*Instant Messenger*
PC–based instant messaging has become very popular among teenage Internet users. Workgroups, enterprises, and family and social groups are also embracing these text-based dialogues.

*Gaming*
By developing the online gaming communities' appetites, interactive communication and entertainment combined with the fact that these games run across IP, it becomes possible for IMS–based architectures to develop new revenue streams from this segment. These games can be downloaded over the IP broadband connection to the wireless or wireline device of choice.

*Voice and Unified Messaging*
Unified messaging supports access to messages of various media types (including voice mail, fax, or e-mail) from a single mailbox. The user can do this via a variety of devices, including wireless or wireline phone, personal digital assistant (PDA) or personal computer (PC) through a Web interface.

*Videoconferencing*
Web conferencing, audioconferencing, and videoconferencing allows participants to view presentation materials, listen to a conference, and hold simultaneous private text conversations, all under the control of a conference moderator.

*Voice and Video Telephony*
IP has increased the availability of video communication between various terminals and PCs. For businesses and geographically dispersed families, this means a videophone solution that does not depend on proprietary networks and equipment and is mobile.

*Presence Services*
The addition of presence services turns a simple handset directory into an availability list, indicating whether a user is available for a voice call or SMS message. The presence server can route calls in the preferred medium, not only to individual users, but also to services and places.

### IMS Value Proposition
*Single Infrastructure and Lower Operational Costs*
The converged network is characterized by the following:

- One network, multiple access technologies
- Open and standards-based
- Single set of services that apply network-wide

*Addresses Carriers' Business Needs*
- Reduced operating costs
  - Lower transmission costs
  - Simplified operations
  - Streamlined provisioning

- Better ability to address subscriber requirements
  - Service consistency for different types of access
  - Ease of service bundling
  - Common billing

*Faster Service Time to Market*
- IP–based application interface
  - Much simpler than old methods
  - Easy to combine voice, data, and multimedia
  - Independent of access details
  - Tailored services to specific markets lowers churn

- The result: fast development of high-value applications
- Directly affects carrier business
  - Adds revenue
  - Builds loyalty

*Enhanced QoS*

IMS provides standardized solution for real-time IP mobile services. Real-time mobile IP communication is difficult due to fluctuating bandwidths, which severely affect the transmission of IP packets through the network. The de facto 'best effort' approach for IP frame communication does not suit the service characteristics required by voice and video. The result is that real-time mobile IP services function poorly or not at all (i.e., voice quality is poor or garbled, video jitter, etc.). The QoS mechanisms were developed to overcome these issues and provide some type of guaranteed level of transmission instead of "best effort." The policy decision function (PDF) contains the intelligence required to enable QoS within a mobile IP network.

# New Service Creation in an IP Environment

## The Advantages of Integrated IMS and Service Delivery Platform Capabilities

## Kyriacos Sabatakakis

*Global Lead of Wireless Networks, Network Service Line*
Accenture

Excitement runs high today about the implications of rapid Internet protocol (IP) and broadband take-up around the world. Operators are migrating as quickly as possible to next-generation networks, confident of the possible cost savings through better aggregation of the pipes that transfer an enterprise's information.

But the business case for IP migration is more significant than the cost savings alone. An IP migration strategy that drives high performance must look at the possibilities created by this new, unified platform that can deliver more powerful services for customers and better user experiences—services that are also faster to create and easier to manage.

Deploying a new infrastructure in a network-driven or service-driven approach gives carriers important efficiencies, but it alone does not help deploy applications and services that allow carriers to realize incremental revenue benefits. What is needed is some way to create, control, and execute services on top of IP, taking advantage of that protocol's unique characteristics.

### The Promise of IMS

The answer to that need is the real growth enabler in the IP world: information management system (IMS), or IP multimedia subsystem. IMS is a standardized architecture—defined by global standards bodies that include the carriers themselves, the equipment manufacturers, and the IP community at large—that can truly commercialize the impact of IP.

How? In part, this is by simplifying access to the network layer through a common standard. In effect, IMS provides the intelligence on top of IP, enabling the creation, control, and execution of new, rich media services. For user-to-user services, as well as user-to-server and multi-user services, IMS is the answer. You want to provide your customers with

not only streaming video to their mobile devices, but also the ability to share it with a friend and watch it together in real time? IMS can make that happen.

In short, the efficiencies delivered by the all–IP infrastructure are great; but creating an impact on revenues is what makes IMS notable.

### Avoiding the Mistakes of the Past

There are many pitfalls ahead, however, for companies moving to the next generation of service creation. Nothing is as simple as it sounds, not even the unifying capabilities of IP network architectures. In fact, one of the possible dangers carriers face is repeating many of the errors made with previous generations of network capabilities.

A number of network vendors in the marketplace are selling rich media services that take advantage of the IP infrastructure. That is, they are selling the services, not necessarily the IMS technology. That is great, but to drive high performance by selling IMS as an enabling platform, carriers must take advantage of the horizontal layer of IMS, which enables offerings to be created without having to develop siloed, vertical services every time a carrier wants something new.

### Application Complexity

Another issue carriers are dealing with is the complexity of application development. This complexity has arisen in part because of efforts on the part of carriers to compensate for the loss of revenue from traditional voice services. To cope with that threat, carriers have increasingly moved over to the data side of service offerings, but that has helped create complexity in the application layer caused by multiple application platforms. Then, to create more innovative applications, carriers found they need to open up those platforms to third-party developers. So on top of everything else, they now have issues with access control and security.

## The Answer: Integrating IMS with Service Delivery Platform Solutions

We believe that an important part of the answer for carriers today—something that not only addresses the access and security issues involved in new service creation, but also prevents companies from repeating past mistakes of siloed application development—lies in bringing IMS together with something they already have: a service delivery platform: an open architecture that supports the creation of service offerings by content developers.

With modular, reusable application programming interfaces (APIs), carrier-grade scalability, and a highly extensible architecture, a service delivery platform gives application developers the means to create new value-added services and content faster, more consistently, and at a lower cost.

The flexibility of a service delivery platform allows developers to innovate rapidly, dynamically responding to changes in market or technology trends. As a result, the seemingly endless search for the "killer application" instead becomes a stepwise, flexible, and iterative process of discovery, with substantially lower business and financial risks for developers and network operators alike.

The point here is, why reinvent the wheel of application development in the IMS world? Again, why repeat past mistakes? Carriers should be doing two things: First, they should be developing services on the horizontal layer, not repeating past mistakes by developing within the vertical silos.

Second, they should be using their existing service delivery platform, eliminating the need to create another development environment for IMS. Carriers that have a service delivery platform in place can get a jumpstart on the competition by augmenting it, instead of starting over. Companies can leverage and enhance their existing service factory infrastructure to accommodate the new, rich media communications that IMS will enable. Rather than having to build a new service creation environment just for IMS, they can bring together all kinds of services in addition to IMS–based offerings.

## Making the Integration of IMS and Service Delivery Platform Happen

IMS functionality might overlap with functionality in a service delivery platform architecture. The challenges of integrating IMS with a service delivery platform architecture will involve identifying such duplicate functions and eliminating the redundancy by implementing a function in either the IMS layer or the service delivery platform layer. The decision where to implement the function will be based on the availability and maturity of technology solutions available from the vendors, the cost of replacing the function, and the complexity of integrating the new function.

An example of a challenge of IMS integration with the service delivery platform is user data consistency. The duplication of IMS functions and user data in existing service delivery environments must be eliminated as much as possible. Overlapping user profile repositories and provisioning flows in IMS components and the service delivery environment might exist, which could compromise the consistency of user data. IMS–based service subscriptions in IMS and the service delivery environments must also be consistently managed.

## How One Company Leveraged Its Service Delivery Platform in the IMS World

The recent experience of one European operator demonstrates how a service delivery platform integrated with IMS is delivering significant incremental revenue. For this operator to expand its presence in the marketplace and grow revenue by retaining its existing customers and attracting new ones, executives knew they had to re-engineer and upgrade the company's network infrastructure and renew its operations support system (OSS) applications to align with a next-generation network model.

The solution involved more than simply the network infrastructure; the company was also looking for strong content and service creation. This was accomplished through a service delivery platform—a horizontal, programmable, multi-service platform based on separating the layers of service, control, and information. Distinctive characteristics of this type of architecture include the adoption of standard protocols and the management of user-related data by a single, logically centralized database.

The company has advanced toward high performance in several areas, including a reduction in operational costs and a decrease in the time needed to roll out new services. Specifically, this initiative has laid the groundwork for significant improvements in business performance in the coming years. Based on its own analysis, this operator is on course to realize the following benefits by 2007:

- A 29 percent increase in revenue from innovative services
- An increase in the innovative handset customer base from 1.8 million to 11 million (from 9 percent of the customer base to 62 percent)
- Growth of value-added services users from 20 to 60 percent

## Overcoming the IMS Challenges

Carriers face a big challenge in the coming months to orchestrate the implementation of IMS in a timely fashion, ensuring high end-to-end service quality. IMS represents a fundamental change in the paradigm for service creation because its adoption involves taking apart the vertically integrated elements of traditional circuit switched networks and reassembling them in a different, more efficient architecture built around a set of standards-based technologies.

However, carriers looking ahead—past just the efficiencies of IMS to its revenue-generating potential—should be planning now to jump-start their service creation capabilities. It is up to the application and services layer on top of IMS to efficiently use IMS service capabilities, combining them into value-added services.

Companies looking to advance more quickly toward high performance should seek a service creation capability that integrates IMS into their existing service creation environment.

# Network Architectures and Design

# On Board IP Architecture

## A New Approach to Computer Telephony Card Design

## Ian Colville

*Product Manager of Marketing*
Aculab

### Executive Summary

As the computer telephony (CT) industry has evolved to deal with the emergence of Internet protocol (IP) and voice over IP (VoIP) telephony, the notion of separate digital signal processing (DSP) resource cards and IP connectivity or gateway cards has evaporated—just like the demise of traditional, single-function digital network access cards, fax cards, and voice processing cards—in favor of combined products. Multipurpose, multifunction cards are now the norm, and the industry is looking at where these latter-day design concepts can be improved and where they can still operate within the industry according to accepted form factor specifications. One reason for this is that, despite the advent of new specifications such as advanced telecom computing architecture (TCA) and associated packet-switched backplane (PSB) technology, in practice the PC server, and hence the CT industry, remains heavily biased in favor of protocol control information (PCI)– or CompactPCI–based designs.

This paper presents an alternative to the traditional PCI–driven parallel bus architecture for the core, on-board design of versatile, multifunction CT resource cards. It demonstrates a "think outside the box" approach that combines adherence to standards with optimum use of prevalent technologies. The unique design hypothesis will illustrate a method of overcoming limitations of proprietary device drivers to present an attractive option for large-scale, distributable CT applications.

### The Traditional Approach

#### From Single- to Multifunction

The traditional approach to CT card design has evolved from the days of single-function integrated services digital network (ISDN), fax, and voice boards—from the likes of Dialogic and Rhetorex in the early days—to the situation where multipurpose, multifunction cards are now commonplace.

Many vendors in the industry now offer combined IP gateway and media processing resource cards. And the competition among manufacturers of universally functional cards in PCI and CompactPCI formats is intensifying, with several new vendors entering the fray.

Most competitive offerings seem to show strengths in either the IP domain or in DSP resources but typically not both. And there are some "tick box" suppliers. Customers mostly want card density and configuration options, so it is desirable for vendors to offer a mix-and-match set of modules, which they balance by cost-effective manufacturing variants, employing high reuse of the same core technology.

#### Conventional Blueprints

With the majority of vendors having migrated from industry standard architecture (ISA) to PCI and/or CompactPCI cards, a common factor in base card design is the PCI bus, a parallel data bus. This has proven to be good enough for first- and second-generation CT applications but presents some limitations for today's third-generation products with their IP telephony cynosure.

A key design point in relation to current-generation PCI and CompactPCI cards is simply that because they are proprietary creations—albeit designed to PCI Special Interest Group (SIG)/PCI Industrial Computer Manufacturers Group (PICMG) bus and form factor specifications—they require vendor-specific device drivers and libraries. These will have been written to be compatible with one or more operating systems (OSs), with different driver software needed for each OS, and more important, the libraries will be able to talk only to the vendor's drivers. This conventional blueprint, which until now was more than adequate, has certain limitations.

The available hardware resources can only be readily controlled from within the platform they are installed in, that is, unless some technology such as MC3 (an H.100/H.110 bus extension option) is employed to scale a solution beyond a single box and some additional middleware similar to the Enterprise Computer Technology Forum's (ECTF's) now-unheralded CT–Media is used to add control of the algorithms or a solution provider/third party is engaged to provide an integrated platform that controls distributed client and server applications and hardware resources. Or perhaps customers—the hard-pressed application developers or systems integrators—have no option but to write their own distribution application programming interface (API).

All of these options are no doubt in widespread use, but they all have one thing in common—they tackle the problem from an API/application perspective, putting the onus on the customer to add something to the mix to achieve the end result. From a low-level, hardware-design, device-driver standpoint, the challenge to the vendor community is, "Why not consider an alternative?" The use of proprietary drivers and libraries precludes the ability to make use of the distribution mechanisms built into today's IP stacks, which come "for free" with the OS. Therein lies a clue.

## The Need for a New Approach

### Does Size Matter?

Well, in truth, if the needs of the application are satisfied by a discrete system in a single chassis and any future scaling is not needed beyond the maximum capacity of the platform, it really does not matter too much. However, if larger systems are envisioned, much larger than the initial deployment, for example, then this could be a significant factor. There are other considerations too, such as the need for the optimum utilization—or sharing—of valuable hardware resources. This is a prime concern of application service providers (ASPs), for instance, for whom capital expenditures (CAPEX) and overheads are key determining factors. If the available resources can only be readily controlled from within the box they are installed in, then sharing of resources cannot be readily achieved. That is not to say it cannot be realized—rather, it becomes more difficult. If the available resources cannot be readily scaled or expanded, then the development of large-scale, distributable CT applications is less easily accomplished. As processor power increases, components of a size offer more per square millimeter and prices are squeezed, which results in customers getting greater value for money than before. And the option of greater channel density in the same form factor, or using even larger card sizes such as AdvancedTCA, also exists for large-scale telco system deployments. That said, addressing the software issues and/or adding bigger product variants to their portfolio is not the panacea and are not the only choices available to vendors. With an investment in PCI and CompactPCI designs and an industry-wide customer base yet to be convinced of the advantages of third-generation PICMG "form and fabric" specifications, the life expectancy for these more mature technologies can be summed up in one word—longevity.

### Captive Cards

So what is the problem we are trying to resolve? In a nutshell, it is the scalability and distribution of system resources. Scalability means expanding the number of resources available to a system or application. Distribution means being able to readily locate those resources independently of the application (i.e., non-co-located), which is ideal for those ASPs or service providers.

Current-generation cards installed in a chassis are essentially captive; they belong to the box they are in. They are locked into the application platform they serve and are not readily accessible to any other application.

### Why Are They Locked In?

A standard PCI card is controlled by device drivers. These are by their nature proprietary, i.e., value-added code designed to control only the vendor's own card or card type(s). The link between these drivers and the application that needs to make use of the card's resource functions is the vendor's API library. The library is then also proprietary and only controls the same card or card type(s).

The library and driver are not designed to control anything else, anywhere else but in the same box. They cannot access devices on another box. There is no method of identifying another device—not even a card of the same type from the same vendor—on another box. The library and driver do not have such a routing mechanism.

There are ways around this, of course, but like with middleware, the key point is that they are not highly efficient.

### To'ing and Fro'ing – an Example

Reasonably sized interactive voice response (IVR) deployments often require the use of more than one host machine to make use of the scalability, reliability, and greater capacity that such systems can offer. An example of this is speech recognition systems, for which the speech recognition engines, and sometimes the applications, are typically on one or more hosts remote from the machine containing the speech resource cards (e.g., time division multiplex [TDM] telephony and display system protocol [DSP] resource cards with record, echo cancellation, dual-tone multifrequency [DTMF] detection, etc.).

A speech recognition engine running on a server may receive recorded speech from a client machine where the speech resource cards needed to process the incoming speech signal are. Similarly, an application running on a server may send speech to the same client machine for replay onto the line. See *Figure 1*.

Using current technology, this client machine will run a "recognition client" process, which interfaces between the speech resource cards and the remotely located speech recognition engines and applications. If a remotely located application wants to play a prompt, it sends a "play_this_file" message, along with the file itself, via the local-area network (LAN) to the "recognition client" process on the client machine. This process opens the file and streams it to the resource card for playback using a "play_this_file" API call. This is inefficient.

Why? Because the file arrives at the client machine's network interface card (NIC), passes over its PCI bus and has to be processed by the client's central processing unit (CPU) before being passed via the API and device drivers over the PCI bus (again) to the speech resource card for replay. Not only does this amount to a "double hop" over the client machine's PCI bus, but you must also write the "recognition client" process application as well as the main application. And you have to write your own distribution mechanism too—a distribution layer above the API—for both machines, because the API only talks to the card driver and card over the PCI bus. Even if the file is stored on the client machine, its CPU still has to get involved to send it up to the card over the PCI bus.

A better way would be to send the file directly to the resource card for replay, if only the card could accept it.

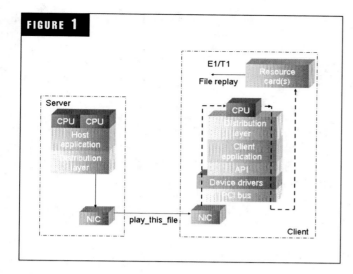

**FIGURE 1**

## The Aculab Concept

### Ubiquitous IP

To achieve scalability and distribution of combined media, VoIP and public switched telephone network (PSTN) system resources, this design concept uniquely employs an on-board and inter-system, all–IP control architecture. The result of this is that multiple hardware resources (cards) can be shared by multiple, independent (non-co-located) application servers.

IP has been deployed on a global scale, and huge investment has made it ubiquitous; it is a stable, proven, reliable technology. For various reasons, controlling a card over IP is considerably lighter on the available host resource than the host performing the equivalent amount of IP telephony. And as IP telephony is achievable now, so IP control of cards is easily achievable.

### How Does It Work?

As has been stated, proprietary drivers and libraries do not have an integral routing function. On the other hand, using a common media access control (MAC)–layer driver and the OS IP stack inherently provides the Ethernet with connectivity and desired routing capability. So why not make use of these distribution mechanisms? After all, they are built in to the IP stacks and so effectively come "for free" with the OS. Using a MAC component on a resource card in place of a more usual PCI interface device means that when the card is plugged in to the host, it is recognized as a NIC and gains an IP identity of its own.

As with current-generation cards, a vendor will continue to provide device drivers and libraries for these "resource NICs," thus enabling applications to communicate with the card(s). The driver software will still have been written to be compatible with one or more OS, but the essential driver technology is available and well proven due to the many deployments of the chosen MAC devices. But the key point is that these drivers and libraries have standard, OS–defined interfaces at the IP stack layer.

This means that the IP stack can readily route data between a library and such a "resource NIC" driver or, indeed, any

regular NIC driver. Therefore, it uses the same mechanism whether routing data between the library and a local card or between the library and a remotely located card (via a regular NIC). As each resource card has its own IP/MAC address and its own Ethernet connection, it is "findable" on the network by any application in any server. In this way, cards can be directly controlled via Ethernet connections on a private LAN.

The efficiency gained through this design feature comes from the removal of the "double hop" previously referred to, which frees up the local CPU to do a lot more work. Apply this technique to the entire control path anatomy of a card and the net result is an all–IP structure. Resources can be shared across both the on-board and external IP networks, enabling not just the card, but the resource functions of the card, to be controlled remotely. This architecture makes it ideal for cards to be controlled either by a local application on the host CPU or from a separate host PC. What is more, the API does not require the addition of a higher layer to achieve distribution, because the IP stack and NIC drivers take care of the routing of messages, whether via the LAN or locally.

### And There Is More

Employing a modular, IP–interconnected design on the card means scaling from low to high densities and is readily achievable by either providing on-board module options for DSP resources and PSTN network connectivity or adding cards in one or more chassis to give intra- or intersystem options.

Moreover, when adding cards in an additional chassis, as previously shown, the control data path need not go via the traditional server-to-client route. The communication data path is directly with the card via its own Ethernet connection, as shown in *Figure 2*. Taken further, this idea means that the use of a "passive" chassis, simply for system expansion, is feasible, with no client CPU requirement at all.

Monitoring distributed chassis and cards is also readily achievable via LAN connectivity using industry standard network management procedures such as simple network

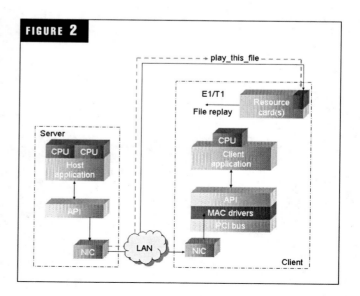

**FIGURE 2**

management protocol (SNMP) to control the remote equipment. The principal of modularity also enables customers to enjoy the flexibility of selecting a given function or not, e.g., keeping the TDM trunk connectivity on a daughter module. Plus, from a regulatory perspective, this separation allows easier achievement and maintenance of type approvals.

## Conclusions

This concept is ideal for products designed to fit into the complex, converged technical space where support for circuit switched digital network access, media processing resources, and IP telephony are all required in a multiplatform, distributed system. And because the interconnection is via Ethernet, the channel count of the distributed, shared resource pool is restricted only by the available bandwidth and the capabilities of the host machine driving it.

The idea of adding modules and then cards to expand a system provides excellent, all-around granularities, from small-scale, low-density, single-card systems up to large-scale systems of virtually unlimited size. Using passive expansion chassis filled with resource cards also contributes to the cost-effectiveness of the end solution. Very high channel counts are possible, with greater value per channel leading to improved margins—music to the ears of the financial director!

The principle of shared resources also leads to optimum efficiency, in contrast to a traditional system where underutilized capacity will be inaccessible to other applications. A crucial factor for customers seeking to migrate is consistency with previous driver APIs. This approach allows customers to reuse existing applications, with a very small number of changes, because the interaction between the API and the driver should be essentially unchanged from an application perspective.

Furthermore, this creative approach can allow open standards platforms and solutions to more readily compete with proprietary enterprise and telco switching systems, enabling bids for projects that were previously out of reach.

# Optimizing Core Networks for IP Transport

## Ori Gerstel

*Technical Leader, Advanced Technology, Core Routing Business Unit*
Cisco Systems, Inc.

## Mallik Tatipamula

*Senior Product Manager, Service Provider Router Technology Group*
Cisco Systems, Inc.

## Kelly Ahuja

*Vice President and General Manager, Core Routing Business Unit*
Cisco Systems, Inc.

## Walid Wakim

*Consulting Systems Engineer, Optical Networking Group*
Cisco Systems, Inc.

## Vik Saxena

*Director of Optical Networking, Advanced Engineering*
Comcast Cable Corporation

## John Leddy

*Vice President, Advanced Engineering*
Comcast Cable Corporation

## Abstract

We introduce a novel architecture that integrates Internet protocol (IP) and dense wavelength division multiplexing (DWDM) technologies from a data plane, control plane, and management plane perspective. Such an architecture provides significant capital and operational cost efficiencies, allowing carriers to reduce the cost incurred per bit while significantly scaling up capacity in the network. This paper also highlights the challenges in today's networks and how the proposed architecture would address those challenges.

## Introduction

This paper describes an IP–over–DWDM (IPoDWDM) architecture, designed for high-speed (10 Gbps and above) IP networks. The first section of this paper describes the drivers that motivate a change in the current network architecture. It also explains why the core of the network should shift from electrical to photonic in order to address the required change and how the IPoDWDM architecture supports this requirement. The second section highlights the benefits of this architecture. The third section discusses the challenges posed by the architecture and how those could be overcome. The fourth section describes the unique role that a control plane plays in this architecture. The fifth section summarizes the paper.

## Drivers for IPoDWDM Networking

The significant growth in IP traffic outpaces any other type of traffic in core networks worldwide. It is largely assumed that this trend will accelerate in coming years because of the widespread deployment of broadband over cable, fiber, and wireless networks, as shown in *Figure 1*. As the last-mile bandwidth to the consumer increases, new bandwidth-hungry applications such as IP video, gaming, and videoconferencing will be introduced, placing new scaling requirements on the current network core. At the same time, service providers (SPs) will continue to drive to lower the cost per bit and optimize the network costs as much as possible—both in terms of the cost of equipment and operational efficiency.

Most SPs historically designed the network as a layered architecture: the design of the transport layer was separate from the design of the IP layer. This approach created many instances of inefficiency in the network, from the perspective of capital expenditures (CAPEX) and operational expenditures (OPEX). This separation of layers should not be taken for granted; the advantages of integration have already been witnessed at the transport layer. For example, the synchronous optical network (SONET)/synchronous digital hierarchy (SDH) layer [1] and the DWDM layer [2] have been integrated together for several years now. This

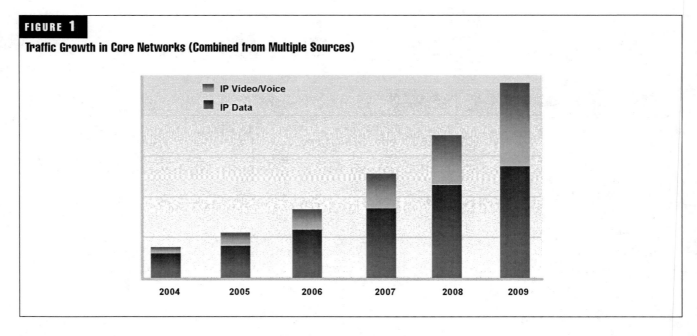

**FIGURE 1**

**Traffic Growth in Core Networks (Combined from Multiple Sources)**

integration manifests itself in the form of native DWDM interfaces on SONET/SDH ADMs, cross-connects and common management systems. We believe that the time has come for a similar level of integration between routers and the DWDM layer in the core of the network, as will be explained in this paper.

A traditional IP over optical network is depicted in *Figure 2*. In such a network, routers rely on Layer-1 devices for bandwidth management and protection through SONET/SDH add/drop multiplexers (ADMs) and cross-connects. These devices use transponders to translate their trunk signals into wavelengths for shipment over a fixed WDM layer, typically consisting of point-to-point links. This architecture worked well when router trunks used much lower bandwidth then the wavelengths that carried them and when the majority of the traffic and services were SONET/SDH.

However, the current drive to migrate most services over IP, coupled with the availability of high-speed trunks (10Gbps and 40Gbps) on core routers, create a strain on the architecture: the traffic between peer routers in a typical core IP network exceeds 2.5 Gbps and sometimes even 10 Gbps. These IP trunk demands can therefore be economically transported through multiple intermediate sites without requiring electrical processing at Layers 1-3. As a result, it is no longer beneficial to insert a layer of SONET/SDH grooming between the router and the DWDM layer. A much lower cost solution is to originate the DWDM wavelengths directly from the router and allow them to propagate through the

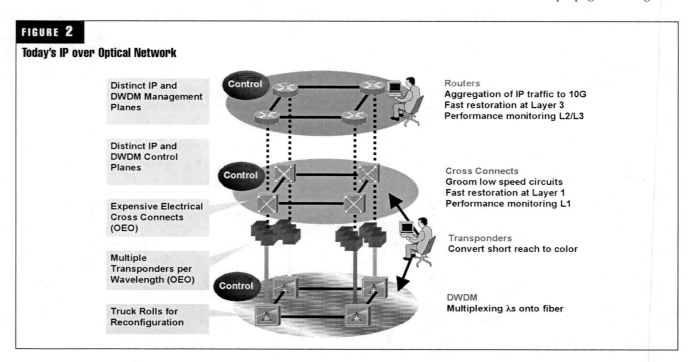

**FIGURE 2**

**Today's IP over Optical Network**

Distinct IP and DWDM Management Planes

Distinct IP and DWDM Control Planes

Expensive Electrical Cross Connects (OEO)

Multiple Transponders per Wavelength (OEO)

Truck Rolls for Reconfiguration

Routers
Aggregation of IP traffic to 10G
Fast restoration at Layer 3
Performance monitoring L2/L3

Cross Connects
Groom low speed circuits
Fast restoration at Layer 1
Performance monitoring L1

Transponders
Convert short reach to color

DWDM
Multiplexing λs onto fiber

core of the network, as shown in *Figure 3*, all without costly conversion between the optical and electrical domains, departing from the common architecture in today's network.

Furthermore, advents in optical switching and transmission technologies allow for optical bulk switching of these trunks and facilitate a flexible optical infrastructure that manages the optical bandwidth in the optical domain.

### The Diminishing Value of Electrical Processing for IP Transport

In the past, some SPs have used SONET/SDH as a transport vehicle for IP traffic to take advantage of features such as performance management and fast protection. Recently, G.709 (digital wrapper [3]) has been proposed as alternative multiplexing scheme, essentially upgrading the basic bandwidth unit to 2.5 Gbps. Such schemes provide two distinct functions in the electrical domain: framing and bandwidth management. The framing function provides overheads for monitoring, error correction, etc., while bandwidth management enables time division multiplexing (TDM), protection and cross-connection at the sub-wavelength level. While we believe the framing function will continue to play an important role over the core, we do not believe that the electrical bandwidth management function will be of value, as explained below.

TDM is very effective if the bandwidth of the client signal is much lower than the bandwidth of the transmission layer. This allows for effective grooming of multiple client signals into a wavelength and switching them automatically as the network evolves. However, today's routers are capable of transmitting at the same bit rate and sometimes even at higher bit rates compared to TDM platforms. For example, 40 G interfaces on routers are being deployed at the same time or earlier that 40 G interfaces on cross-connects. As a result, TDM no longer provides value in the core. Note that this observation holds not just for SONET/SDH but also for the emerging G.709 multiplexing hierarchy (2.5 G into 10 G into 40 G).

Another historic value of electrical bandwidth management was fast and reliable protection against network failures such as fiber cuts. However, with the advent of reliable core routers (supporting the 99.999 percent availability that transport equipment is famous for) and reliable and fast (less than 50 ms) restoration in the packet domain, such as multiprotocol label switching (MPLS) Fast ReRoute (FRR [4]), the value of protecting traffic at the TDM layer becomes questionable. Moreover, in multiservice networks, not all traffic has the same protection needs. For example, an SP may choose to degrade the performance of peer-to-peer traffic under failure conditions while keeping the performance of business traffic intact. This is very hard to achieve at the TDM layer, as no visibility into packets is available. Therefore, network protection would largely be replaced by IP restoration.

Additionally, the reasons against an electrical transport layer are deeper and far-reaching: electronic processing is intrinsically more sensitive to the traffic format than photonic processing, as demonstrated by the following examples:

*Bit-rate robustness*: Today, much of the core traffic is at nx10 G level, we will see 40 G router interconnects deployed in next-generation networks and believe that their use will be important for core networks. In the photonic domain, 40 G transmission can be handled through novel modulation formats that do not require significant changes in the existing transmission network—effectively quadrupling the available network capacity—while processing 40 G traffic in the electrical domain does not offer such benefits. It simply uses up the switching resources four times faster.

*Future protocol robustness*: An electrical network element is unable to support even small deviations from the bit rates and protocols it was designed for without significant work. For example, if an electronic cross-connect (EXC) is designed for 10 Gbps traffic; modifying it to support 10 GE LAN PHY is not possible without major redesign. The photonic layer is insensitive to fairly significant protocol changes.

*Functional robustness*: Even an electrical system that supports protocols transparently requires complex hardware designs to accommodate unforeseen limitations on the traffic. For example, fiber channel (FC) traffic requires extremely low

### FIGURE 3

#### Proposed of IP over DWDM Architecture

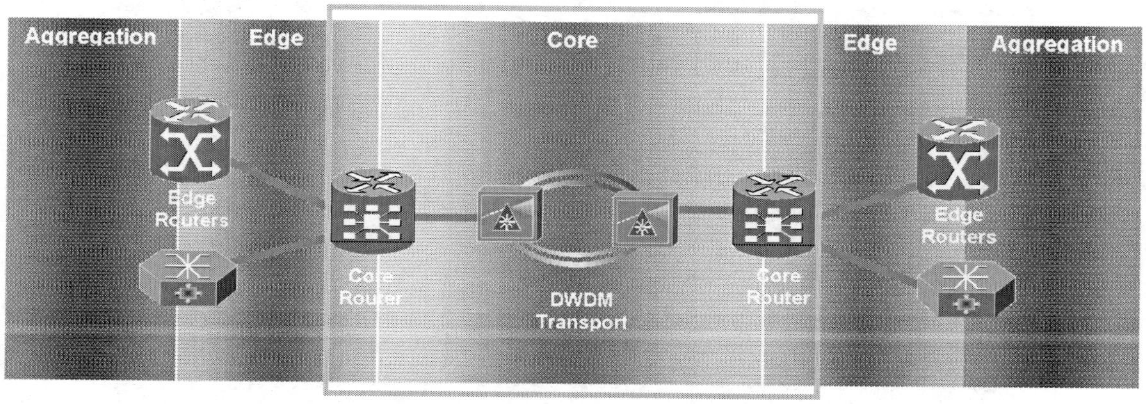

latency. A system that was not designed with this in mind is likely to fall short. In the photonic domain, the signals go through a minimal number of elements that do not significantly interact with the protocol and therefore are likely to meet such demands. Specifically, the latency is very low, supporting the FC example.

In each of the above examples, one can come up with an electrical system that will take the specific requirements into account, but this does not alter the fact that the solution is sensitive to the next set of requirements that were not anticipated at the time of deployment. Greater care is required in planning core networks that are being deployed for longer periods of time—often 10 to 20 years. Judging from past experience, how many new protocols will come up over such a life span?

### An Integrated IPoDWDM Core Transport Architecture

The proposed architecture recognizes the need to provide more integration of IP and DWDM networks in the core as well as more automation in the optical layer to bring it to the level of flexibility that packet networks have been expecting. It is based on the following three building blocks:

- *Element integration*: The integration of DWDM interfaces into core routers, as well as the integration of photonic switching into the DWDM layer, allows bypassing of transit points optically when there is no need for Layer-3 processing.

- *Management integration*: The integration of management functions from both the IP layer and DWDM layer into a single streamlined solution—still allowing different users to have specialized views based on the role.

- *Control integration*: Integrating control plane functionality, building on existing generalized MPLS (GMPLS) standards, and extending them to the DWDM layer.

The three building blocks of the architecture are depicted in *Figure 4* and will be described in more detail in subsequent sections.

This architecture leads to the following transformation in the network as shown in *Figure 5*. This network is greatly simplified from the traditional architecture depicted in *Figure 2*.

### Benefits of the IPoDWDM Architecture

The IPoDWDM architecture provides tremendous value over traditional transponder-based architectures and over architectures based on next-generation EXCs. Some of these values are as follows:

- *CAPEX savings*: Perhaps the most obvious advantage is removing transponders, the shelves that house them and the short-reach interconnects between router interfaces and Layer-1 devices. Existing architectures imply many more optical-to-electrical (O–E) conversions and separate systems that have a nontrivial price tag. This is discussed in greater depth in the next subsection.

- *Labor savings*: The only solution that economically provides all the switching resources at intermediate sites from the onset is integrated photonic switching in the DWDM layer. While they in theory provide automation, solutions based on EXCs still require considerable manual visits to install new DWDM transmitters/receivers, since these interfaces are expensive and cannot be fully pre-deployed at initial deployment.

- *Footprint savings*: Transponders and the active optics and electronics that implement them require space in COs and PoPs. Based on today's transponder technology, for a router supporting 64x10 Gbps interfaces in one rack, four to six shelves of transponders would be required to convert the 10 G short-reach interfaces from the router into DWDM signals. This represents about 30 to 50 percent of the footprint required by the IP and the optical layers combined. IPoDWDM allows more modest use of precious CO and PoP real estate.

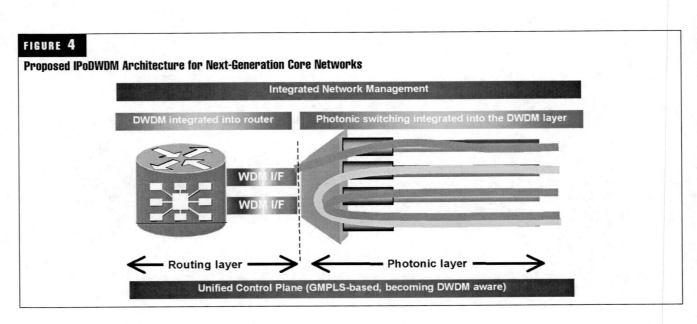

**FIGURE 4**

**Proposed IPoDWDM Architecture for Next-Generation Core Networks**

Integrated Network Management

DWDM integrated into router

Photonic switching integrated into the DWDM layer

WDM I/F

WDM I/F

◄— Routing layer —► ◄— Photonic layer —►

Unified Control Plane (GMPLS-based, becoming DWDM aware)

**FIGURE 5**

**Network Architecture with IPoDWDM**

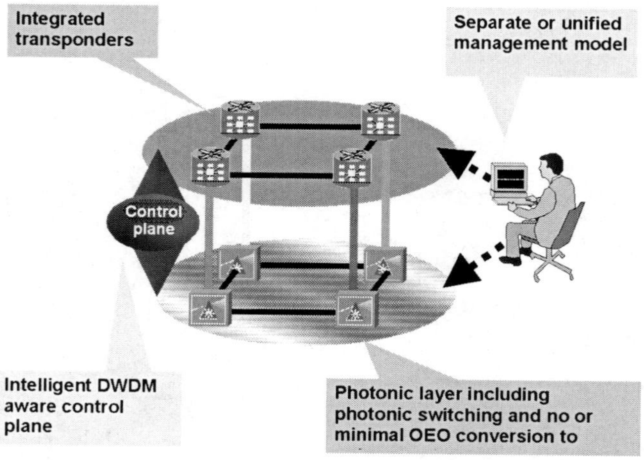

Integrated transponders

Separate or unified management model

Control plane

Intelligent DWDM aware control plane

Photonic layer including photonic switching and no or minimal OEO conversion to

- *Power/cooling savings*: Transponders consume significant power and require cooling. Adding them into core routers does not necessarily translate into added power and cooling requirements in core routers, but does reduce the power and cooling in the central office by eliminating the transponder shelves. Today's configuration requires short-reach optics in the router feeding a transponder at the transport layer, we now replace the short-reach with the integrated optics and eliminate the transponders, effectively reducing power and cooling requirements.

- *10GE with SONET/SDH–like operations, administration, maintenance, and provisioning (OAM&P)*: The integration of G.709 technology and 10 GE directly on a router enables key performance and alarm management functions that used to only be available via SONET/SDH. This represents a new Ethernet interface, which we call WDMPHY.

- *Proactive protection*: In a traditional IP–over–optical architecture, the router only learns about degraded optical signals after the fact—when the signal is failed. This is because the transponder shields the degradation from the router, predominantly because of the behavior of the forward error correction (FEC) mechanism. While the router may implement mechanisms to perform fast restoration, the overall recovery time is dependent on the failure detection mechanism. With integrated G.709 framers on the router, the router now has visibility into the actual optical signal and can switch away from a deteriorating connection ahead of a failure, without any hit on traffic.

- *Streamlined operations*: SPs today have a model for network operations that is segmented for each layer—transport and IP, for example. Each set of elements in a layer of the network has its own element-management system (EMS) and often times a dedicated network-management system (NMS). This results in increased operational burden. The architecture aims at removing these barriers for operators who can benefit from a sin-

gle integrated management and control system for both the optical layer and packet layer.

Of these values, the most readily quantifiable one is the CAPEX savings. This is the topic of the next subsection.

***Quantifying the Network Cost for Different Architectures***
A comparison of the overall CAPEX of a network for the IPoDWDM architecture to other prominent architectures for the core was conducted and results are demonstrated in *Figure 6*. The following three architectures were considered and are described below and depicted in *Figure 6A*:

- A solution in which transponders are physically patched to the router or to each other for optical pass-through of traffic that does not need to go to the router ("TXP"). This often represents today's core network architecture.
- A proposed alternative architecture, in which the bandwidth is managed via an EXC—typically SONET/SDH today, and moving to G.709 OTU1/OTU2 granularity. The figure shows two variants of this approach—one using existing POS interfaces on the cross-connect and router ("EXC w POS"), the other based on next-generation cross-connects, allowing for 10 GE interfaces to the router ("EXC w 10 GE").
- The next-generation IPoDWDM architecture that we propose in this paper, in which the traffic is manipulated in the photonic domain, via photonic switches. Again, two variants are shown: one using 10 GE DWDM ("10 GE WDMPHY") interfaces and the other using 40G DWDM POS interfaces ("40 G WDMPOS").

*Figure 6B* shows a cost comparison between these three architectures. The cost comparison was conducted using a real core network from a large PTT in Europe representing traffic projections for the next three years. As noted in the caption, the cost does not include components that are identical between the various architectures such as amplifiers, service interfaces on routers, etc.

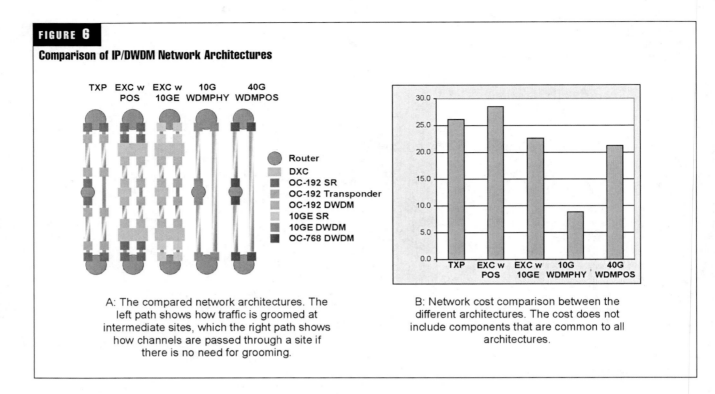

**FIGURE 6**

**Comparison of IP/DWDM Network Architectures**

A: The compared network architectures. The left path shows how traffic is groomed at intermediate sites, which the right path shows how channels are passed through a site if there is no need for grooming.

B: Network cost comparison between the different architectures. The cost does not include components that are common to all architectures.

We have seen similar cost savings with other realistic network models, representing long-haul U.S. networks, regional U.S. networks, and other national networks in Europe.

When assessing the value of a technology, it is important to consider the evolution of the network beyond its initial state. In *Figure 7*, we have quantified the evolution of CAPEX and OPEX for the same network over five years. While CAPEX growth is straightforward, the modeling of operational costs involves many factors and merits further explanation. It is also more sensitive to customer-specific assumptions and therefore should be viewed as a typical model rather than an accurate assessment of the costs incurred in this specific network. The OPEX model includes many factors such as the cost of leasing footprint in offices, cost of labor, power consumption and management overhead per shelf, as well as the cost of site visits, including truck rolls. As expected, these costs are similar for the two EXC models as they require the same amount of manual involvement as well as consume the same amount of space and power. The same holds for the 10 G WDMPHY and 40 G WDMPOS models.

**FIGURE 7**

**CAPEX and OPEX Evolution over Five Years**

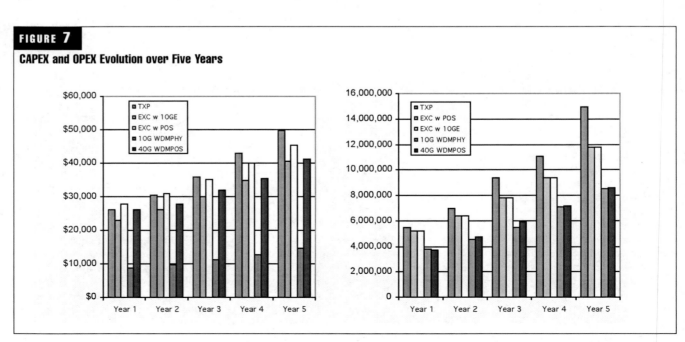

## What Are the Challenges of the Architecture, and How Can They Be Overcome?

The following challenges would need to be addressed in the implementation of the proposed architecture:

- *Enhancing the usability of all-optical networks*: The removal of active electronics from the photonic layer implies that this layer has less visibility into the bits encoded in the signal. This loss of functionality must be recovered through less costly means. Therefore, next-generation DWDM systems should have heavy instrumentation, allowing network management the visibility into what is changing over time. In addition, "analog control plane" technology inside next-generation DWDM systems should provide "plug-and-play" behavior in the optical layer.

- *Supporting a multivendor environment*: Since service provider networks are often built on best-of-breed elements, the proposed architecture must support an open approach. This affects the design of the DWDM layer, which must facilitate convergence by instrumenting it to condition an alien wavelength coming from a third-party system and use standards-based control plane features. It is also required to design the DWDM interface on routers to ease interoperability with a third-party DWDM layer. Note that keeping the solution open requires that it be somewhat overbuilt so as to ensure that the DWDM interfaces on the router can work over a less intelligent DWDM system and likewise add intelligence to the DWDM system to compensate for less sophisticated alien DWDM interfaces (such as DWDM XFPs). We believe, however, that we have achieved such an overlap in functionality without burdening the system with extra cost or complexity.

- *Separating administrative domains between transport and packet domains*: Several SPs may have separate groups for managing their data and transport networks. Each group is responsible for the operation of its part of the network and sometimes even has dedicated teams within the network operations center (NOC). The IPoDWDM architecture will accommodate multiple options for the management of the network. Three key options would include integrated IPoDWDM management, physically separated DWDM and IP management and virtually separated DWDM and IP management. These options will be addressed in detail below in Section 3.3.

Some more details on how we solve the above challenges are discussed in the sections below.

### Enhancing the Usability of All-Optical Networks

Next-generation DWDM systems should provide unprecedented flexibility and ease of use to provide the OPEX savings that SPs need. The following enhancements are required to accommodate next-generation DWDM systems in IPoDWDM networks:

- *Ubiquitous photonic switching*: One of the most prominent features of next-generation DWDM systems is full automation. This trend has started with reconfig-

urable optical add/drop multiplexers (ROADMs), which support ring-based switching. Leveraging highly integrated photonic devices, ROADMs come with essentially no premiums. More important, with proper planning tools, ROADMs eliminate the traditional dependency on accurate traffic forecasts and allows for any wavelength to any destination any time without affecting existing traffic. New technologies, such as wavelength selective switches (WSS) provide the same level of automation from ring networks to arbitrary mesh topologies, which are more common for core IP networks.

- *Making the DWDM layer operationally friendly*: Next-generation DWDM system must be designed to match and exceed the ease of use of SONET/SDH network elements. To this end, extensive and accurate monitoring is necessary. Every card is equipped with multiple photo diodes for accurate troubleshooting. The advent of photonic integration allows for multiple monitoring points, even at the individual wavelength level, without significant added cost. To reduce manual provisioning, DWDM systems must have access to data from network planning tools in order to automate the provisioning of optical set point. Finally, analog as well as digital control planes must be used to enable plug-and-play operation. For example, the analog control plane must adjust the working conditions of the network based on actual fiber and device losses as well as detect and compensate for impairments, such as fiber plant aging, without human intervention. A digital control plane (likely based on GMPLS) will automate the discovery of network resources and ensures the EMS accurately reflects the network conditions. The ramifications of such increased usability can be quantified as shown in *Figure 8*. It should be noted that this model was based on a network with a single ROADM. Multiple ROADMs at all nodes will create an even wider gap between the systems.

*Streamlining OAM&P in the digital domain*: The framing of client signals in the DWDM layer is achieved via a G.709 frame (digital wrapper). This frame not only provides operational features similar to those in SONET/SDH, but also supports forward error correction (FEC). FEC is used to overcome impairments in the photonic layer and allows for a significant increase in reach.

As shown in *Figure 9B*, the value of integrating G.709 framing in the router extends beyond the obvious CAPEX and OPEX savings. Non-integrated architectures cannot provide a common end-to-end framing structure from router to router. These traditional architectures rely on a nested framing structure: G.709 framing across the optical layer and SONET/SDH framing end-to-end, as depicted in *Figure 9A*.

As stated in Section 2, the non-integrated approach does not allow the router to have visibility into the optical domain. As a result, it extends the time it takes for a router to detect link failures and prevents the router from switching away from failures before they affect traffic.

In addition, this approach only works for POS interfaces. Since 10 GE interfaces do not have a framing structure, they must rely on embedded OAM&P frames (known as

FIGURE 8

**Work Hours Needed to Install an Eight-Wavelength System and Upgrade it to Thirty-Two Wavelengths: Traditional DWDM versus Next-Generation DWDM System**

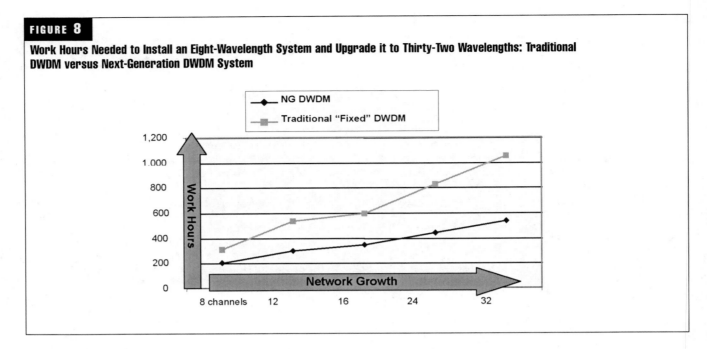

Ethernet OAM&P), an approach that is not yet widely adopted. Having a 10 GE signal embedded in a G.709 frame allows the system to provide OAM&P identical to that of SONET/SDH.

Finally, having two structures is cumbersome and results in unnecessary overhead, both from a bandwidth perspective and a network management perspective.

*Supporting a Multivendor Environment*
While long-haul transport systems have been extensively deployed in the past, they have been typically confined to point-to-point links with fixed wavelength add/drop in intermediate sites. These systems have been expensive and hard to deploy. They have also been closed systems, typically not allowing for multi-vendor interoperability. Standards bodies have lagged in defining analog attributes

FIGURE 9

**OAM&P Capability in Non-Integrated and Integrated Networks**

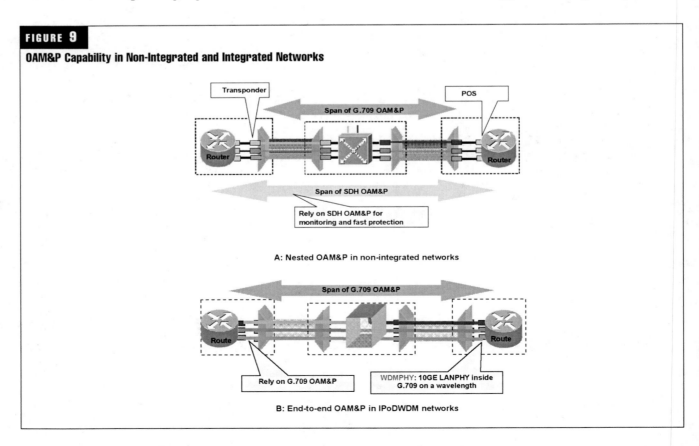

A: Nested OAM&P in non-integrated networks

B: End-to-end OAM&P in IPoDWDM networks

for DWDM systems, which would allow for multivendor interop. Until standards are defined, DWDM systems must rely on de facto standards based on off-the-shelf components such as the 300-pin multi-supplier agreement (MSA) modules or XFPs for DWDM transmitters/receivers. Another important example would include off-the-shelf G.709 framers supporting FEC, which enables systems to reach up to 1000 km. Ultra-long-haul (ULH) systems, such as needed across the United States, are typically closed solutions, though inexpensive regenerators will allow such systems to be open. The steep cost reduction that standard transmitters/receivers are enjoying due to their larger user base makes such a regenerated solution a cost-effective proposition for ULH networks.

### Administrative Domain Separation between the Transport and the Packet Domains

Network operators can gain significant operational simplicity by merging and streamlining the operations of the packet and transport layers. Integrated element management systems are available today that can be used to manage both the transport elements—next-generation DWDM systems in particular—and the routers.

While integrated network management represents the easiest way to manage an IPoDWDM network, we recognize the need in certain organizations to be able to manage the IP and optical layers separately. To this end, next-generation integrated network management should support different modes such as the following:

- *Integrated IPoDWDM management*: Allowing for combined performance and alarm management as well as

wavelength provisioning from router port to a peer router port, through an optical layer.

- *Physically separate DWDM and IP management*: Allowing management of the router layer and the optical layer via separate management systems, using separate DCNs and often separate management protocols (TL-1 for the optical layer; SNMP or XML for the router layer).

- *Virtually separate DWDM and IP management*: Allowing management of the router layer and the optical layer via separate views of a common database and a common management infrastructure (servers, data communications network [DCN], etc). Each user is restricted to functions as defined by their role.

These three modes are depicted in *Figure 10*.

The second mode may be implemented in two ways: allow the network management instance of the optical layer to directly communicate with the router to manage the optical attributes of its interface; or the optical layer management systems will only communicate with the optical network elements, which in turn communicate with the router to implement some of the requests. This latter communication should be achieved via a control plane, through a protocol called link management protocol (LMP).

For example, consider provisioning the wavelength on the router interface to some frequency X. According to the first approach, the optical layer management system will communicate to the router and provision the value directly on

**FIGURE 10**

**Different Network Management Models**

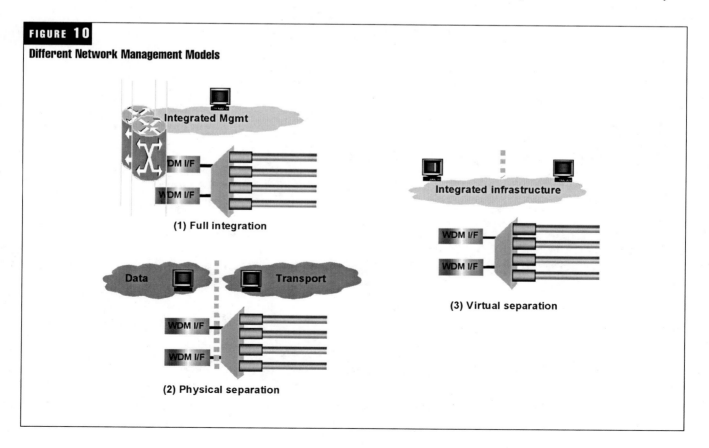

(1) Full integration

(2) Physical separation

(3) Virtual separation

the router interface. In the second approach, the optical layer management system will provision the optical interface with allowable wavelength values. The router management system may also provide some constraints on allowable wavelength values. LMP will facilitate negotiation between the optical layer and the router, arriving at a wavelength that is allowable for both layers.

These two approaches are depicted in the *Figure 11*.

## The Role of an Integrated Control Plane for IPoDWDM

Much has been written about the merits of an end-to-end control plane for fast connection setup, restoration, and similar functions [5]. In addition, the complexities of adapting the control plane to the DWDM layer have been discussed [6]. In the context of IPoDWDM, the control plane takes on unique functions, which are the focus of this section.

Our approach to control plane integration is tied to the requirements of the management solution. It is based on the following phasing approach:

- *Segmentation*: Signaling to accommodate the administrative boundaries between the IP and optical layers, allowing the management of these domains to remain separate and autonomous.

- *Collaboration*: Tighter coordination between the IP and optical layers to simplify operations, by reducing the amount of manual provisioning and alarms, leading to enablement of future applications such as combined IPoDWDM protection and restoration.

Segmentation allows service providers to maintain their current operational boundaries, or "business as usual," while optimizing the network hardware. The collaboration requirement takes the next step: improving the operations and network robustness.

*Figure 12* demonstrates the role of a control plane to achieve the segmentation requirement. It explains how a failure in the optical domain propagates to the IP domain to allow both management systems to detect and put an alarm on it. The example has two parts: The present mode of operation, with a SR handoff between the router and the transponder; and the new paradigm, in which the transponder has been eliminated.

In the IPoDWDM case, without a control plane, there is no way for the optical layer management system to know that certain problems have occurred. The instrumentation in the optical layer will detect hard failures that will cause a loss of light on a wavelength. However, since the optical layer has no visibility into the payload, there is no way for it to detect soft failures, such as increased bit error rate.

This challenge does not exist in the present method of operation (PMO), since the optical layer includes transponders. Once the transponders are removed, the network must rely on a control plane—specifically the link management protocol (LMP) to regain this capability. LMP will enable the router to notify the adjacent optical node that an interface experiences problems, via a backward defect indicator (BDI). This allows the optical layer to report the alarm to its management system.

## Summary

We have shown the various elements of next-generation IPoDWDM architecture in the above sections. We have delved into some of the technical challenges in implementing the architecture and have shown how they can be overcome to deliver a streamlined solution that has both better CAPEX than today's solution, as well as an easier and more streamlined operation, thereby reducing the OPEX for the core.

The following table summarizes the values of this architecture:

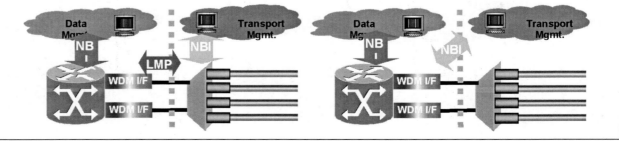

**FIGURE 11**

**Managing the Optical Interface on the Router from the Optical Layer Management System**

Signaling:
- The optical NMS talks to the DWDM layer
- The DWDM layer acts as a proxy to the router
- LMP used to interact between the DWDM layer and router
- LMP negotiation allows to restrict what the optical layer can do

Direct management interface:
- The optical NMS talks directly to the router (CLI, XML)
- Router security features can be used to limit the capabilities of certain users

ORI GERSTEL, MALLIK TATIPAMULA, KELLY AHUJA, WALID WAKIM, VIK SAXENA, AND JOHN LEDDY

**FIGURE 12**

**Fault Reporting in PMO and New Approach**

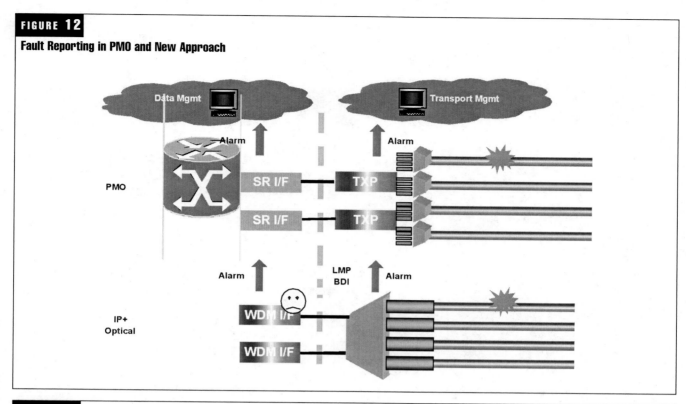

**TABLE 1**

| Feature | Benefit |
|---|---|
| DWDM interfaces on router into all-optical layer—minimize OEO conversion in the optical layer | Lower CAPEX: transponderless network<br>Lower OPEX: reduce footprint and power<br>Higher reliability: fewer components |
| 10 GE in end-to-end G.709 wrapper (WDMPHY) | Ethernet price point with better than POS OAM&P features (such as hitless protection switch) |
| Photonic switching for automated network configuration | Reduce human labor and human errors (OPEX) |
| Open and transparent photonic layer | Future-proof: no need to overhaul network for new service types (CAPEX) |
| Operational usability comparable to SONET/SDH at lower costs | OTN incorporates technologies for easy management and troubleshooting (OPEX/CAPEX) |
| Scalable to more WLs and higher bit rates (40G) | Future-proof: long-term use for investment |

Acknowledgement: We would like to thank Rob Batchellor and Jeff Meek for their network modeling work resulting in the graphs in Section 2.

## References

1. www.iec.org/online/tutorials/sonet.
2. R. Ramaswami and KN Sivarajan. Optical networks: a practical perspec- tive. Morgan Kaufmann Publishers, 1998.
3. ITU-T G.709/Y.1331 recommendation "Network node interface for the optical transport network (OTN)", www.itu.int/ITU-T/studygroups/com15/otn/transport.html.
4. Pan, P. et al., "Fast Reroute Extensions to RSVP-TE for LSP Tunnels," IETF RFC 4090. www.ietf.org/rfc/rfc4090.txt?number=4090.
5. www.iec.org/online/tutorials/gmpls.
6. J. Strand, A. Chiu and R. Tkach, "Issues for Routing in The Optical Layer," IEEE Communication Magazine, February 2001.

# A Functional View of Service Management for Complex IP Networks

Bruce Stewart

*Chief Architect, Network & Service Providers, Consulting and Integration*
Hewlett-Packard

## Introduction

Complex Internet protocol (IP) networks are becoming very common today as more and more service providers look to IP technology to simplify their networks and offer the opportunity for consolidation and convergence of services onto a single communications platform. This paper is aimed at looking at the service assurance part of managing complex IP networks and defining an architecture platform for service management.

### Background

The TeleManagement Forum (TMF) introduced the telecom operations map (TOM) in 1995, and this evolved to enchanced TOM (eTOM) in 2001. eTOM has become the de facto standard for defining the building blocks for operating a telecommunications business and is universally accepted by most telecom operators. The operations component of eTOM is shown in *Figure 1*. This shows the three main direct operations functions—fulfillment, assurance, and billing (FAB)—with the support function of operations, support, and readiness. Across these functions are the following four groupings:

- Customer relationship management, which deals with all aspects of the interaction and relationship with customers

- Service management and operations, which deals with aspects associated with the services offered to customers

- Resource management and operations, which deals with the actual equipment or network elements that are combined to provide the customer services

- Supplier and partner relationship management, which deals with providing services within the wider framework beyond the specific service provider

*Figure 2* shows a further breakdown of the eTOM model, providing the high-level view of the operations processes required in each area. This paper is concerned with the management and operations area for services and resources with a specific focus on assurance together with the fulfillment aspects that relate to assurance.

The primary function of fulfillment is to streamline the service provisioning process: accepting a service order, determining which inventory resources to use, and provisioning the actual network equipment that enables the service. However, the provisioning of a service does not stop at just creating the service. It is also important that the proper service-level measurement, monitoring, and assurance management be provisioned for this service as well to ensure the overall customer experience. The telecom market is hyper-competitive these days; it is not only important to fulfill an initial service but also to ensure ongoing customer satisfaction. This maintains customer loyalty and allows further upsell opportunities. Service fulfillment and assurance therefore must work hand-in-hand to ensure business success.

The functions of problem management and quality analysis, action, and reporting from a resource and service perspective provide the capability to monitor and report on the operation of the components that make up a service. The separation of these processes into a layer directly for the resources or equipment and a layer for services is important to allow the resource layer to focus on the equipment status and the associated equipment failure/diagnose/fix life cycle and leave the service layer to focus on the service aspects that directly affect the customer.

A key requirement for the service assurance function is the facility to be able to monitor and report on problems and status from a customer-service perspective. To achieve this objective, it is necessary for the service assurance system to receive up-to-date information on the service configuration

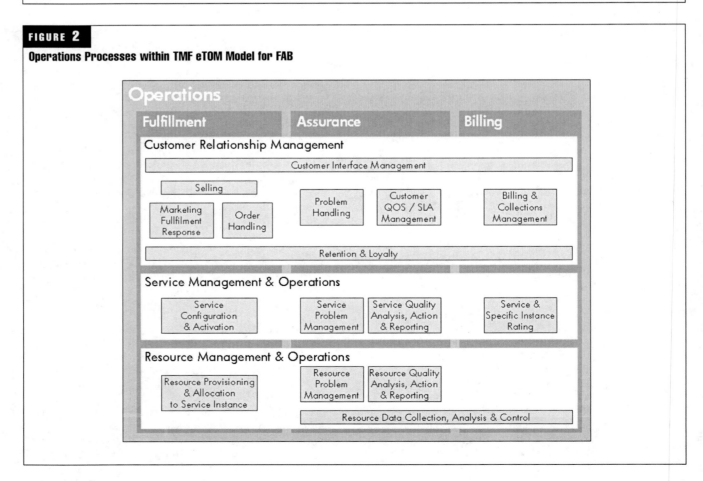

**FIGURE 1**

**TeleManagement Forum eTOM Model for Operations**

Operations

| Operations Support & Readiness | Fulfillment | Assurance | Billing |
|---|---|---|---|

Customer Relationship Management

Service Management & Operations

Resource Management & Operations
(Application, Computing and Network)

Supplier/Partner Relationship Management

**FIGURE 2**

**Operations Processes within TMF eTOM Model for FAB**

Operations

| Fulfillment | Assurance | Billing |
|---|---|---|

Customer Relationship Management

Customer Interface Management

Selling

Marketing Fullfilment Response · Order Handling · Problem Handling · Customer QOS / SLA Management · Billing & Collections Management

Retention & Loyalty

Service Management & Operations

Service Configuration & Activation · Service Problem Management · Service Quality Analysis, Action & Reporting · Service & Specific Instance Rating

Resource Management & Operations

Resource Provisioning & Allocation to Service Instance · Resource Problem Management · Resource Quality Analysis, Action & Reporting

Resource Data Collection, Analysis & Control

and how this relates to the underlying network, IT, and application components. The service configuration that is created and maintained as part of the service delivery components of service activation and inventory and design needs to be linked to the service assurance components as will be discussed below.

Before going on to discuss service assurance for complex IP networks in more detail, it is necessary to start with a discussion on what we mean by complex IP networks.

## Complex IP Networks

The trend to use IP as the convergence platform for all communication functions in a network and service provider environment is gathering pace. While it will still be many years before the plain old telephone service (POTS) or the public switched telephone network (PSTN) will be completely retired, network operators are already looking at moving toward a pure IP–based network for all voice, data, and video services. *Figure 3* illustrates the likely future communications scenario.

Currently, most IP networks consist of an underlying synchronous digital hierarchy (SDH) or a synchronous optical network (SONET) core network connected together on a fiber backbone, often using wavelength division multiplexing (WDM)/dense wave division multiplexing (DWDM) technology. The IP network is on top of this SDH backbone. For IP networks, the trend is now to use multiprotocol label switching (MPLS) technology for the IP overlay. This consists of high-speed switch routers connected together in a mesh arrangement with the paths between each router provided by SDH, WDM/DWDM, or even Ethernet technologies.

One of the key benefits of MPLS is that it provides a superior and flexible virtual private network (VPN) service infrastructure, where traffic is switched efficiently, allowing a wide variety of services to be provisioned dynamically and economically. MPLS enables the following three main types of MPLS VPN service:

- In a Layer-3 (L3) service, each customer site is connected to a provider edge (PE) router in the provider's network via a customer edge (CE) router. The PE routers in the provider's network work together so sites of the customer's network can exchange routing information and IP traffic but are shielded from traffic in VPNs belonging to other customers or even multiple separate VPNs belonging to the same customer.

- The second type of service is the Layer-2 (L2) virtual private local-area network service (VPLS), where each customer site is attached to an Ethernet port on a PE device, again typically via a unique CE switch as the interfacing device on each customer site. The provider network then emulates a single bridge interconnecting all the customer Ethernet segments to effectively form a single LAN. This service leaves the choice of L3 protocol (e.g., IP, Internet packet exchange (IPX), NetBIOS, systems network architecture (SNA), or a combination of protocols) to the customer. In the terminology of the Metro Ethernet Forum, this type of service is called a multipoint-to-multipoint Ethernet virtual circuit.

- As an alternative to multipoint service across the provider's network, point-to-point connections at L2 are known as virtual private wire service (VPWS). These often replace physical circuits or asynchronous transfer mode (ATM)/frame relay (FR) virtual circuits.

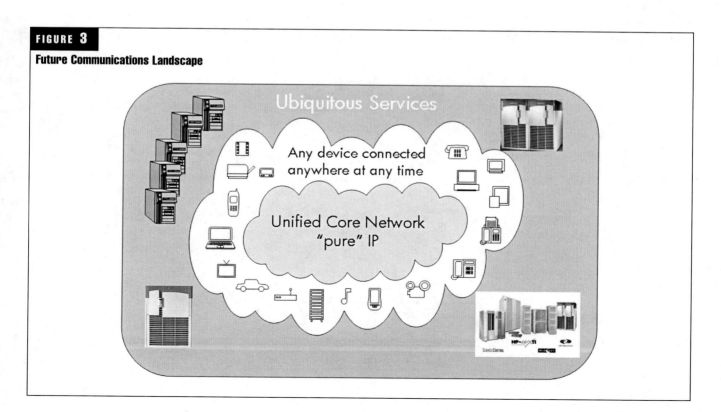

**FIGURE 3**

**Future Communications Landscape**

In addition to VPN services, MPLS also offers advance traffic engineering capabilities, allowing the service provider to specify, on a per-customer per-service basis, bandwidth requirements, class of service, and even recovery priority in case of a path failure.

MPLS networks today are fully capable of supporting time-sensitive voice and video traffic as well as the traditional data traffic. This gives service providers the opportunity to offer truly integrated services. For example, a service provider may start by offering classical VPN service to interconnect the customer offices and provide these offices with Internet access. Once this VPN service is established with the customer, the service provider can offer managed voice over IP (VoIP) and videoconferencing services over the same infrastructure. It is also feasible to provide gateway points within or outside the cloud that allow separate VPNs to connect to other revenue-generating services such as hosting, off-site storage, rich media, and other IP–based services. Over time, the MPLS VPN can become an enhanced revenue-generation platform for a service provider, offering an integrated platform for delivering multiple IP–based services.

*Figure 4* shows a typical service arrangement for an enterprise. From a management perspective, the example shows the service provider managing the core network, all the connections to the sites and the customer-premises equipment (CPE) for some of the sites (colored). One of the sites is shown where the service provider manages additional equipment within the site (shown as darkened triangle). There is also the notion of value-added services provided by the service provider (e.g., Web hosting, collocation, hosted mail, virus checking).

The management problem faced by the service provider in managing this type of network is complex. Usually many monitoring systems are connected to the multiple technologies that contribute to the services offered to the customer. Typically, an alarm surveillance group monitors problems and events that occur in the network. Their task is to try to fix the problem immediately; however, often the problem is too complex, and a handoff to an expert group focusing on a particular technology type occurs. For this to be successful, the alarm surveillance group needs to have analyzed the problem correctly to identify the root cause and to have selected the correct group to fix the problem. When this does not occur, the problem may end up being passed from group to group, with successive analysis slowly revealing more aspects of the problem, until its root cause is finally identified and the problem is resolved. All of this activity is usually focused on the technology and equipment side without the operators being aware of the impact of the problem on the customer services that are being provided. Taking a customer-service view adds another layer of complexity. The establishment of the service impact of a problem in the network requires a complete knowledge of how the service is constructed from its individual components. This service configuration information needs to be accurate and kept up-to-date as changes in the network occur.

The communication regime summarized in *Figure 4* is seemingly quite simple, however the underlying technology required to support these services is much more complex than is shown here. *Figure 5* is an example of a typical asymmetric digital subscriber line (ADSL) service. This complex arrangement maps to just the single line "a" to "b" shown in middle-right of *Figure 4*, connecting the PE point at the net-

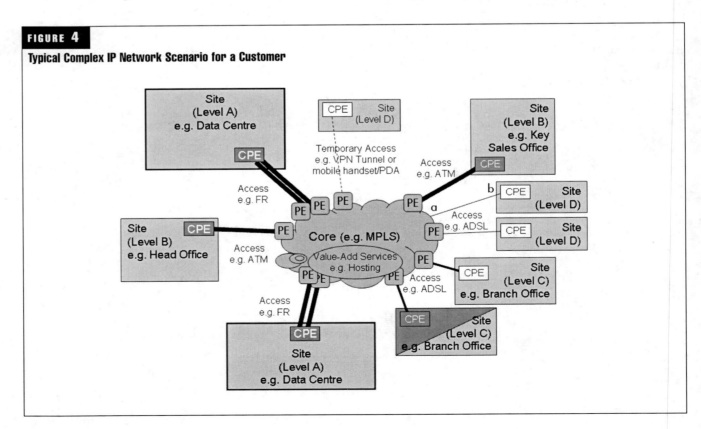

FIGURE 4

Typical Complex IP Network Scenario for a Customer

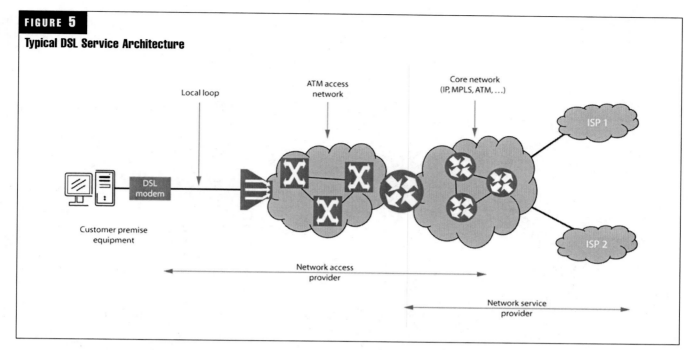

**FIGURE 5**

**Typical DSL Service Architecture**

work service provider (NSP) side and the customer site (level D). The DSL modem maps to the CPE. A similarly complex connection scheme applies to the other connection types, for example, FR or ATM.

Managing this complex IP network is not easy. From a customer perspective, the customer would like to just focus on the services provided and negotiate service-level agreements (SLAs) with the service provider for each of these services. Provided all the services are delivered within the agreed service-level objectives (SLOs) that are used to define the SLAs, the customer is happy. When a problem occurs, the customer would ideally like the service provider to already be aware of the problem and to have identified the root cause of the problem and initiated actions to rectify it. This could be in the form of an interactive voice response (IVR) message that the customer listens to when they call the service provider or a Web page indicating live status of the customer's services. If SLAs are violated or are in jeopardy of being violated, again some proactive communication with the customer is preferred. At the end of each month, the customer wants to see summary SLA reports showing the performance of all the customers' services over the entire month.

The service provider needs to put together a range of support systems to provide the service-oriented support to the customer. As indicated in *Figure 2*, this is the topic of service assurance.

## Service Assurance for Complex IP Networks

Network service providers have a difficult task of monitoring and managing the technology and infrastructure they use to provide services to customers. In the early POTS days, most of the infrastructure was designed to be highly robust with a substantial amount of redundancy in the infrastructure beyond the customer's local loop connection to an exchange. Most users today just expect to get a dial tone when they pick up the phone to make a call. The pic-

ture is very different with more complex networks that are now prevalent, an ever-increasing range of services being offered, and waves of technology standards being implemented to meet particular service needs. It is common for a network operator to have a range of ways in which a particular service is offered and have a problem retiring the old technologies and moving the services across to the most recent standards. In the example shown in *Figure 4*, frame relay is an old technology that would today be in the process of being phased out; nevertheless, it will still be around for many years. In addition, some of these services have been introduced at a lower grade/quality of service than the POTS approach, yet the customer still expects to get a "carrier-grade" level of service from network operators.

The eTOM model for operations as shown in *Figure 2* shows the high-level view of the processes required in an operations environment. This model has been adapted as shown in *Figure 6* below to indicate a functional view of the components required. The model will be used in this paper as the basis for defining the building blocks required for assurance management of complex IP networks.

Network operators usually make a distinction between two categories of support systems. These are operations support systems (OSS) for management of the network infrastructure—resources and business support systems (BSS) for customer-related activities associated with running the business. The separation of the TMF eTOM model functions into OSS and BSS components is open to interpretation. Functions such as order management, customer relationship management (CRM), and billing are always considered as BSS functions. The lower-level functions of problem management and quality analysis, action, and reporting for services and resources are usually considered OSS functions.

*Figure 6* shows three distinct levels of operation as indicated in the eTOM model. At the top is the CRM level, consisting of systems that are focused on direct interaction with customers. In the middle are the service management and oper-

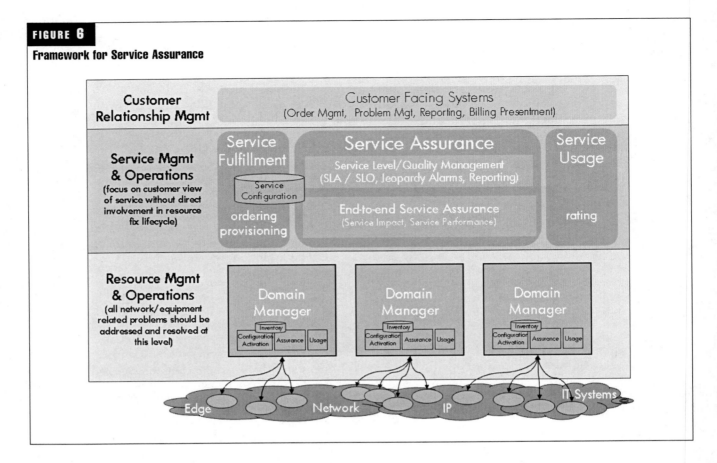

**FIGURE 6**

**Framework for Service Assurance**

ations level, which focuses on providing a customer service based operations model for customers, and at the bottom is the resource management and operations level, which provides management support for the actual equipment infrastructure.

First, we need to recognize that the infrastructure and equipment level itself straddles many technology domains. At the edge of the network, next to the customer, is the access network, which takes the signal from the customer over the last-mile connection and combines it through a multiplexer (MUX) onto higher bandwidth connections. Behind the access network is the core network—switching systems that aggregate up signal paths and connect them together across large geographical areas and across the world, using protocols such as ATM, SDH, or MPLS over optical fiber, radio, or satellite. There is also a trend toward many of the services provided to include components hosted on IT computing servers.

The infrastructure or equipment (resource) level consists of many technology domains connected together in a complex fashion. The management of these technology domains has to take into account the specific aspects of the particular technology. At this level, it is useful to establish the concept of a domain manager that addresses the OSS requirements for a particular domain. Each domain manager would need to include four major components:

- An inventory module that keeps a record of all the equipment elements in the domain and how they connect or relate to each other

- A configuration/activation module to provide the facility to configure the equipment elements to establish the services that the particular domain provides as part of the overall service offered to customers

- An assurance module that provides the facility to monitor the status of all the equipment elements to ensure that they are functioning correctly and to support the process of identifying and fixing problems when they occur

- A usage component that provides the facility to collect usage information on services provided for reporting to higher-level systems (this component is often treated separately as part of a usage collection/mediation/billing function)

All of these components need to combine to provide summary information to the service level above.

The service level is an important layer that needs to be considered separately from the resources level. Its function is to establish an end-to-end view of the services offered to customers and provide the facility to monitor, report, and manage the health of these services. The introduction of this level as a distinctly separate functional layer is key to providing customers with higher quality of service.

### The Distinction between Service and Resource Management
The separation of the OSS function into two levels reflects an important distinction between the notion of service man-

agement which is concerned with the offering of services to customers and resource management which is concerned with handling and restoring problems associated with the infrastructure and equipment used to provide the customer services.

Ideally, the process of identifying and fixing equipment/network problems should occur at the resource level as part of the fault/performance management function. Aspects such as having a correlated view of how the equipment/network components relate to each other and using this information to determine the root cause of a problem need to be addressed at this level. With this separation the service level can focus on the status and performance of the customer's service from an end-to-end perspective and only have to be aware of the problem/fix life cycle rather than being directly involved in this process.

This separation needs to be reflected in the trouble-ticketing systems used to track and manage problems and events. Ideally, there should be two trouble-ticket systems, one focused on handling equipment/network/infrastructure (resource) problems, and the other focused on service- and customer-related problems.

### Layering of Services

Another important concept to introduce is the layering or abstraction of services. In any reasonably sized network, the number of equipment elements and associated services is very large, consisting of many millions of objects. All of the

equipment elements and services are related to each other in some way, and this relationship changes over time. Keeping an accurate reference model (centralized or distributed) of all the services and equipment objects in the network in an accurate and up-to-date form is a very difficult task. The layering and abstraction of services can go a long way to alleviating this problem if it is implemented correctly.

If we take the ADSL architecture example from *Figure 4* and apply the abstraction process, we get the result as shown in *Figure 7*.

At the top, we have the ADSL service as reported to the service level from the domain manager for ADSL services. This is shown as a line between the ADSL modem for Customer A and the point of presence (PoP) for the Internet service provider (ISP) for the particular customer. If there are no problems with the overall service, then ADSL service for Customer A will be reported with an "OK" status. Any problems with the service are reported with the associated message indicating the problem together with an indication of the severity—in jeopardy, degraded, out-of-service, etc.

There are a number of aspects to this abstraction and reporting process, summarized as follows:

- The end point on the customer side from a management perspective is the ADSL modem and does not include customer equipment on the other side of this modem.

**FIGURE 7**

**Example of Layering of Services, ADSL Service**

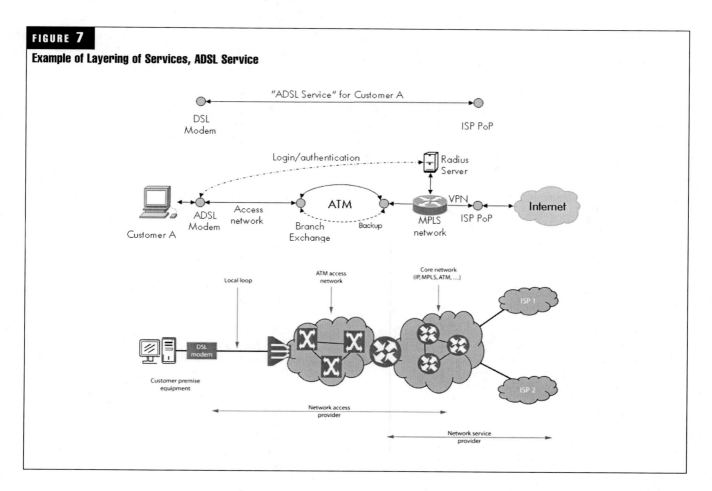

- The path from the ADSL modem all the way to the ISP PoP goes through a variety of technologies, all or part of which may have failures or degradation of service.

- The ATM portion is showing a "backed up" path that will take over in the event that the primary path fails. From an ADSL service perspective, the service is still available but potentially reported as "in jeopardy" when the backed-up path is providing the link.

- The MPLS network is a fully meshed network, and it may not be possible to know whether the service is in jeopardy because of the dynamic nature of the VPN path through the MPLS network. The reporting from the MPLS network will be more around quality of service (QoS) and performance metrics.

- The login/authentication process is part of the service. The ADSL service may only report that there is a valid path from the customer modem to the radius server and that the radius server as an IT application is operating within measured parameters. This will be reported as "authentication service OK." However, the user may still have problems logging in when outside the scope of the ADSL service monitoring (e.g., incorrect password).

- The login/authentication process is shown here as being provided by the service provider. Some customers may have further authentication provided by the ISP. While this is out of scope from an ADSL service monitoring perspective, the customer will not know this and needs to know whether to call the service provider or ISP when there are problems. The handoff issues associated with dealing with this scenario are non-trivial.

- Any of the communication paths may be subject to temporary bandwidth constraints and provide a degraded service for a period. Performance-monitoring systems with defined thresholds on performance levels need to be included to monitor this aspect and report performance violations to the ADSL service monitoring system.

- Some service providers install "probes" at particular points in their network and use these "probes" to initiate and monitor dummy transactions that are injected into the network path to determine if the service is working satisfactorily. This type of measure can monitor transit delays and end-to-end connectivity. The results of these transactions need to be included in the ADSL service.

- There are many varied measurements that can be made across the overall technologies and networks that combine together to provide the overall service. Aggregating up these measures and assigning the associated priority and "weighting factor" to be applied toward the overall ADSL service measure is a complex task.

- Keeping track of the changes in the network and equipment that provide the overall service is a challenging task. Decisions need to be made about

whether the information is kept in a centralized location or contained within each domain. In most cases some form of hierarchy of inventory/network model/service configuration and the associated challenge of keeping this in sync is inevitable.

Automation of the process of configuring and changing the network/service configuration provides the opportunity to have a more accurate inventory. Integration between the inventory system(s) and other components is required. This may be a challenging task in very large networks where there are millions of services and many more underlying elements that contribute to the services.

Keeping the network/service configuration accurate and up-to-date is crucial to achieving a high quality of service for the customers. Wherever possible automated discovery techniques need to be applied to continually maintain the inventory/configuration information.

The abstraction of the services is key to allowing a separation between the fault management side, with its problem-diagnose-fix life cycle, and the service management side, which is solely concerned with the status and quality of the overall service being offered to customers.

***Building Blocks for an Effective Service Assurance Platform***
Having set the scene with an overall model for service assurance for complex IP networks, we now go on to look at the building blocks required for an effective service assurance platform. Note that the discussion here is focused on the functional requirements and does not indicate what and how specific products will be used to provide these functions. As an example, the element manager (EM) functions may be implemented by multiple separate products, each providing a portion of the required solution in an integrated manner.

The discussion will be divided into two areas covering the resource level and service level.

The following six key building blocks are required:

- Resource-level management

  1. Access domain manager to manage and aggregate up all the services associated with access to the customer from the service edge. There are typically legacy systems still in place, but this is likely to be largely focused on DSL/broadband services.
  2. IP domain manager to address all IP network–related aspects. At the highest level, this will be the MPLS service.
  3. IT domain manager for management of the IT applications and associated resources that provide IT value-added services (VAS).
  4. EMs for underlying technology domains. The most important of these are likely to be access EM (ADSL multiplexer), ATM EM, SDH EM and IP EM. The IP domain management aspect may be addressed as part of the IP domain management above.

- Service-level management

  5. Service configuration and inventory
  6. Service quality management

**FIGURE 8**

**Building Blocks for Resource Level**

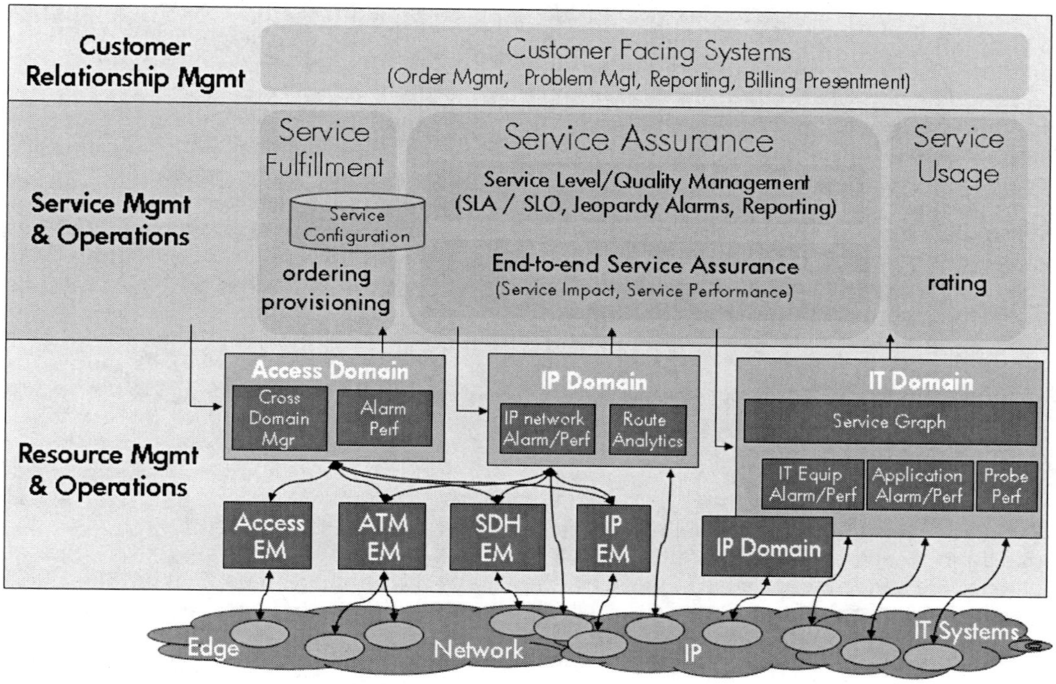

*Building Blocks for Resource Level Management*
The key building blocks for the resource level management are shown in *Figure 8*. The following three domains are shown:

- An access domain manager to aggregate the connection points to the customer
- An IP domain manager to aggregate the IP network components
- An IT domain manager for the IT components

The domain managers also need to be integrated with the underlying EMs that provide for the direct management of each of the technologies used.

*Element Managers*
Each EM is concerned with the management of the specific technology domain. Ideally, the EMs should provide the following functions:

- Complete configuration and activation of network elements (NEs) within the domain. Support for an external application programming interface (API) to provision the services within the domain from a high-level perspective, e.g., "set up a path with xxx characteristics from point A to point B."

- Keep an automated, up-to-date, accurate inventory/topology model of all the NEs within the domain together with the connection points (services) and inter-relationships between the NEs.

- Support for correlation of alarm/events within the domain to show a hierarchy of alarm/events.

- Provide a fault/alarm management function for the domain. Provide support for the identification of the root cause of a problem.

- Aggregation of alarm/events to indicate the status (to the service level) of the service elements provided by the domain, e.g., "status of a path from point A to point B from a service perspective."

- Provide a northbound interface to report status/alarm information to higher-level systems. This interface should include support for filtering so that service-affecting alarms and equipment fault alarms can be reported separately and sent to different destinations.

- The operator graphical user interface (GUI) for the system should allow integration of the client GUI with a GUI from a higher-level system to support context-sensitive "drill-down" from the higher-level system to interrogate and diagnose a problem. This avoids the need for all alarm/event information having to be sent to higher-level systems; only the summary information needs to be reported up. There are significant advantages with this approach in hiding as much complexity as possible and the associated management of this complexity within the domain.

- Provide support for synchronizing the alarm management lifecycle for specific alarms with a higher-level system.

*Access Domain*
The access domain is required on top of EMs because of the complexity of providing access to customers. The ADSL

example summarized above illustrates this complexity. *Figure 8* shows a cross-domain manager (CDM) in addition to the alarm/performance function. This is required to allow for automated setup of access services for customers using paths provided by underlying technologies. In the ADSL example, this includes multiplexers at the edge of the network connecting directly to customers, an ATM path, and an IP–based MPLS path. The CDM receives provisioning commands from the service level and then sets up the connection points to provide the service.

All the functions previously outlined for the element managers also apply to the access domain except that there is an abstraction of services provided by underlying EMs and their associated domain. For example, a required path may be from a customer connected to an exchange in one city, across to an exchange in another city, and then to some IT equipment sitting in a data center. This would be provisioned and managed from a CDM perspective as a path from point A to point B. The actual path may include a variety of technologies, e.g., ADSL, ATM, and SDH to set up the link from the customer to the data center.

Over time it is expected that the access domain will become a lot simpler as the IP domain, with its inherently simpler provisioning model, is extended to the edge of the network close to the customer.

*IP Domain*
The IP domain does not require the complexity of a cross-domain manager because of the inherent simplicity of the provisioning model. The connectionless nature of IP and the notion of tagging or labeling packets provides for a very simple provisioning model. With MPLS, complex VPNs through a very large and diverse IP network spanning multiple cities and even countries can be set up using an MPLS EM. All the issues around setting up the path and maintaining it during equipment failure situations are automatically handled by the routers and the associated protocols.

The IP domain still requires monitoring from an alarm/performance perspective. The functional requirements are as follows:

- The configuration and activation of services would be handled directly by the MPLS domain manager.

- IP domain managers have adopted the concept of automated discovery of inventory/topology information since the inception of IP networks. Most systems have sophisticated techniques to discover L2 and L3 topology and display it on graphical maps. This involves connecting to the routers in the network and extracting information from the management information base (MIB) within the routers. With very large networks, this is becoming a problem where the background traffic over the network to discover and maintain the status of the network is becoming a problem. Some IP domain managers have adopted the concept of an active problem analyzer (APA) where only high-level information is continuously monitored from the network. When a problem occurs, the APA works out where the problem is in the network and initiates a more detailed gathering of information from

the routers close to the problem. This is then analyzed and the root cause of the problem identified.

- A northbound interface for the service level to report on status/alarm information is required, but the inherent resilience of a fully meshed MPLS network simplifies this reporting substantially.

- The fault/alarm function is required. However, the focus is more on performance and QoS rather than the actual path being provided through the network. An IP domain manager needs to include an integrated alarm/performance monitoring system that supports the myriad protocols that control the network. This includes the following:

  ○ *Advanced routing*: Intelligently diagnoses dynamic networks for IPv6, open shortest path first (OSPF) and Cisco hot standby router protocol (HSRP)
  ○ *LAN/WAN edge*: Monitors frame relay connectivity between an enterprise and its service provider
  ○ *MPLS VPN*: Monitors MPLS service availability
  ○ *IP multicast*: Monitors availability and performance of the IP multicast service

- Performance reporting on jitter, delay, and related measures for time-sensitive traffic.

- A new facility that has been recently introduced into IP network monitoring is the concept of route analytics. This works by placing routers in the network that listen to and participate in the routing protocol exchanges between routers as they talk to each other. A key advantage of the route analytics approach is that it does not impose an additional traffic load on the network (just listens to the protocol exchange) and it provides information immediately as it happens. This provides a new perspective on the status of the network to do the following:

- Provide computing of a real-time, network-wide routing map (similar to the task performed by individual routers to create their forwarding tables, but computed for all routers)

- Support monitoring of changes as they happen rather than after the event

- Support detecting and alerting on routing events or failures as they happen

- Support analyzing the impact of routing changes on the "as-running" network before they happen

- The IP domain manager should also support integration of the GUI to allow drill-down from higher level systems and synchronization of the alarm management lifecycle.

*IT Domain*
More than ever, service providers are offering new application-based services to customers to create new revenue streams. These value-added services (Vass) rely on an IT infrastructure. These VASs need to be monitored for their

availability, performance, and QoS and need to be included as part of the services offered to customers.

Managing an IT infrastructure is very different from managing a network infrastructure. In communication networks a particular type of technology is usually established (e.g., router, switch) and then used as a repeatable building block many times—often with millions of instances of the component. The equipment for the technology type is manufactured using a rigorous set of standards defined by international standards bodies (e.g., International Telecommunication Union [ITU], Internet Engineering Task Force [IETF]). This produces a substantial amount of repeatability and standardization of services across the equipment.

IT infrastructure, on the other hand, is much more random and less repeatable in the way it is implemented and used. It consists of the computing hardware and operating system, network infrastructure to connect the various computing elements together, and the applications that are loaded on the computing elements that interact together to provide a service function for customers.

The management of changes in a network infrastructure is usually strictly controlled and subject to a rigorous change process that is repeatable and well defined. This is not the case for IT infrastructures, particularly with respect to the applications that provide the services.

As a result of the complexity and changing nature of an IT infrastructure, it is important to view it as a separate domain and use the IT domain manager concept as a tool to hide the complexity of the services provided and offer a consistent service-oriented alarm/event reporting mechanism for the higher service-level management system.

The high-level functional requirements for the IT domain as summarized in *Figure 8* are as follows:

- *IT equipment alarm/performance system*: This needs to support the monitoring of IT equipment (central processing unit [CPU] load, memory usage, disc storage, etc.), the status of the operating system and other fault/performance measures.

- *Application alarm/performance system to monitor and report on the health of applications that are running on the IT equipment*: The problem here is that each application is different, so efficient tools to simplify the reporting of application health are required. The notion of predefined monitoring packages for popular applications that are maintained and delivered as a product is important.

- A probe-based transaction monitoring system that supports sending dummy signals to systems in the network to establish that network paths are open, applications are responding to requests, and to measure round-trip times for the transactions.

- *An IP domain monitoring system for the network paths that connect the IT systems together*: In this case the need is only to report on the status of the path and leave the

actual management of the IP network to the network domain.

A service graph monitoring facility that allows all the various monitored items to be combined together in a service hierarchy with appropriate propagation mechanisms to indicate the overall status of the "IT service"—The service graph and associated utilities used to set up the service monitoring need to be sufficiently open and flexible to support the unstructured and varying nature of IT systems.

### Building Blocks for Service-Level Management
The key building blocks for the service level management are shown in *Figure 9*. This diagram also includes the resource-level components and shows the associated linkages (high-level only) between the components.

#### Service Configuration and Inventory
The service configuration and inventory system provides the central repository for all the services provided to customers. The system usually includes a provisioning process for ordering, designing, configuring and activating these services as well.

The decision as to where to place an inventory and provisioning system in an overall telecommunications operator environment is open to much debate. A fully centralized approach is only really feasible for second-tier level operators or for a portion of the services where the overall network scope is not too large. This discussion has indicated a cross-domain manager, which includes a configuration and inventory system within the access domain component. This is appropriate and will depend on historical factors as to how a particular operator's network has evolved over time together with the associated OSS systems.

From this discussion paper's perspective, it is important that the service configuration for the complex IP network services is centralized at the top level. How the top level propagates down to lower levels and how these configuration stores are kept consistent and up-to-date will vary depending on the local circumstance. A fully centralized approach is best for data consistency but requires that all the underlying systems have an interface into the centralized system and may have significant constraints with respect to performance. Local copies of data will be required for performance purposes, and these data stores will need to be kept synchronized. A distributed approach where no one system contains all the information is more practical and more likely to be the case.

The approach suggested in this paper of keeping as much information as possible local to the lower-level systems and setting up an abstraction mechanism for services is important. This includes providing access via GUIs and other programmatic systems to lower-level systems when information needs to be accessed. Most of the large volumes of data tend to be at the lower levels, so if this can be contained to that level with appropriate abstraction and "drill-down," then there are substantial advantages in simplifying the data-sharing regime. Containing the complexity to the lower-level domain managers has the added advantage that the rate of change of the service information for the higher level is likely to be much less as well.

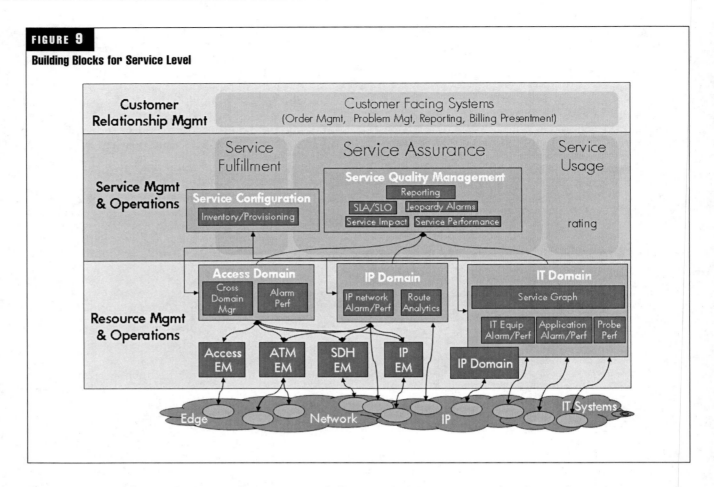

**FIGURE 9**

**Building Blocks for Service Level**

The maintenance of an accurate up-to-date inventory of all equipment and services is a very difficult task to accomplish, particularly if there are manual steps involved. As discussed earlier, the use of automatic discovery mechanisms can simplify this task substantially. The question then arises as to whether the underlying EMs do the discovery or it is done by the centralized provisioning/inventory system. This will tend to vary depending on the local service provider situation.

*Service Quality Management*
The final building block for service assurance for complex IP networks is service quality management. This is the process whereby all the services offered to a particular customer are collected together and service-level objectives (SLOs) are assigned for each of the service components provided to the customer. The services provided to the customer are then monitored on a continuous basis to assure that the SLOs for the constituent services remain within the assigned service level agreement settings that have been contracted with the customer. Summary reports are also provided to validate the service levels provided to the customer.

The bottom of the diagram shows the various components of a service provider's business, covering the network and IT infrastructure, VAS platforms, business processes, etc. Associated with these components are functional groupings of network operations, IT, planning, and so forth. A wide variety of information can be extracted from all of these areas, ranging from fault and performance data to process status information and other information. The approach is to define key

performance indicators (KPIs) from the various data sources and establish an interface link to these data sources. The service operations function is used to link these KPIs and aggregate them up for each customer grouping. The system needs to provide the function of service-level management and provide a service health and SLA compliance view for each customer. This information can be used across a broad part of the service providers business, for example, sales, ordering, customer care, product marketing, and so forth.

It is also important to emphasize that the service-quality management process goes through a life cycle as shown in *Figure 11* and that the service model itself will be subject to adaptation over time. The approach involves the development of a service object model that represents the relationship between the services that need to monitored and reported on for internal and external customers. Once this has been established, there is a repeatable cycle that is carried out to define the services and their associated SLAs, to configure these and then perform real-time monitoring to assure that the there is compliance with the SLA parameters. The life cycle also recognizes that SLAs are not static and, over time, an assessment process will indicate a need to refine and adapt the services and service levels to improve the service-quality management process.

The key functional requirements for a service quality management system are as follows:

• Separate function from fault/alarm management systems

## FIGURE 10

**Vision for Service Quality Management**

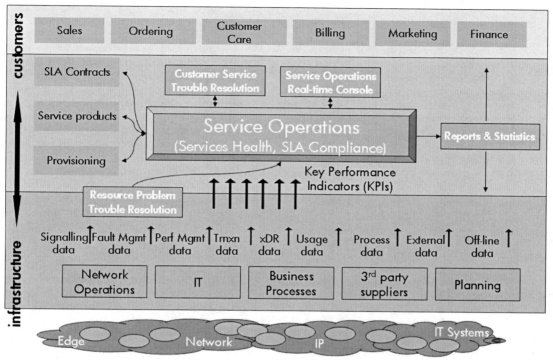

## FIGURE 11

**Service Quality Management Lifestyle**

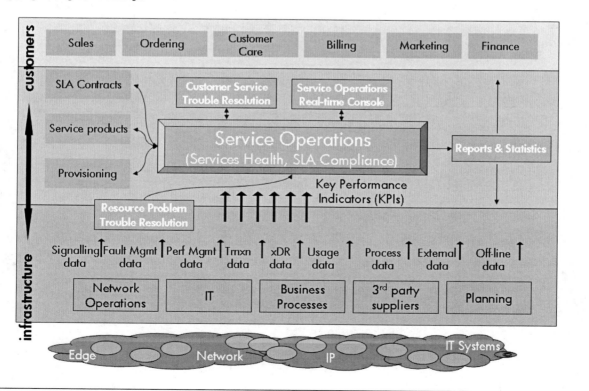

- Facility to model the customers' view of service and incorporate key performance indicators (KPIs) from multiple sources and aggregate these into a propagation model with flexibility in setting the weighting factors for each KPI
- Interface to service configuration/inventory system to maintain the relationship of services over time
- Support for managing the life cycle of developing and maintaining the service model as required by the customer from the customer's perspective
- Real-time operations console of the status of customer's service and compliance with SLOs; jeopardy alarming of potential service violation situations
- Reporting of service compliance for internal users and customers
- Interface to customer/service trouble resolution and equipment/problem trouble resolution systems

## Management Scenario

The following hypothetical scenario provides an example of the myriad management systems and management scenarios that are required in providing a complex IP–based solution to customers. This is by no means complete, and the actual scenario would typically be even more complex than is outlined here. Nevertheless, this example does illustrate many of the key issues that need to be addressed.

The example scenario, outlined in *Figure 12*, consists of the following:

- All the communication services for a company are provided by the service provider. A series of service contracts have been defined and agreed upon for providing the services to the customer. The service areas directly managed by the service provider are shown in the darker color in *Figure 11*. The service provider also manages the central cloud and all of its components.

- The services offered consist of voice and data services, a Web hosting service, and management of the network infrastructure for a series of branch offices. Services to some of these branch offices are still being rolled out.

- All the customer sites are connected together via a VPN that uses mixed technology. The VPN service was originally provided to the customer as a point-to-point service using frame-relay technology. Over time this has evolved to a fully meshed MPLS core for some of the services, but older versions of IP–based core technologies still exist.

- The customer has two large data centers that are critical to the operation of the entire company. These are connected by redundant frame-relay connections.

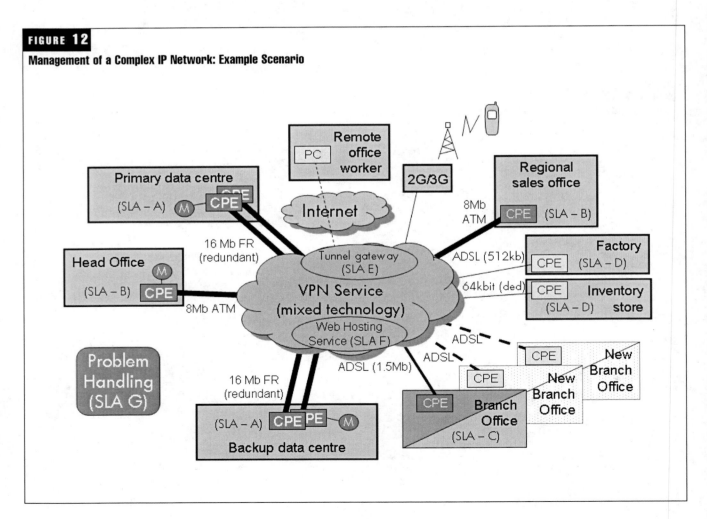

**FIGURE 12**

**Management of a Complex IP Network: Example Scenario**

- The company has ATM connections to the head office and a regional sales office and voice traffic using VoIP is transferred between these two sites. QoS on this connection is monitored.

- A managed service with ADSL connections and local managed router network is provided to branch offices. Further branch offices are being commissioned and a contract with associated SLAs has been signed for service levels associated with the provisioning and commissioning process.

- The customer also has other sites connected to the VPN, e.g., a factory connected via ADSL and a warehouse connected by 64 kbit dedicated digital line connection—an old technology that is being phased out but still widely used by customers.

- The customer also uses the remote access service where mobile workers can connect to the customer VPN using secure tunneling technology. The SLA for this service extends to the gateway that connects to the Internet; however, customers can call the service provider for help if problems occur connecting in. Customers also connect to the VPN through mobile devices and personal digital assistants (PDAs).

- The service provider has provided service status Web pages that the customer can view to monitor the status of the services provided. This is accessed from the head office. The two data centers also have a dedicated monitoring console to view the status of the services for these sites.

- The customer has recently signed a problem-handling SLA where service guarantees are defined for reporting and fixing faults within prescribed limits.

- The customer would like to agree to an overall set of SLAs for the services provided and is willing to pay a premium for this, provided the service provider can guarantee to support them and report on them accurately on a periodic basis.

For service providers, this is a typical situation that is becoming increasingly common. While this example is for a single customer, in the actual situation there would be many hundreds of customers fitting this profile and the customer indicated here would probably be of mid-range complexity, with some customers having a much larger and more complex network to manage.

The service provider will offer the services to the customer based on a range of product offerings that have been developed over time, where the level of integration between the operations systems that provision and manage these services are likely to not be well integrated. This example will assume the following factors:

- The frame relay service is managed from a custom-built system that was developed specifically for the FR service and supports about 50,000 customers and 70,000 connection points.

- The 64 kbit dedicated line (digital network service) is managed by a different custom-built system that is difficult to collect information from and is no longer being developed, with only a small maintenance team supporting it. There are still 30,000 customers using this service.

- The ATM system is managed by a domain manager provided by the equipment vendor.

- There are a number of domain management systems for the VPN service. These are managed separately by specialist groups.

Consider now how the architecture of products outlined in this paper can be used to support the effective service-based assurance of this complex environment. Each domain manager will need to provide complete monitoring and reporting of services within the domain. When a problem occurs in a domain it is quickly analyzed by the domain manager, the root cause is identified and trouble ticketing is initiated to marshal the required people to fix the problem as quickly as possible. The service-level monitoring system is aware of the problem in real time and its effect on customer service, and it can initiate reporting to customers to warn them about potential problems. In many cases, the problem does not have a direct effect on customer service and may only be reported as "degraded" or "in jeopardy."

The older technologies that are being phased out will typically have their own bespoke management system. The concept of abstraction can be used to get the bespoke monitoring system to just report on the high-level health of each service. The service operations system needs to be able to incorporate these monitoring points into the customer's service model directly without having to integrate them with any other system.

The access domain management is contained to this domain so that the complexity of having multiple paths to provide the service is localized. The service-level management is only concerned with the state of the overall end-to-end service. Reporting to the service level is not based on specific equipment failures or the priority of these. Ideally, the service operations group should be able to be aware of a failure of a particular service and then see from the trouble-ticketing system that the root cause of the problem has been identified and an expected time to repair indicated.

The IP network, with its inherent ability to maintain connection points even under equipment failure conditions, presents a different problem. The complex protocols that occur within the IP domain can sometimes create subtle problems that are transient in nature and very difficult to analyze. The route analytics facility can help significantly in resolving these problems. QoS and performance reporting are very important as often the degradation of service is actually a bottleneck somewhere in the network. Aspects such as jitter and delays for time-sensitive traffic need to be addressed.

In this scenario, the IT domain components are related to Web hosting, which is relatively simple to monitor. Using the IT domain management components and the overall health of the Web hosting service for an individual customer or group of customers can be aggregated up from all the IT/network components that contribute to this service. The "Web hosting status" at the top of the service graph view can be passed on to the service operations system hiding the complexity of the underlying components. A similar arrangement can be used to monitor other IT service components that are increasingly becoming an integral part of the services offered to customers. Examples are the domain naming servers (DNS) and radius servers, media and gateway servers, and softswitches.

The service operations system provides the focal point for managing customer service based on real-time monitoring of all the constituent elements that make up the service and how they relate together to contribute to the SLA entered into with the customer. The information is monitored, tracked, and reported in the customer's language. The service operations system can also have a generic SLA agreement for most customers and then tailor this agreement for specific high-value customers. Reports are produced for each customer on say a monthly basis to prove to the customer that the agreed quality of service is being delivered. A Web-based interface can be provided for customer to monitor the status of their service via a portal.

The provisioning of services for the new branch offices, and the associated SLAs around service delivery, need to be monitored by the service operations team. There may be a situation where a service has been delivered late to a specific branch office and the overall SLA for the customer is being reported as in jeopardy. The service operations team can initiate a notice to the field provisioning group to ensure that other branch office rollouts for this customer have higher priority. When a network problem occurs, the service operations team can direct that the part of the network affecting the specific customer to be repaired first. This can be assessed on the basis of the SLA contracts for the specific customer, e.g., platinum customers before gold. All of this contributes toward meeting the overall SLA for the customer and having visibility of the weighting factors assigned to each of the services and how they contribute to the overall SLA for the customer.

## Summary

The costs of integrating a potentially large range of products from multiple suppliers together and maintaining these integrations over time is becoming prohibitive. One of the biggest issues for a service provider in deploying OSS systems is the cost of the integration between the components. A contributor to the overall cost and complexity of an OSS system is the lack of a clear architecture and vision so that individual solutions get implemented without a clear view as to how they contribute toward the overall solution. This paper outlines an architecture and approach to enable effective service assurance of the overall environment. Weaknesses in one area of the solution often end up propagating to other areas of the solution and creating an inconsistent and inefficient outcome.

To provide an effective service assurance regime for complex IP networks, there are six key building blocks that need to be established and integrated in a modular fashion. These are as follows:

- Resource-level management

  1. Access domain manager to manage and aggregate up all the services associated with access to the customer from the service edge. There are typically legacy systems still in place, but this is likely to be largely focused on DSL/broadband services
  2. IP domain manager to address all IP network–related aspects. At the highest level, this will be the MPLS service.
  3. IT domain manager for management of the IT applications and associated resources that provide IT value-added services (VAS)
  4. EMs for underlying technology domains. The most important of these are likely to be access EM (ADSL multiplexer), ATM EM, SDH EM, and IP EM. The IP domain management aspect may be addressed as part of the IP domain management above.

- Service-level management

  5. Service configuration and inventory
  6. Service quality management

The definition of a clear set of building blocks and agreement on what each building block needs to provide for the complete solution to work effectively is very important. Even if parts of the solution do not meet the desired criteria, they can be adapted over time to align to the overall vision.

# Design Considerations for Next-Generation IPv6–Based Virtual Private Networks

## Mallik Tatipamula

*Senior Product Manager, Service Provider Router Technology Group*
Cisco Systems, Inc.

## Tejas Suthar

*Network Consulting Engineer, Advanced Services*
Cisco Systems, Inc.

## Jim Guichard

*Principal Engineer, MPLS Engineering*
Cisco Systems, Inc.

## Khalid Raza

*Distinguished Engineer, Customer Advocacy*
Cisco Systems, Inc.

## Abstract

Businesses across all industries are facing the challenge of integrating data, voice, and video traffic—the so-called triple play—in easily manageable, scalable, economical, and flexible networks. Network-based virtual private networks (VPNs) have emerged as a de facto solution for meeting this challenge, and business customers are now looking to service providers for value-added, cost-effective Internet protocol (IP) VPN services. By outsourcing VPNs to a service provider, business customers gain the advantages of reduced capital and personnel costs, simpler network management, and pay-as-you-grow scalability. [1]IP VPN technology, as described in RFC2547bis, enables a service provider to sell scalable VPN service to customers. This scalability is achieved using a peer IP connectivity model, whereby each customer site peers with the service provider network from a routing perspective (as opposed to peering directly with other sites). Today, end users of some service providers are starting to deploy IPv6 in their network. There are several reasons for this, though the benefits of IPv6 are predominantly derivatives of its much larger addressing space that is required to cope with the Internet expansion and the explosion of Internet-capable appliances such as mobile phones and personal digital assistants. These end users are requesting from their service provider that IPv6 support be added to the IP VPN service with the same look and feel as the current IPv4 IP VPN service. The purpose of this paper is to present a detailed architecture and design implementation for transport of emerging IPv6 services and applications over next-generation IP, multiprotocol label switching (MPLS), or generalized MPLS VPNs.

## Introduction

A virtual private network (VPN) is defined by a set of administrative policies that control connectivity and service policies among sites. These policies dictate routing (whether unicast, multicast or a combination of the two), IP addressing, security, and perhaps quality of service (QoS). The network is virtual because a common infrastructure is used to support multiple VPN customers. The network is private because administrative policies of one VPN customer could be completely independent of other VPN customers' policies, and connectivity between VPNs is prevented implicitly.

IP VPN services for IPv4 traffic have been widely accepted and deployed by service providers and the customer adoption of such services has been widespread. Such services are commercially available globally today from most national and global service providers.

Salient points of the IP VPN for IPv4 architecture are illustrated in *Figure 1*. The customer edge (CE) routers are connected to the provider's backbone (referred to in the architecture as the P network) via provider edge (PE) routers. The routers in the service provider's core are called provider (P) routers. Only PE routers perform VPN specific tasks. The key mechanisms used to build a VPN network are as follows:

- Constrained distribution of VPN customer routing information among different customer sites through the P-network backbone. This is achieved using multi-protocol border gateway protocol (MPBGP) using appropriate route export policies at ingress PE router (PE–1) and appropriate route import policies at egress PE router (PE–2).

- Multiple VPN routing and forwarding tables are maintained at PE routers. These tables are called virtual routing forwarding (VRF) instances.

- Use of a specific address family (namely VPNv4). This is accomplished by prepending a 64-bit field called a route distinguisher (RD) to the customer routes.

- MPLS forwarding over a label switched path (LSP) within the VPN backbone. The basic idea is that the ingress PE router imposes two MPLS labels on entry of traffic into the P network as shown in *Figure 1* below. The outer label (L2) is used to label-switch packets from ingress PE to egress PE. The inner label (L1) is used to determine a forwarding operation toward the appropriate remote CE router from the egress PE. In this paper, we will refer to the outer label and inner label as LDP–imposed label and VPN label, respectively.

The key benefits of the above VPN model are the following:

- The backbone network is easily managed and configured because of the nature of the peer-to-peer connectivity model, where a CE router is a routing peer of one or more PE routers. This applies both to the service provider and the VPN customer.
- Because of private IPv4 addressing capability, addresses within a customer private network are not constrained to be globally unique.
- Complete isolation across VPN customers means secure data transmission is provided in each VPN.
- Only PE routers hold the relevant VPN routes that isolate the core (P routers) from any VPN state.

MPLS is widely accepted as a core technology for next-generation networks (NGNs) that provides a different packet-forwarding paradigm that is easily adaptable for the deployment of value-added services. Service providers who offer IP VPN services to their IPv4 customers are now looking to add IPv6 VPN services to their portfolio so that they are ready to serve their customers when widespread IPv6 deployment becomes a reality. Service providers typically want to support IPv6 in traditional ways but have until now had few options—tunneling or IPv6 native/dual-stack MPLS. Tunneling methods (e.g., manual, tunnel broker, generic route encapsulation [GRE] or intra-site automatic tunnel addressing protocol [ISATAP]) have scalability problems because of the nature of their overlay connectivity model. IPv6 native/dual-stack MPLS cores also have issues that need some consideration:

- For IP VPN services, service providers made a significant investment in building the IPv4/MPLS backbone. The return on investment (ROI) thresholds has, in most cases, yet to be met.

- Backbone stability is paramount for service providers who wish to offer reliable services, especially when it comes to support of voice over MPLS. Stabilization of IPv4 infrastructure is for the most part complete, and therefore service providers are hesitant to make another significant move when it comes to supporting IPv6.

With the emerging adoption of IPv6 in enterprise networks, a similar requirement to that of IPv4 IP VPN service is surfacing so as to support IPv6 customers. The 6VPE technique, which will be described in more detail in this paper, allows the service provider to add such an IPv6 VPN service with no upgrade/reconfiguration of the IPv4 MPLS core (i.e., the P routers) and with configuration/operations that are virtually identical to their current IPv4 VPN service.

With the 6VPE technique, only PE routers (i.e., routers at the edge of the MPLS cloud) need to support IPv6. The PE routers run a dual stack—IPv4 and IPv6. Such a PE router is

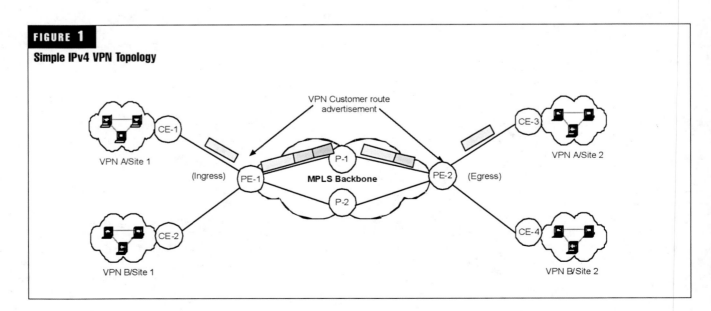

**FIGURE 1**

**Simple IPv4 VPN Topology**

referred to as a 6VPE. IPv6 traffic received from customer IPv6 networks is transported to the destination IPv6 network using the IPv4 MPLS infrastructure in the core. In the MPLS core, all control protocols, including label distribution and interior gateway protocols (IGPs) remain IPv4.

The key differences between 6VPE operations over IPv4/MPLS cores and IPv4 VPN operations are as follows:

- Use of a different address family, which is referred to as the VPN–IPv6 address family
- Encoding in MPBGP advertisements the IPv4 address of the egress PE in the next-hop field using an IPv4–mapped address
- Forwarding at the ingress PE (and at the egress PE in the case of aggregate routes) based on IPv6 addresses

Other elements of operations (e.g., concept of VRFs, use of MPBGP to exchange VPN reachability and constrain distribution of such reachability within/across VPNs , MPLS forwarding in the core from ingress PE to egress PE, use of a second label advertised through MPBGP) are common to both IPv4 and IPv6 VPNs.

### IPv6 MPLS VPN Architecture

Before we look in detail into 6VPE, it is important to define "dual stack." The dual stack technique allows IPv4 and IPv6 to co-exist on the same router interface, as illustrated in *Figure 2*. Today, IPv4 has roots in most of the hosts that run applications. Moreover, stability and reliability of new applications over IPv6 is maturing. So, the co-existence of IPv4 and IPv6 is a requirement for initial deployment and expected to last for some time to ensure a smooth migration to IPv6 through various phases, including business case, funding, design strategies, training, and developments of support tools. For support of IPv6 on an MPLS network, the following are two important aspects of the network to examine:

- *Core*: The 6VPE technique allows for the carrying of IPv6 in a VPN fashion over a non–IPv6–aware MPLS

core. It also allows IPv4 and/or IPv6 communities to communicate with each other over an IPv4 MPLS backbone without modifying the core infrastructure. By avoiding dual stacking on the core routers, the resources can be dedicated for their primary function and avoid any complexity from an operational standpoint. The transition and integration in respect to the current state of networks is also seamless.

- *Access*: To achieve native IPv6 support, the access network that connects to IPv4/IPv6 domains needs to be IPv6–aware. Service provider edge elements (PE routers) can exchange routing information with end users. Therefore, dual stacking is a mandatory requirement on the access layer, as shown in *Figure 2*.

The IPv6 VPN solution defined in this document offers many benefits, especially where a coexistence of IPv4 and IPv6 is concerned, as the same MPLS infrastructure can be used without putting additional stress on the P routers. Also, the same set of MPBGP peering relationships can be leveraged. This is independent of whether the core runs IPv4 or IPv6. This means that the IPv6 VPN service can be supported before and after a migration of the core to IPv6, and this is independent of the customer VPN.

Within the multiservice MPLS core, the backbone IGP (ISIS or OSPF) populates the global routing table (v4) on all PE and P routers. Using the label distribution protocol (LDP), it also populates the label forwarding tables on all PE and P routers. 6VPE routers maintain separate routing tables for logical separation. VRF tables associated with one or more directly connected sites (CEs) form a closed IPv6 and/or IPv4 speaking community. The VRFs are associated to physical or logical interfaces. Interfaces may share the same VRF if the connected sites share the same routing information. MPLS nodes forward packets based on the top label. IPv6 packets and IPv4 packets share the same common set of forwarding characteristics or attributes also known as forwarding equivalence class (FEC) within the MPLS core. If config-

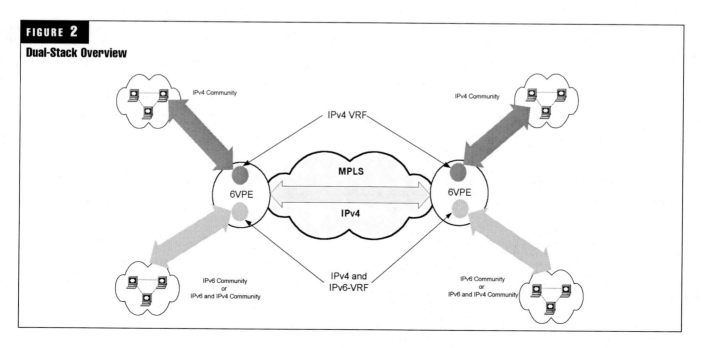

**FIGURE 2**

**Dual-Stack Overview**

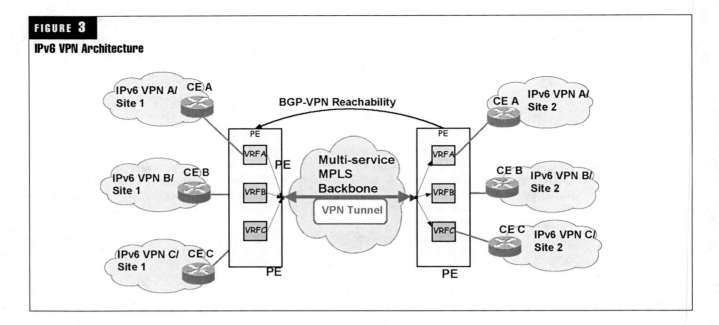

**FIGURE 3**

**IPv6 VPN Architecture**

ured for dual stacking, the interface belongs to multiple VRF instances, IPv4, and IPv6. Each instance maintains its own routing information base (RIB) and forwarding information base (FIB).

RFC2547bis has address family identifier (AFI) concepts as well as MPBGP to carry the VPN information across the MPLS network. This enables the formation of a full-mesh traffic matrix between customer sites. The PE routers advertise their VPN membership (identified by their VRF route-target import/export policies) to other PE routers via direct MPBGP relationships or through route reflectors. There are

new address families introduced to support IPv6 within a VPN, IPv6 and VPNv6. MPBGP is now capable of handling the VPNv6 address family to advertise the IPv6 prefix information across the VPN.

*6VPE Control-Plane and Data-Plane Operation*
When IPv6 is enabled on an interface that is participating in a VPN, it becomes an IPv6 VPN. The CE-PE link may run IPv6 and/or IPv4 natively. The addition of IPv6, and the ability to segment VPN control-plane state on a PE router, turns the PE into a 6VPE–capable device. This enables the service provider to support IPv6 over the MPLS network.

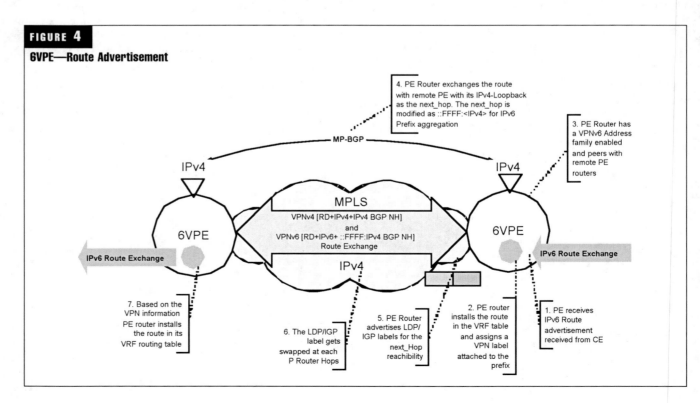

**FIGURE 4**

**6VPE—Route Advertisement**

As outlined in *Figure 4*, PE routers use VRF to maintain the segregated reachability and forwarding information of each IPv6 VPN. MPBGP, with its IPv6 extensions, distributes the routes from 6VPE to other 6VPEs via a direct MPBGP session or through VPNv6–capable route reflectors. Within the MPBGP update, the next hop of the advertising PE router still remains the IPv4 address (normally it's a loopback interface), but with the addition of IPv6, a value of "*::FFFF:*" gets prepended to the IPv4 next hop (NH). The technique can be best described as automatic tunneling of the IPv6 packets through the IPv4 backbone. The MPBGP relationships remain the same as for VPNv4 traffic, with an additional capability of VPNv6. Where both IPv4 and IPv6 are supported, the same set of MPBGP peering relationships may be leveraged.

MPBGP is enhanced to carry IPv6 in a VPN fashion that is known as VPNv6. This uses a new VPNv6 address family. The VPNv6 address family consists of 8 bytes—an RD followed by 16 bytes IPv6 prefix. The combination forms a unique VPNv6 identifier of 24 bytes. The RD value has a local significance on the router and the route target (RT) advertises the membership of the VPN to other PE routers just as in the VPNv4 case.

As illustrated in *Figure 5*, once the ingress 6VPE router receives an IPv6 packet, a destination lookup of the IPv6 address is performed in the VRF table. This destination prefix is either local to the 6VPE (which is another interface participating in the VPN) or learned from a remote egress 6VPE router. For the prefix learned via a remote 6VPE router, the ingress router does a lookup in the VPNv6 forwarding table associating with the incoming interface of the IPv6 packet. The VPNv6 route will have an associated MPLS label and an associated BGP next hop. This MPLS label is imposed on the IPv6 packet. The ingress 6VPE router performs a "push" action, which is a top-label bind by LDP/IGPv4 to the IPv4 address of the BGP next hop to reach the egress 6VPE router through the MPLS cloud. This topmost imposed label corresponds to the LSP between 6VPEs. The bottom label is bound to the IPv6 VPN prefix via MPBGP, and the top label

is bound by LDP/IGP. The IPv6 packet now with two labels gets label-switched through the IPv4/MPLS core (P routers) using the top label only (referred to as the IGP label). Since only the top label is of significance to the P core, it is unaware of the IPv6 information in the bottom label.

The egress PE router receives the labeled IPv6 VPN packet (with the top label already removed) and performs a lookup on the second label, which uniquely identifies the target VRF instance and the egress interface. A further Layer-3 lookup is typically performed in the target VRF instance and the IPv6 packet is sent toward the proper CE router in the IPv6 domain.

In summary, from a control-plane perspective, the prefixes are signaled across the backbone in the same way as they are for regular IPv4 VPN prefixes advertisement. The top label represents the IGP information that remains the same as for IPv4 MPLS. The bottom label represents the VPN information that the packet belongs to. As described earlier, in addition to this, the MPBGP next hop is updated to make it IPv6–compliant. The forwarding or data-plane function remains the same as is deployed for the IPv4 IP VPN. The packet forwarding of IPv4 on the current IP VPN remains intact.

### Building Blocks for IPv6 MPLS VPN
For a service provider to deploy IPv6 VPN service, the 6VPE technique depends on many building blocks/components, which are explained in the following subsections.

*VRF for IPv6 and Its Association to Interfaces*
To support the IP VPN architecture, the concept of VRF was introduced. This concept allows for the use of either global or private address space in each VPN and allows the service provider to provide selective reachability to each VPN. Each customer site is attached to a particular VRF instance. All VPN sites that share the same routing information and communicate directly with each other can be placed in a common VRF instance. A VRF instance is associated with the following:

**FIGURE 5**

**6VPE—Packet Forwarding**

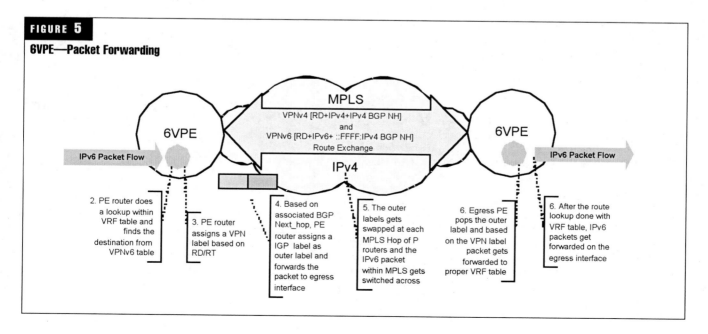

- An instance of RIB that is different from the global RIB (i.e., the RIB that is used for all the interfaces that has not been associated with a VRF instance)
- An instance of FIB that is derived from the RIB instance corresponding to the VRF instance
- A set of interfaces that use the above-mentioned instance of FIB
- Rules that control the import and export of VPN routes to and from the RIB corresponding to the VRF instance
- Routing protocols through which a PE router learns VPN routes from a CE connected to it
- An RD used to make exported IPv6 addresses of VPN customers unique in the backbone
- RT(s) used to export/import and identify policy for IPv6 addresses of VPN customers

6VPE supports IPv6 VRF in the same manner as IPv4 VRF is supported for IPv4 VPNs. Once a VRF instance is configured on a PE router, physical or logical interfaces between this PE and the connected CE routers can be associated with the VRF instance. When the interface is associated with a particular VRF instance, its IPv6 address is removed from the global RIB.

RFC2547bis for IPv4 recommends only one VRF instance per interface. When running dual stack on a 6VPE, multiple VRF configurations on a single physical or logical interface are required (IPv4 and IPv6). Each VRF instance configuration on a dual-stack interface forms IPv4 and IPv6 address families. Each address family within the VRF runs VRF–aware routing protocols; BGP (BGP with IPv6 enhancements for IPv6), open shortest path first (OSPF) (OSPFv3 for IPv6) or routing information protocol (RIP) (RIPing for IPv6). Static routing (static IPv6 unicast routing for IPv6) is also possible.

*VPN IPv6 Address Family and Route Distinguisher*
Another key concept of IPv6 IP VPN is the formation of VPNv6 addresses out of the standard IPv6 address. 6VPE supports the VPN-IPv6 address family as defined in [2].

A VPNv6 address is a 24-byte entity, beginning with an 8-byte RD and ending with a 16-byte IPv6 address. If two VPNs use the same IPv6 address prefix (effectively denoting different physical systems), the PEs translate these into unique VPNv6 address prefixes using different RDs. This ensures that if the same address is used in two VPNs, it is possible to install two completely different routes to that address—one for each VPN.

The purpose of the RD is solely to allow one to create distinct routes to a common IPv6 address prefix, similarly to the purpose of the RD defined in [3]. As it is possible per [3], the RD can also be used to create multiple routes to the same system. This can be achieved by creating two different VPNv6 routes that have the same IPv6 part but different RDs. This allows BGP to install multiple routes to the same system and allows policy to be used to decide which packets use which route.

Note that VPNv6 addresses and IPv6 addresses are always considered by MPBGP to be incomparable.

A VRF may have multiple equal-cost VPNv6 routes for a single IPv6 address prefix. When a packet's destination address is matched against a VPNv6 route, only the IPv6 part is actually matched.

Each VRF configured on a PE router must have an associated configured RD. The RD associated with the VRF is the one pre-pended to the IPv6 prefix in that VRF when advertising VPNv6 reachability to other PEs. The same RD value can be used for both IPv4 VPN and IPv6 VPN.

The RD encoding is specified in [3] with the following format, where the interpretation of the VALUE field depends on the value of the TYPE field:

- TYPE field: 2 bytes
- VALUE field: 6 bytes

As it is the case in [3], three encodings can be used:

- TYPE field = 0: the VALUE field consists of the following subfields:
  - Administrator subfield: 2 bytes, it contains an autonomous system number
  - Assigned number subfield: 4 bytes
- TYPE field = 1: the VALUE field consists of the following subfields:
  - Administrator subfield: 4 bytes, it contains a global IPv4 address
  - Assigned number subfield: 2 bytes
- TYPE field = 2: the VALUE field consists of the following subfields:
  - Administrator subfield: 4 bytes, it contains a 4-byte autonomous system number
  - Assigned number subfield: 2 bytes

*Route Target*
Each VRF instance is associated with an RT import and/or export rule. 6VPE uses the following RT import and export rules exactly as IPv4 VPNs:

- Associated with each VRF is an "export RT" list. When a VPN route is exported into BGP and advertised to other PEs in MPBGP, all the RTs in the export RT list of the corresponding VRF are included in the BGP advertisement. The egress PE performs this task.

- Associated with each VRF is an "import RT" list. This list defines the values it should be matched against to decide whether a route is eligible to be imported into that VRF. The import rule is that all routes tagged with at least one RT associated with a given VRF will be imported into that VRF. The ingress PE performs such filtering during route import.

When advertised in MPBGP, the list of RTs is included in the optional BGP extended community attributes which are described in [2].

Each extended community attribute has a community type of 16 and is encoded as follows:

- 2-byte attribute type (for extended community of extended type, as is the case for RT)
- 6-byte attribute value

The attribute value has two components—administrator subfield and assigned number subfield. Depending on the attribute type, the attribute value is encoded as follows:

| Encoding RT Extended Community | | |
|---|---|---|
| Attribute Type | Attribute Value | |
| | Administrator Sub-field | Assigned Number Sub-field |
| 0x0002 | 2 byte (ASN) | 4 byte |
| 0x0102 | 4 byte (IPV4 address) | 2 byte |
| 0x0202 | 4 byte (ASN) | 2 byte |

6VPE supports the 2-byte ASN and the IPv4 address format.

*BGP Capability Negotiation*
For two PEs to exchange labeled IPv6 VPN network layer routing information (NLRI), they must use BGP capabilities negotiation to ensure that they are capable of properly processing such NLRI. This is done as specified in [5] and [10], by using capability code 1 (MPBGP), with AFI=2 (for IPv6) and SAFI=128 (MPLS–labeled VPNv6). 6VPE supports this BGP capability negotiation for labeled IPv6 VPN NLRI.

The address family within MPBGP is modular in nature to facilitate distinct peering relationships. The regular BGP capabilities are exchanged once the peering sessions are turned on. For two PE routers to exchange labeled IPv6 VPN prefixes, they must use BGP capabilities negotiation to ensure that they both are capable of processing such information. When the service provider network is running VPNv4 peering sessions with other respective elements in the network, it exchanges the VPNv4 AFI capabilities with others. Once the VPNv6 peering sessions are turned on, it renegotiates the capabilities, and fresh peering sessions are established. The peering sessions are established based on common features if either of the peers does not agree on any of the capabilities.

In *Figure 6*, there are three PE routers, two of which need to exchange VPNv6 routes. All three PE routers need to maintain their existing VPNv4 capabilities. This is possible with the address family configuration feature. This makes the migration steps very smooth, and service providers can mix and match VPNv4 and VPNv6 PE routers as required. 6VPE

functionality can be turned on when and where required.

*VPN IPv6 NLRI Encoding*
The advertising PE router assigns and distributes MPLS labels with the IPv6 VPN routes. Essentially, PE routers do not distribute IPv6 VPN routes; they label IPv6 VPN routes [7]. When a PE router learns a local route to a connected site, it must allocate a label before distributing the route to other PE routers. When the advertising PE receives a packet that has this particular advertised label, the PE will perform a lookup on this and process the packet appropriately (i.e., forward it directly based on the label or perform a lookup in the corresponding IPv6–VPN VRF).

The BGP multiprotocol extensions [5] and extensions to encoded labels in BGP [7] are used to encode the MP_REACH NLRI and MP_UNREACH_NLRI. The AFI and subsequence AFI (SAFI) fields are set as follows:

- AFI: 2, for IPv6
- SAFI: 128, for MPLS–labeled VPNv6

The labeled VPNv6 MP_REACH_NLRI itself is encoded as specified in [7]. The prefix belongs to the VPN–IPv6 address family and thus consists of an 8-byte RD followed by an IPv6 prefix.

BGP must provide VPNv6 address-specific processing. It must be able to create, modify, and/or delete VPNv6 network entries in the BGP tables based on the NLRI in the VPNv6 update. Furthermore, BGP must be able to perform best-path calculations for prefixes obtained via VPNv6 updates and update the corresponding routing tables. Note that VRF contains IPv6 prefixes. Thus, before updating VRF instances, BGP has to translate VPNv6 prefixes into IPv6 prefixes.

*BGP Next-Hop Encoding*
MPBGP has the constraint that the "BGP Next Hop" field in the MP_REACH_NLRI attribute needs to be of the same address family as the NLRI encoded in the MP_REACH_NLRI attribute. In the case of VPNv6 NLRI advertisement, this means that the BGP next-hop field must belong to the VPN–IPv6 address family.

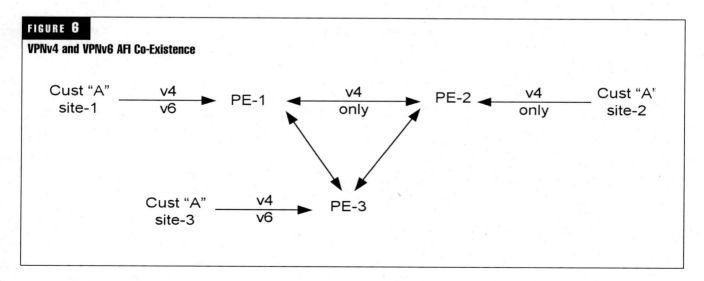

**FIGURE 6**

**VPNv4 and VPNv6 AFI Co-Existence**

Since the 6VPE feature supports IPv6 VPN service over an IPv4 backbone, the BGP next hop is encoded with a VPNv6 address that has an RD that is set to zero and a 16-byte IPv6 address that is encoded as an IPv4–mapped IPv6 address [11] containing the IPv4 address of the advertising PE.

*Encapsulation*
The ingress PE router will tunnel IPv6 VPN data over the backbone toward the egress PE router identified as the BGP Next Hop for the corresponding IPv6 VPN prefix. The ingress PE uses the IPv4 address contained in the 16-byte v4–mapped IPv6 address contained in the BGP Next Hop field, to perform BGP recursion and to determine the IPv4 LSP to use to tunnel the IPv6 VPN packet through the IPv4 MPLS core.

Note that the ingress PE router directly pushes the LSP tunnel label on the label stack of the labeled IPv6 VPN packet (i.e., without pre-pending any IPv4 header). The topmost label imposed corresponds to the LSP starting on the ingress PE router and ending on the egress PE router. The bottom label is the label bound to the IPv6 VPN prefix via BGP.

*Automatic Inbound Route Filtering*
Unless it is a BGP route reflector (RR), a PE router should discard any VPNv6 route with an RT that does not match any of the import targets of any VRF configured on that router. This feature should be available by default, and no extra configuration is needed. 6VPE supports this automatic inbound route filtering for IPv6–VPN address families.

When the policy of a PE router changes, such as when a new VRF instance is added or a new import RT is added to an existing VRF instance, the PE router must acquire the routes it may previously have discarded. This is done using BGP's "route-refresh" capability as described in RFC 2918 (route refresh capability for BGP–4). Note that route-refresh messages defined in RFC 2918 should have an AFI value of 2 and a SAFI value of 128 for requesting a refresh for VPNv6 routes. 6VPE automatically triggers route-refresh requests on relevant PE changes.

Manual refresh of VPNv6 routes is possible by clearing BGP session(s). BGP sessions can be cleared on a per-neighbor basis or on a per-address family basis. 6VPE allows for the clearing of the entire VPN–IPv6 address family. If a particular import RT is no longer present in any of a PE's VRF instances, the PE may discard all routes imported using that RT. When an export RT is deleted or modified, the PE router should withdraw all previously advertised routes from its MP-BGP peers. A router not attached to any VPN and is not a route reflector (i.e., a P router), never installs any VPNv6 routes at all.

*Leverage Route Reflectors for BGP Scalability*
For advertising VPN membership, PE routers peer with VPNv4 route reflectors for scalability. This avoids the need of full-mesh MPBGP sessions among all PE routers. The same concept is supported for VPNv6. The same VPNv4 route reflectors can be upgraded to support VPNv6 address families. Route reflectors can also make the addition or removal of a PE router from a network simple and flexible. Alternatively, the BGP confederation option can also be deployed to provide MPBGP peering sessions among PE routers.

*Link-Local Addresses on PE–CE*
The IPv6 address architecture [11] defines the concept of scope for IPv6 addresses and defines the different unicast address scopes: link-local, site-local and global.

The link-local addresses are embedded on the IPv6–enabled interface for certain local tasks. Since link-local scope addresses are defined as uniquely identifying interfaces within (i.e., attached to) a single link only, the 6VPE feature allows link-local addresses to be used on the PE–CE link but those cannot be used for reachability across IPv6 VPN sites and will never be advertised via MPBGP to remote PEs.

Global unique aggregately IPv6 prefixes are defined as uniquely identifying interfaces anywhere in the network. These addresses are expected for common use within and across IPv6 VPN sites. They are obviously supported by this IPv6 VPN solution for reachability across IPv6 VPN sites and advertised via MPBGP to remote PEs.

*Handling Site-Local Addresses as Global Addresses*
The 6VPE feature handles site-local addresses in the same way as global addresses (i.e., without any consideration for their scope). Site-local addresses are defined as uniquely identifying interfaces within a single site only. Quoting from [11], "a 'site' is, by intent, not rigorously defined."' However, it is anticipated that, when used in an IPv6 VPN network, the concept of site-local scope will typically be used in either of the following scenarios:

• All IPv6 VPN sites of a given VPN are within the same site-local zone. Thus, site-local addresses may be used for reachability across all IPv6 VPN sites and corresponding site-local routes may be advertised via MPBGP to remote PEs and installed in VRF instances. In this scenario, site-local routes MUST effectively be treated by the PE in exactly the same way as global routes. The following is an example of this scenario:

| VPN | VPN site | Site-local zone | Site-local addresses within |
|---|---|---|---|
| Yellow | 1 | 1 | FEC0::CAFE1/64 |
| Yellow | 2 | 1 | FEC0::CAFE2/64 |
| Yellow | 3 | 1 | FEC0::CAFE3/64 |

This scenario is directly supported by the 6VPE feature since the required behavior for site-local addresses is the same as for global addresses.

• Each IPv6 VPN site of a given VPN is in a different site-local zone. Site-local addresses can not be used for reachability across IPv6 VPN sites, and corresponding site-local routes should not be advertised via MPBGP to remote PEs. The following is an example of this scenario:

| VPN | VPN site | Site-local zone | Site-local addresses within |
|---|---|---|---|
| Green | 1 | 1 | FEC0::BEEF1/64 |
| Green | 2 | 2 | FEC0::BEEF2/64 |
| Green | 3 | 3 | FEC0::BEEF3/64 |

This scenario can also be used by the 6VPE feature by effectively hiding the site-local routes to the PE. This can be seen as placing the site-local zone boundary on the CE (as opposed to on the PE) and thus locating the PE outside the site-local zone. Then, when dynamic IPv6 routing is used between the PE and CE (v6 IGP, MPBGP), the CE will not distribute site-local routes to the PE. Thus, no special handling of site-local routes is required on the PE, since there are none.

*Global Addresses in VPN*
Global-scope addresses are defined as uniquely identifying interfaces anywhere on the Internet. Global addresses are expected to be commonly used within and across IPv6 VPN sites. They are obviously supported by the 6VPE feature for reachability across IPv6 VPN sites and advertised via MPBGP to remote PEs and processed without any specific considerations to their global scope.

*QoS for IPv6 VPN*
QoS and queuing of important application traffic requires distinct policies for IPv4 and IPv6. This may require additional operational tasks where IPv4 and IPv6 networks co-exist. Both IPv4 and IPv6 have a commonality, which is the 3-bit IP precedence (or type of service [ToS]) field within the IP headers. Alternatively, the differentiated services (DiffServ)–compliant QoS models can also be employed. The following are two methods for QoS from a 6VPE perspective:

- DiffServ QoS on ingress PE
  - Ability to classify IPv6 packets received on the PE from the CE based on the fact that it is IPv6 traffic and based on the DiffServ Code Point (DSCP).
  - Ability to meter classified traffic against rate/burst (conform/exceed).
  - Ability to mark/remark the DSCP of the classified/metered packets and the EXP field of the pushed label-stack entries. This supports the uniform mode of MPLS DiffServ tunneling. [12]
  - Ability to mark/remark the EXP field of the pushed label-stacked entries of the classified/metered packets without modifying the DSCP. This supports both the pipe mode and the short pipe mode of MPLS DiffServ tunneling. [12]
  - Ability to schedule/drop MPLS–encapsulated IPv6 VPN traffic on the PE–P link based on the EXP field of imposed outer label (this includes the ability to schedule/drop the IPv6 traffic jointly with MPLS–encapsulated IPv4 traffic with same EXP value).

- DiffServ QoS on egress PE
  - Ability to classify the IPv6 traffic on the PE-CE link, since it is an IPv6 packet and based on DSCP in the IPv6 header.
  - Ability to remark-DSCP/schedule/drop/shape based on this classification (this includes the ability to schedule/drop/shape the classified IPv6 traffic separately or jointly with some classified IPv4 traffic). This supports the short pipe mode and the uniform mode of MPLS DiffServ tunneling [12].
  - Ability to classify IPv6 traffic on the PE–CE link based on the EXP field of the packet as it was received by the egress PE before label pop. Ability

to meter/mark DSCP/schedule/drop/shape based on this classification. This supports the pipe mode of MPLS DiffServ tunneling [12].

Irrespective of the marking technique—whether it is IP precedence—or DSCP–based, the QoS is such an important factor when low-speed links are concerned. However, there is no additional advantage of QoS on IPv6 versus IPv4, but at some point, IPV6 can be different by using the flow label in IPv6 header. The QoS within the MPLS core remains based in MPLS Experimental Value (MPLS_EXP) and is untouched but still is effective for the IPv6 addition.

*PE–PE MPLS Traffic Engineering Tunnels*
The 6VPE feature supports operations where IPv4 reachability of the egress PE from the ingress PE is achieved through an MPLS traffic engineering (TE) tunnel rather than LDP. In this case, the 6VPE feature on the ingress PE will push the following two labels:

- the BGP label bound to the IPv6 VPN NLRI as the inner label
- the MPLS TE tunnel label as the outer label

The criteria for the 6VPE to use a given TE tunnel is purely whether the IPv4 address of the egress PE advertised as the BGP next hop is routed onto that TE tunnel. The use of IPv6 packets will not dictate the need for tunnels. So, in typical designs, a given TE tunnel is either used by both v4 VPN and v6 VPN traffic (from same ingress PE to same egress PE) or used by neither.

It is conceivable that the network administrator may wish to steer IPv6 VPN traffic in a PE–PE tunnel and steer IPv4 VPN on other paths than this tunnel. This will be allowed by the 6VPE feature but through the use of more complex designs. One example design would be where the IPv6 VPN NLRI is advertised with a different IPv4 PE address than the IPv4 VPN NLRI and where static routes are used to route the IPv4 PE address used for IPv6 VPN to go down the desired TE tunnel while the IPv4 PE address used for IPv4 VPN is routed on another path.

*Device Management*
Finally, the device management is another important aspect that service providers must consider. The device management in a dual-stack network can be done through an IPv4 and/or IPv6 address. Where the IPv6 VPN service is supported over an IPv4 backbone, and where the service provider manages the CE, the service provider may elect to use IPv6 for communication between the management tool and the CE for such management purposes. The management systems, including OSS servers, need to be aware of IPv6 and must run proper simple network-management protocol (SNMP) stacks to perform IPv6–based management. From the VPN perspective, it still remains transparent how the device and services are managed.

**Simultaneous Operations of IPv4 VPN and IPv6 VPN**
As stated earlier, RFC2547bis introduced AFI concepts as well as MPBGP to carry the VPN information across the MPLS network. This enables the formation of a full-mesh traffic matrix between customer sites. The PE routers advertise their VPN membership to other PE routers via direct MPBGP or through route reflectors based on the configured "VRF route-target import" value(s).

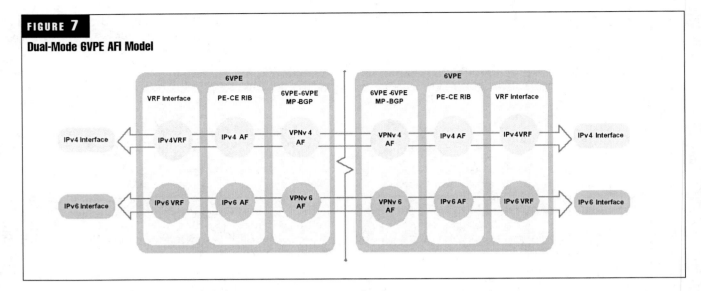

**FIGURE 7**

**Dual-Mode 6VPE AFI Model**

As highlighted in *Figure 7*, there are new address families introduced to support IPv6 within VPN, IPv6, and VPNv6. If configured for a dual stack, the interface belongs to multiple VRF instances, IPv4 and IPv6. Each instance maintains own RIB. The MPBGP is now capable of handling a VPNv6 address family to advertise the IPv6 prefix across the VPN.

### Enhancements to the Draft

The IP VPN services that service providers have implemented are based on RFC2547bis. The "BGP-MPLS VPN extension for IPv6 VPN" is the current draft that addresses the need for IPv6 support over MPLS networks within a VPN. Also, to avoid an extra layer of signaling, the draft addresses the scalable automatic tunneling of VPN–based IPv6 prefixes. The basic functionality remains the same as is outlined in RFC2547bis. Some of the extensions outlined will require additional work on the PE routers to be effective in the service provider network.

### Conclusions

In this new era where explosive use of IPv6 is envisioned, it is extremely important for the network administrators to have a simplified, automated, fail-proof and cost-effective network. The Internet draft discussed drives the capabilities to achieve these and becomes very important to spread the knowledge so that a practical approach is taken in supporting IPv6 for customer's next-generation applications. In a nutshell, this highlights a solution and recommendations to bring the IPv4–to–IPv6 transition one step closer with a simple, cleaner, cheaper, and scalable solution. Supporting large-scale and wide-ranging VPN services increases the value added by service providers and decreases operational cost.

### References

"IPv6 – Service provider view in advancing MPLS networks," Internet Protocol Journal- Vol. 8, No. 2, June 2005.

BGP-MPLS VPN extension for IPv6 VPN, draft-ietf-l3vpn-bgp-ipv6-07.txt.

Rosen et al., "BGP/MPLS VPNs," draft-ietf-l3vpn-rfc2547bis.

Mallik Tatipamula, Patrick Grossetete and Hiroshi Esaki, "IPv6 integration and co-existence strategies for next generation networks," IEEE communication Mag., Jan. 2004.

Bates, Chandra, Katz, and Rekhter, "Multiprotocol extensions for BGP4," June 2000, RFC2858.

Deering, S. and R. Hinden, "Internet protocol, version 6 (IPv6) specification," RFC2460.

Rekhter and Rosen, "Carrying label information in BGP4," RFC3107.

MPLS Technology and Applications, Bruce Davie and Yakov Rekhter, Morgan Kaufmann Publishers, ISBN 1-55860-656-4.

MPLS and VPN Architectures, Ivan Pepelnjak, Jim Guichard, Cisco Press ISBN 1-57805-002-1.

Chandra, R., Scudder, J., "Capabilities advertisement with BGP-4," draft-ietf-idr-rfc2842bis-02.txt, April 2002, work in progress.

Deering, S., and Hinden, R., "IP version 6 addressing architecture," draft-ietf-ipngwg-addr-arch-v3-11.txt, work in progress.

Black, D., "Differentiated services and tunnels," RFC 2983, Oct. 2000.

### Note

1. This paper presents IP VPN services deployed using MPLS label switching as the transport mechanism. However, the same service can be provided using an IP encapsulation, such as L2TPv3, as the transport in the core of the network.

# Quality of Service

# Next-Generation Services: Co-Opting Disruption

## T. S. Balaji

*Group Head, COE – NGN Solutions and Services*
Tech Mahindra Limited

## Ajit Kalele

*Solution Architect, COE – NGN Solutions and Services*
Tech Mahindra Limited

## Naveen Sharma

*Solution Architect, COE – NGN Solutions and Services*
Tech Mahindra Limited

## Manik Seth

*Business Analyst, COE – NGN Solutions and Services*
Tech Mahindra Limited

## Abstract

Telecommunications service providers (SPs) have been focusing on evolving their networks toward an access-agnostic Internet protocol (IP) core that could expose application programming interfaces (APIs) to in-house and third-party application SPs (ASPs). While this is likely to provide a platform for quickly providing new value-added services (VASs) in the short run, , they will have to contend with business models that are prevalent on the Internet—including the Web 2.0 concepts enshrined in the business processes of players such as Skype and Google—in the long run. This paper analyzes how telecom SPs could evolve their networks and applications to co-opt these otherwise disruptive business models.

## Introduction

The telecom industry is undergoing seismic shifts in the manner in which services are packaged and delivered to their end users. Deregulation and licensing processes have created competitive pressures on the revenues of traditional telephony SPs, driving them to address the need to provide more value to their end users than the newly emerging telephony SPs.

Toward this, SPs have taken recourse to engineering their networks, service delivery mechanisms, and associated IT infrastructures in a quest to provide new customized services to end users. Typically, the transformed network architectures involve an IP–based core network with various forms of access networks that provide relatively larger bandwidths. Standardization processes in various telecom forums have been focusing on defining universally acceptable models and frameworks to ensure interoperability.

While the new infrastructure is likely to address quick development of services in the short term, the following questions remain:

- Does the transformed infrastructure address all the threats that confront an SP today?
- If all SPs embrace similar philosophies of transformation, how would service differentiation be maintained?
- What criteria would be employed to decide on a successful catalog of user-specific services that should be introduced?
- Would the new services be priced on the basis of value or as cost-plus, and would the subscribers be willing to pay that price?

The purposes of presenting this paper are as follows:

- To address the above questions and propose possible architectures suitable for next-generation services delivery. The threats confronting telecom SPs and how the current transformation trends may not serve to address the same are presented. A potential service/business model is described, and a service example is also presented.
- To throw open new paradigms that telecom SPs would need to investigate and explore. Subscriber behaviors vary across cultures and geographic regions. However, the model that is ultimately likely to succeed is one that allows the subscribers a greater choice in the kind of services they require rather than a greater choice in the kind of service they could subscribe to.

FIGURE 1

**Network Evolution Drivers**

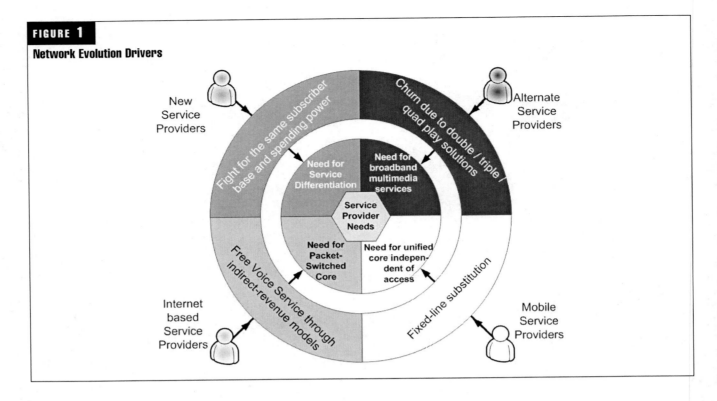

## Telecom Network Evolution

### Threats Faced by Telcos

To appreciate the evolution of the network, it is important to understand the sources of threats that are faced by SPs that provide fixed telephony services. Primarily, the threat for incumbent operators comes from emerging and newly licensed telephony SPs, other SPs foraging into telephony, Internet-based telephony SPs, and Internet SPs (see *Figure 1*).

### Considerations for a Next-Generation Network (NGN)

New SPs include emerging SPs that have been recently licensed and have commenced operations. The issues arising from this market reality is one of struggling to tap into the same subscriber base and their spending power. To stave off this threat, it would be necessary for existing SPs to maintain service differentiation. This essentially implies the need for a mechanism through which services could be designed and delivered quickly. Most network evolutions today include a service delivery mechanism that supports application development and management.

Alternate SPs such as cable operators and broadband SPs have also started offering voice services. These SPs have thus put pressure on traditional telcos to provide multimedia services through seemingly large bandwidth connections. The result is that the evolving network typically includes support for one or more VAS aimed at exploiting the broadband access capacity such as IP television (IPTV) and grid computing.

The convenience of mobility as experienced by end users has come to stay. According to a 2005 survey of enterprise networks and their usage, reported by IDC, cellular phones are fast becoming the primary phone of users and are

already ranked second behind digital desktop phones [1]. To counter this threat of mobiles substituting for fixed lines, it would be essential for NGNs to be realized with capability to connect through any access network.

Lastly, Internet-based SPs that provide instant messaging (IM) and voice services through voice over IP (VoIP) have emerged. These players, having to continue the existing business models of providing free e-mail and IM services, provide voice services for free. To compete effectively, SPs are forced to subsidize voice services and will have to minimize their operational expenses (OPEX) to minimize their losses. Thus, most NGN designs incorporate a core network that uses packet-switching technology. Packet-switching technology reduces complexity of the network, thus reducing the OPEX of maintaining and operating the network.

*Figure 2* depicts a simplified structure of an NGN based on the requirements as explained above and tallies with the definition of NGN frameworks such as the 3GPP IP multimedia subsystem (IMS) architecture [2].

### Potential New Threats and Way Forward

The generic NGN addresses most of the threats to SPs that were identified in *Figure 1*. However, the following points may also need to be noted about the architecture.

The packet-switched network, typically implemented using IP technology, would expose end users to services that are available over the Internet more and more. Like pressures on speech services, any service introduced by SPs directly or through third-party ASPs would be susceptible to threats from services provided on the open Internet, thus reducing the NGNs to mere IP pipes and reducing the SPs to mere connectivity providers.

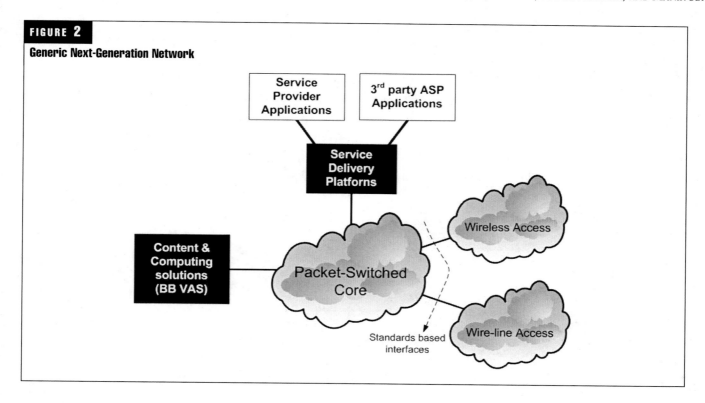

**FIGURE 2**

**Generic Next-Generation Network**

Further, with abundance of choice in look, feel, and flavor of a service available on the Internet, SPs would be hard-pressed to make justifiable business cases and choose which service is likely to be successful. This could lead to an analysis paralysis in designing and launching new services that exploit the next-generation infrastructure created.

The way forward could be one that co-opts the successful models of service delivery and businesses on the Internet and enable a similar one through the next-generation infrastructure. One such successful design of service delivery and associated business process named "Web 2.0" provides an ideal framework for designing the new services [4].

## What is Web 2.0?

"Web 2.0" is defined on Wikipedia as "a phrase coined by O'Reilly Media in 2004 and refers to a supposed second generation of Internet-based services—such as social networking sites, wikis, communication tools, and folksonomies—that let people collaborate and share information on-line in previously unavailable ways" [5].

Web 2.0 is based on the following principles related to the design of the service and business process associated with a particular service [4]:

- *The long tail:* It is observed that the majority of the Internet content or applications accessed are those of small communities of users. This also implies the need for self-service mechanisms that would allow users to access the niche applications they need.

- *Data is the driving force:* Applications are increasingly data-driven, so for competitive advantage, it would be

important to own a unique, hard–to–re-create source of data. The subscription profiles of end users and other dynamic information regarding subscriber availability, location, etc., available with telcos could be advantageous.

- *Users add value:* The key to competitive advantage in Internet applications is the extent to which users add their own data to what is provided. Therefore, users should be involved both implicitly and explicitly in adding value to service applications. For example, while subscriber profiles are available with SPs, business interests and hobbies of subscribers could be added through a self-service portal by the user-enabling future communications centered on the business interest or hobby of the subscriber. Tagging and rating third-party or SP content are other examples of users adding value.

- *Some rights reserved:* Intellectual property protection limits reuse and prevents experimentation. Since the Web 2.0 model is based on benefits coming from collective adoption, it should be ensured that barriers to adoption are low by following standards, and it should use licenses with as few restrictions as possible. Telecom transformations, by virtue of having adopted a standards-based approach, are better placed to align with Web 2.0 than are Internet-based SPs, which typically employ proprietary interfaces.

- *The perpetual beta:* When devices and programs are connected to the Internet, applications are no longer software artifacts; they become part of the service offerings. Hence, the applications could be offered as beta software with incremental releases instead of periodic monolithic releases. This also enables users to partici- pate in real-life testing, which leads to better attune- ment to user requirements and makes the service development risk-free.

- *Cooperate, do not control:* Web 2.0 applications are net- works of cooperating data services. Therefore, the offer of Web service interfaces and content syndication and the reuse of the data services of others would need to be supported.

- *Software above the level of a single device:* The personal computer (PC) is no longer the only access device for Internet applications, and applications that are limited to a single device are less valuable than those that are connected. Thus applications built around the core of the network will need to have front-end support across personal digital assistants (PDAs), mobiles, PCs, and specialist devices.

An easier way to understand Web 2.0 would be through examples of what services on the Internet are considered Web 2.0 [6]. The following are some examples of applica- tions identified to be offering services and conducting busi- ness in a manner aligned with Web 2.0:

- *FlickR:* FlickR [7] is a photo-sharing Web site and an on-line community platform. Users can store photos on-line and tag keywords to them for searching later on. FlickR has created a platform out of the Web wherein users can network with each other, resulting in increased involvement.

- *del.icio.us:* del.icio.us [8] is a social site that allows users to bookmark their favorite sites, news, and pictures in an on-line repository, and allows them to share their bookmarks with people around the world along with metadata tagging of the information.

- *Digg:* Digg [9] is a community-moderated news Web site that combines the ideas of news dissemination through social bookmarking and democratic moder- ation. It is rapidly becoming a preferred location to find relevant news items than news Web sites them- selves.

- *Wikipedia:* Wikipedia [10] is a Web-based, multilingual open encyclopedia that anyone can edit. It rests the responsibility of creating and managing entries on the users.

## Architecture for Service Delivery Aligned with Web 2.0

### Features of Web 2.0 Services

The features of popular services on the Internet that are aligned with Web 2.0 could be summarized as indicated in *Figure 3*.

End users can get connected to the service through a client application. The client application, in its simplest form, could be a browser available on a user device (PDA, PC, or special device such as a music player).

End users would engage with a front-end portal that authenticates the user and presents the user with the screens of choice. The screen of choice would also include asynchro- nous information that the user had subscribed to from the community of users registered with the same service.

This interface would also allow users to deposit content into the public store and describe the same through a collection of metadata strings. Three types of stores could exist: for applications (called "gadget repository" in the figure), tar- geted advertisements, and generic user information.

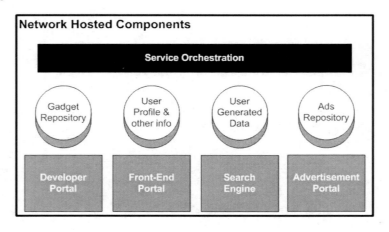

**FIGURE 3**

**Features of Web 2.0 Services**

Network Hosted Components

Service Orchestration

Gadget Repository | User Profile & other info | User Generated Data | Ads Repository

Developer Portal | Front-End Portal | Search Engine | Advertisement Portal

Client Application

While front-end portals could be used to provide the necessary user interface for end users, separate portals may be maintained for getting advertisements to be served as featured links when users search for the specific information. Similarly, a separate portal may be maintained to download gadget development kits and upload finished gadgets by end users.

A search engine with an algorithm to search the metadata information and allow end users to access the information or service gadget of their choice by a typing in a search string is also typically supported.

The service orchestration component provides an interface for the gadgets to interact with other applications on the Internet or access other relevant information available on the network for delivering a service. The service orchestration layer provides an execution mechanism that is typically shrink-wrapped and supplied as part of the gadget development kit.

### Next-Generation Services Model

*Figure 4* depicts the unification of the Web 2.0 model and generic NGN model described earlier.

The service orchestration layer integrates with the service delivery platform through Web services for capabilities of the telephone network and the Internet for services requested by end users.

End users log on to the network and update the client application with the gadgets of their choice. Accessing the service happens in the following manner (see *Figure 5*):

- The user initiates the service request toward the gadget.
- The service request is interpreted by the network elements and SDP and appropriately relayed to the relevant gadget.
- Upon receiving the service request, the gadget invokes the back-end procedures through the service orchestration layer. This may also involve connecting with various other services supported by the SP or third-party applications registered with the SP's network.
- The results of the operations in the previous step are communicated back to the gadget.
- The gadget then uses the network/SDP to relay the end result to the end-user application.

### A Simple Service Example

The example given in this section assumes that the NGN infrastructure implemented by the SP is based on an IP–based core network that supports session initiation protocol (SIP) [11] for realizing an IM service based on Internet Engineering Task Force (IETF) SIP/SIP for Instant Messaging and Presence Leveraging Extensions (SIMPLE) principles [12].

The SDP is assumed to expose session control, presence, and messaging capabilities through Web services that could be

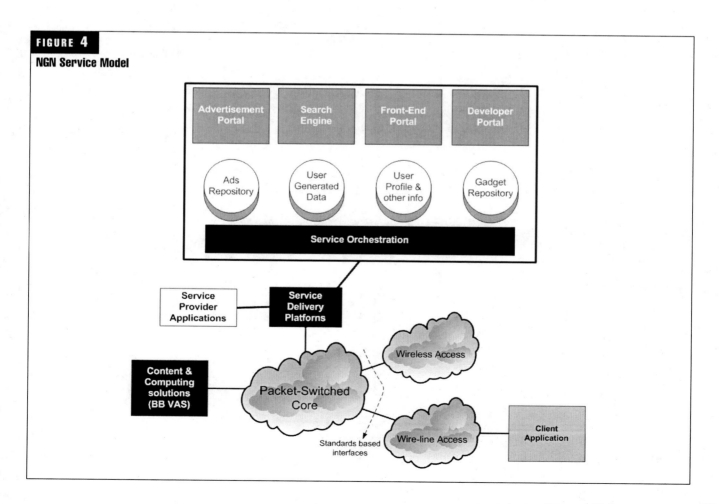

**FIGURE 4**

**NGN Service Model**

## FIGURE 5

**Features of Web 2.0 Services**

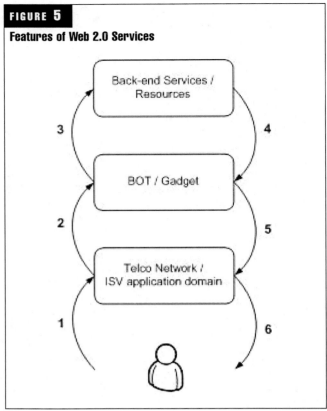

invoked from an IM robot (BOT) developed using the gadget development kit. For example, a BOT developed by a cinema multiplex that uses Web services for ticket purchases can be uploaded through the developer portal. An end user who wants to see the latest Harry Potter movie only has to go to the movies tab on the IM client and key in the search string "Harry Potter."

The search returns published BOTs registered with the SP and a few sponsored links, as shown in *Figure 6*. The results of the search could be designed to consider the location of the user as identified in the network and present choices appropriately.

The user could have the options to right-click on the BOT and send an e-mail or make a call. If the user wishes to make a call, the same would be achieved through a third-party call control mechanism that uses the published telephone number of the cinema multiplex and the phone number associated with the user profile.

The user could use some text commands, as shown in *Figure 7*.

As seen in *Figure 7*, a default "help" command, included as part of the gadget development kit for the BOT, allows for valid commands to be listed. The user could use any of the valid commands to get fulfil his or her requirements. There could another gadget to provide a graphical interface to

## FIGURE 6

**Invoking Network Features through a Contributed BOT**

## FIGURE 7

**Enabling Services through Contributed BOT**

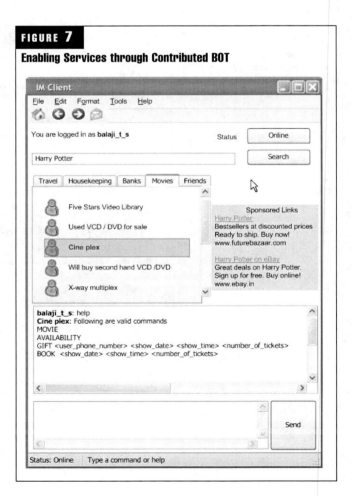

access the services, instead of the cumbersome command line interface.

While performing this task, the user could click on the friends tab and check out the availability of friends who could accompany the user to the movie. Again, the information of the availability of a friend could be ascertained through a call, an IM (if on-line), an SMS (if the user is reachable through the wireless mechanism), or a BOT command that fetches the current location of the friend based on network registered information.

It may be noted that these transactions would take place only after the user has validated his or her own identity through authentication mechanisms supported.

While this service is being offered by the SP, it is assumed that the user is not expected to pay for anything other than any tickets to the show. The page gets sponsored through the advertised link on the IM client, the owner of which could be charged on a per-impression or per-click basis.

## Conclusion

The disruptive nature of Web 2.0 on traditional telecom services is easily witnessed in the number of minutes of voice traffic lost to Internet-based SPs. The NGN infrastructure being created by telecom SPs could be enhanced to co-opt these models and create a richer end-user experience.

## References

[1] IDC's Technology Assessment, "Extending the Enterprise PBX to Mobile Communications: Single Number Dialing" – Nora Freedman, Abner Germanow (IDC #201463, May 2006)
[2] "Service requirements for the Internet Protocol (IP) multimedia core network subsystem; Stage 1" – Third Generation Partnership Project Technical Specification (3GPP TS 22.228 V7.3.0 [2005–12])
[3] "Embracing the Future: BT's vision for a 21st Century Network" (www.btplc.com/21CN)
[4] "What Is Web 2.0: Design Patterns and Business Models for the Next Generation of Software" by Tim O'Reilly (www.oreillynet.com/go/web2)
[5] "Web 2.0" (en.wikipedia.org/wiki/Web20)
[6] "Levels of the Game: The Hierarchy of Web 2.0 Applications" by Tim O'Reilly (radar.oreilly.com/archives/2006/07/levels_of_the_game.html)
[7] FlickR – (http://www.flickr.com)
[8] Del.icio.us – (del.icio.us)
[9] Digg – (www.digg.com)
[10] Wikipedia – (www.wikipedia.org)
[11] SIP: Session Initiation Protocol (www.ietf.org/rfc/rfc3261.txt)
[12] SIP for Instant Messaging and Presence Leveraging Extensions (SIMPLE) (www.ietf.org/html.charters/simple-charter.html)

# Beyond QoS: The VoIP User Experience

## Tom Flanagan

*Director, Technical Strategy, DSP Systems*
Texas Instruments

## Debbie Greenstreet

*Director, Service Provider Marketing, DSP Systems*
Texas Instruments

Today, most legacy network service providers have just begun to roll out voice over Internet protocol (VoIP) services. At this early stage, many provide only a pipe for VoIP traffic, and management or provisioning capability for VoIP services is rudimentary at best. And when problems occur, they lack the tools and know-how required to resolve them. Even next-generation VoIP providers offering services that compete with the incumbent local-exchange carriers (ILECs) have minimal ability to quickly determine the cause of a problem and fix it.

As a result, customer service representatives who are lacking effective tools have little choice but to suspect that the VoIP customer-premises equipment (CPE) is defective. Users are often asked to return their voice gateway for a replacement when relatively simple diagnostic and management techniques could be used to resolve the problem remotely and without the need to replace the CPE.

The good news is that service providers are beginning to embrace and understand the scope of the issues that can have a negative impact on the quality of IP–based services as perceived by the end user. They also are realizing that what consumers expect from a telecommunications service in terms of quality involves more than just voice.

The emergence of equipment that incorporates voice signal quality monitoring is a step in the right direction toward providing operators with the information they need to better understand the consumer experience with VoIP. In fact, systems now are coming to market that are able to calculate the mean opinion score (MOS) for every VoIP call on the fly. This paper explores several additional areas that are often overlooked when considering the consumer VoIP experience.

## More than VoIP

As communications providers migrate from legacy circuit-switched networks to IP–based networks, consumers will expect their VoIP phones and "phone lines" to enable them to do, as a minimum, the same things they do today with their telephony lines and devices. Of course, they will use their VoIP connections to talk, but they also will use them to send and receive faxes, connect to home security systems that use embedded modems, and much more. Even their TiVo may need a telephone line connection for program guide updates.

In short, end users quite justifiably will expect a VoIP voice service to be a full-featured telecommunications service capable of handling any service and any device they have plugged in to an RJ-11 jack in the past. Equipment manufacturers and service providers alike need to anticipate these requirements when creating their products.

Today, much is being discussed and written about VoIP quality and the concept of quality of service (QoS) in VoIP services, networks, and devices. Typically, these discussions are centered on the network performance characteristics of bandwidth allocation and reservation and the priority queuing of voice packets. While these characteristics are important to IP service quality, they are by no means the only factors that affect IP performance.

Network-based QoS cannot overcome a poor implementation of a system to manage network impairments such as jitter, packet loss, and delay/latency in the end equipment. Some impairments may be overcome by using networking techniques, and some are best addressed within the voice processing domain, but often a solution that makes sense to the voice engineer may exacerbate the problem due to its effect on the network. (The reverse is also true.) For example, a common solution to overcoming lost packets is to retransmit redundant packets. But, if this is all that is done, the resulting network loading may actually have a negative impact on the end user's experience due to increasing congestion and delay.

The challenge is to compensate for network impairments with a balance of voice and network processing for maximum effectiveness. Overcoming these impairments is the "black art" of VoIP. The best systems are based on products created by designers who have learned through years of experience what it takes to reliably deliver voice over a hostile packet network. For this reason, equipment manufacturers and service providers need to understand the pedigree of their core technology providers.

Today, service providers and enterprise managers have very few tools to help them determine the causes of poor IP–based service quality. This ultimately leads to subpar end-user experiences with VoIP, because troubles often reside a layer deeper than can be revealed by taking measurements that reveal the level of jitter, packet loss, or delay.

It will be many years, if ever, before all calls and communications originate on IP devices and travel exclusively across pure IP networks. In the meantime, issues affecting the end-user experience will occur at points in the network where the legacy network and the IP network interconnect. This includes media gateways, line cards, and voice gateway devices used in home networks.

## Tone Detection

It is important to remember that the legacy network of devices that facilitate telephony calls will be with us for many years to come. Voice- and tone-activated private branch exchanges (PBXs) and virtual operator systems are everywhere, and they are unlikely to disappear anytime soon. Our banks and credit card companies, voice-mail systems and nearly all customer-service systems use the familiar "press 1 for English, press 2 for Spanish" dual-tone multifrequency (DTMF) signaling for navigation, authentication, etc.

Consumers have become accustomed to communicating with automated attendants and voice-mail interfaces using touch tones. Legacy telephones and cellular phones generate standard tones that are transmitted across the network and are received by a wide variety of tone-driven interfaces.

VoIP phones mimic legacy phones by sending IP–based "tones" to the media gateway. At the media gateway, the tones are detected and repackaged in real time and then transmitted to their destination. Many things can go astray in this process.

Working together, the VoIP CPE device and the media gateway must detect and re-create tones, including dialed digits, fax detection, modem detection, and call progress tones that have the same amplitude, frequency, and timing as those generated by the legacy telephony equipment to which they are connecting. For example, if tones sent by an IP phone to access voice mail are not accurately detected, translated, transmitted, or received, access will be denied and the end user will not be

able to check his or her voice mail. A good VoIP implementation will exceed International Telecommunication Union (ITU) Q.24 tone detection standards.

Today, end users, service providers, or enterprise managers would be hard-pressed to pinpoint which part of the process was causing a problem in any particular instance of trouble. This is because this type of quality issue cannot be diagnosed and pinpointed to the source by measuring the elements that contribute to a mean opinion score (MOS).

Rather, this type of trouble relates to in-band dual-tone multifrequency (DTMF) transmission using the G711 encoding scheme (G711 is used to encode voice), or it relates to DTMF relay and transmission of tones in packages or separately to the receiving end of a call. Or, separately these problems could be exacerbated by, or be simply due to, poor user equipment such as a cordless handset that is not transmitting the tones properly. Furthermore, there are so many tone types generated and tone standards worldwide that if a service provider chooses to offer global services, the VoIP equipment must be robust enough in design to operate in such diverse environments.

## More DTMF Possibilities

There are many opportunities for call failures that relate to DTMF–related signaling issues. Ultimately, how well IP–based solutions are implemented and how robustly they are executed with this in mind will determine end-user experience. When switching to IP phones and devices, end-user experience should be equal to, or better than, end-user experience with legacy telephones and devices.

Another important consideration that IP device designers and carriers must take into account is that media and telephony gateways used in different parts of the world have different tone parameters. Making matters even more complex, the DTMF issue can extend outside the network. There have actually been cases in which VoIP devices have mistaken end users' voices for actual DTMF tones. When this happens, the equipment abruptly halts the voice transmission and transmits an erroneous tone.

In addition, how well the wide variety of IP network interfaces adhere to existing DTMF standards will factor in to a network element or device's ability to deliver a quality experience to end users. The challenge for IP equipment makers is to design robust systems that can reliably reproduce true tones; quickly, and with 100 percent accuracy,

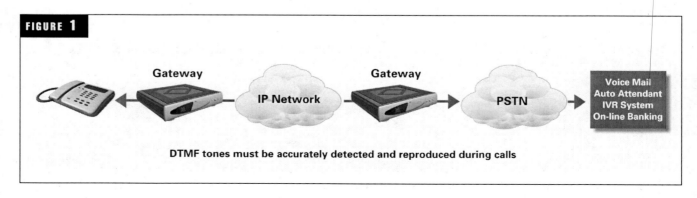

**FIGURE 1**

Gateway      Gateway

IP Network      PSTN      Voice Mail / Auto Attendant / IVR System / On-line Banking

**DTMF tones must be accurately detected and reproduced during calls**

detect a true tone; and meet 0 percent "false detects," meaning that the device will never mistakenly characterize an end user's voice as a tone. The challenge for service providers is being able to determine if, and when, DTMF–related issues are to blame for poor user experience.

Fortunately, there are metrics in VoIP gateway and end devices that can be collected and sorted through to help troubleshoot and resolve DTMF–related issues. Equipment is emerging that will take full advantage of these metrics. Eventually, all communications will be IP–based, but this may take several decades. When this finally occurs, user interfaces will be voice- and message-based and the need for tone-based signaling will disappear. Until then, accurate detection and replication of DTMF-based signals will be a key parameter affecting the customer experience.

## Echo Cancellation

Echo cancellers are used throughout the legacy public switched telephone network (PSTN). Line echo occurs any time a four-wire to two-wire interface is encountered in the network. In the PSTN, where interoffice trunks meet local loops are four-wire to two-wire interface junctions called hybrids. When signals travel from the four-wire network to the two-wire network, their energy is reflected back onto the four-wire network.

In legacy time division multiplex (TDM) networks, the reflections typically occur quickly. PSTN-based echo cancellation equipment is calibrated for these quick echoes. Packet networks, with longer delays, produce reflections beyond the time threshold within which existing echo cancellers' work today. Therefore, VoIP end equipment has to be equipped with embedded echo cancellers capable of handling long echo tails in order to eradicate echo caused by packet-based traffic flows.

The process of locating echo in speech is called convergence. Poor echo canceller designs can take a "long time" to converge. When echo is heard early in a VoIP call, it is because the echo cancellers have a hard time pinpointing the echo and nullifying it. This is because they have not converged on the echo. In normal circumstances, once the echo canceller does converge on the echo, it performs adequately for the duration of the call. There are situations where convergence is lost during a call and the echo canceller must restart.

It is important to test VoIP equipment's echo cancellation ability in challenging conditions to determine if its designers have made the right decisions regarding cost versus quality. With the ubiquity of mobile phones, it is even more likely that one party to a call will be in a noisy environment such as a car or airport. High background noise, particularly when the background noise changes suddenly, is especially challenging to echo cancellation.

While not common in polite face-to-face company, double talk, which occurs when both speakers talk at the same time, occurs more frequently in phone conversations due to network latency. Like background noise, double talk challenges echo cancellers. Line echo induced by hybrids is not a problem on pure VoIP calls in which both parties are using IP phones. This is because the IP phones are connected directly to the packet network from the start and packets never touch any legacy telephony interfaces.

Service providers need to be aware that if they opt for the cheapest VoIP phones or media gateways, echo cancellation is the one thing they are most likely to sacrifice in the process. This cost/performance tradeoff could end up being more costly, especially when one considers that echo is the problem most often cited by end users as the cause of low-quality experience.

Yet another form of echo, acoustic echo, is of concern. Speakerphones experience acoustic echo from the reflections of the end users' voices as their voices bounce off the walls, desks, windows, etc., in the rooms in which the speakerphones are used. The design of the speakerphone itself may also induce acoustic echo, which is caused by the placement of the speaker and microphone. A quality echo cancellation implementation can, to a certain extent, be used to "tune" a phone's acoustic attributes.

Video IP phones will likely feature speakerphone functionality, so acoustic echo cancellation will be a critical element influencing end users' experiences with them. Consumers are likely to pay a premium for video phones, and they also are likely to be the provider's premier customers. Poor VoIP implementation in a video phone, especially in the tricky area of acoustic echo cancellation, could result in the consumer blaming the service provider for quality issues caused by poor acoustic echo cancellation on their video IP phone.

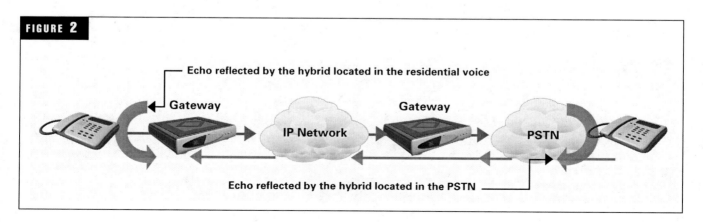

**FIGURE 2**

Echo reflected by the hybrid located in the residential voice

Gateway            Gateway

IP Network          PSTN

Echo reflected by the hybrid located in the PSTN

All echo cancellers are not created equal, and adherence to the G.168 standard for cancellation is not a "seal of approval" for echo cancellation. Simply put, it is not easy to implement good echo cancellation. It takes a lot of digital signal processor (DSP) power and memory to do it well. Quality echo cancellation will have a marginal cost impact associated with the additional millions of instructions per second (MIPS) and memory resources when compared with bare-bones solutions. This cost can be minimized when the developers have control over all system resources. Thus, an integrated echo canceller that was conceived as part of the whole voice subsystem rather than a third-party add-on is likely to be more resource-efficient.

## More than Voice

As mentioned earlier, voice is not the only application for which telephony lines are used. Modem-based devices are still quite common, with fax being the most common of these. There are a tremendous number of issues to contend with when sending a fax, but the first is simply detecting that the call is a fax call and not a voice call or some other kind of modem call. False detection of fax tones can result in frustrating failure of the call. Once detected, fax tones can be transmitted using G.711 encoding, but the scheme is not as robust as the T.38 fax relay, which breaks a fax call down and sends the fax data across in a packet format. The result is greatly improved reliability and call completion rates.

Fax is designed to operate between two machines directly connected via the PSTN. This is a nearly optimal connection from the perspective of delay, and therefore fax machines are very intolerant of "unusual" delays. Connecting fax machines via a packet network virtually assures that the delays will be outside the fax machines' operating parameters. Fax transmissions implement a fax protocol between two machines involved in the transmission of a fax. This protocol can be "spoofed" to compensate for the delays. If packets are lost during negotiation between the machines, or even in the middle of the fax call, problems will occur.

## Knowledge Is Power

There are ways to mitigate many of the issues described above, but service providers need to do more than adjust their networks for jitter, packet loss, and delay. As legacy networks and IP networks unite, knowledge is power for service providers.

The better the tools, the better the chance of resolving issues that many end users will have difficulty describing simply because they will be new to them. Having access to calls in real time to resolve problems, being able to measure a call while speaking with the consumer, and/or the ability to (with consumer approval) play back a call will help service providers troubleshoot problems.

At a higher level, it will be useful to correlate the similar complaints, or troubles, that take place at the same time or on the same day. In addition, correlating troubles on calls running through the same pieces or types of equipment will help service providers establish trends and identify problem areas in the very-distributed and multi-network environment in which IP–based calls will be placed.

Just as there were systems and tools created to manage TDM networks effectively, tools are emerging that will examine the full spectrum of characteristics that can have a negative impact on the voice QoS. They will do the same for the telephony applications that originate on packet networks and in places where packet networks merge with TDM networks. Until then, tools that examine the end points and places of demarcation between packet networks and telephony networks will be of great value to service providers. VoIP phones and IP PBXs at the enterprise can provide metrics and statistics that help service providers and enterprise managers better understand the true causes of end users' complaints.

Service providers that are spending more than $3,000 to acquire new customers will see the wisdom of investing in technology that will enable them to provide those customers with the high-quality experience they have come to expect in the telephony environment, in the IP environment. Based on their strategic plans, even ILECs, like next-generation voice service providers, are moving toward complete IP–based networks as a foundation for their IP communications. A large reason for this is that service providers have been convinced that an IP–based infrastructure will reduce their overall operational expenditures (OPEX), offering a more competitive business model.

However, service providers must remember to consider the operational costs of managing subscriber issues. The cost of poor management and the resulting subscriber churn can offset the OPEX benefits of a common infrastructure. Rather, if consumer satisfaction is met early on in new services by employing accurate and proactive quality management, the OPEX model is improved further.

And, as they roll out IPTV, the challenges of providing quality IP video will be even more critical to delivering a quality experience. This is because the eye is even less forgiving than the ear. Fortunately, techniques and technology used to troubleshoot IP–based telephony QoS issues will be ramped up to meet the coming IPTV QoS challenges.

# 10 Critical Questions to Ask Before Launching into VoIP

## Mission-Critical Considerations for Every Service Provider

## Rich Grange
*President and Chief Executive Officer*
New Global Telecom

### Market and Business Focus

#### What Market(s) Should I Target?
Addressable market segments for voice over Internet protocol (VoIP) services include the following: residential, small office/home office (SOHO), small and mid-size businesses (SMBs), and enterprises. However, VoIP penetration is extremely low in all these segments today—we are, in effect, just at the leading edge of the VoIP industry growth curve. At the same time, VoIP line shipments to the business sector will actually exceed time division multiplex (TDM) line shipments in 2005, indicating an enormous gain in interest in and acceptance of VoIP solutions. In addition, 50 million homes and 55 percent of all adults have broadband Internet access (up 60 percent in the past year) and 90 percent of SMBs have DSL or faster Internet connectivity—important enablers for VoIP service penetration.

The market segment(s) chosen by any service provider will drive the development of a service value or sales proposition having the most appeal to that marketplace—for instance, there is a vastly different proposition for primary line residential service than for secondary line service, and a very different proposition for hosted PBX service to a business than for trunk replacement service (i.e., connectivity between an onsite PBX and the public network). Clarity around your target market is central to effective product definition and development.

#### What Delivery Models Should I Consider?
Service providers can opt for a "build-my-own" (i.e., self-provisioned) hosted VoIP services solution to address their markets, or they can use a managed wholesale option. This is the classic "build versus buy" decision. This chart reflects some of the most important factors (though by no means all the factors) that figure into this decision.

Capital cost avoidance (for instance, in relation to application servers) and operating cost savings (such as technical skill sets), in conjunction with access to IP (and legacy TDM) experience, are sound reasons to consider a wholesale solution. Also compelling, given the complexity and effort

required to create an end-to-end service, is increased speed-to-market and access to tested systems for smooth delivery. At the same time, a managed wholesale solution may not offer the same degree of network control and flexibility in voice product packaging as the "build-my-own" alternative. A thorough assessment would consider these and other important aspects of product development, deployment and support for each particular service provider.

Whichever delivery model is chosen, service providers will want to ensure they have a solid position in the market, relative to competitive alternatives. In the business market, for instance, where you will compete head-to-head with IP private branch exchanges (PBXs) and other hosted services, your solution needs to effectively counter the following IP PBX limitations:

- *Technology obsolescence*: PBXes require ongoing investment to maintain state-of-the-art features and technical capabilities; indeed, most IP PBX manufacturers issue "dot" releases monthly and major releases every six to 12 months (sometimes requiring expensive hardware upgrades).

- *Technology limitations*: IP PBXes cannot generally interface to third-party software or other vendor equipment, and the PBX hardware often has limited resale value after purchase and often uses outdated technology (such as H.323).

- *Challenging interface*: The system interface is often not user-friendly and may require 10 to 15 days' training for your customers' system administrator.

- *Security issues*: Any breach of security from external sources jeopardizes the integrity of your customers' local-area network (LAN).

#### How Will I Maximize Customer Acquisition?
Creating a compelling sales proposition that is sustainable in the long term is, of course, the key to customer acquisition in any competitive environment. While creating appeal

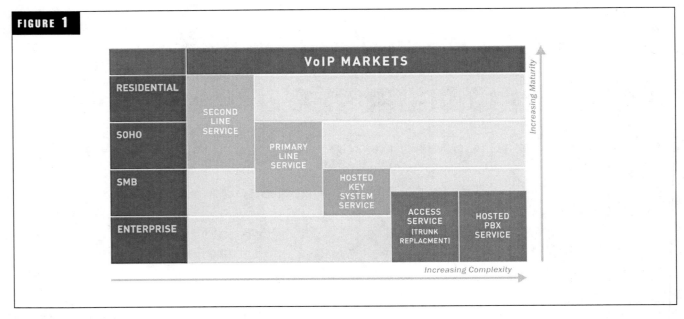

**FIGURE 1**

**FIGURE 2**

| VoIP Service Provider Options | Capital Cost | Operating Expense | Access to IP Experience | Speed-to-Market | Market-Ready OSS/BSS | Network Control | Access to IP Experience Flexibility |
|---|---|---|---|---|---|---|---|
| Build-My-Own Hosted Solution (**Build**) | ☐ | ☐ | ☐ | ☐ | ☐ | ■ | ■ |
| Managed, Hosted Wholesale Solution (**Buy**) | ■ | ■ | ■ | ■ | ■ | ☐ | ☐ |

☐ Potential Area of Concern          ■ Relative Advantage

for the "early-adopter" user can lead to some short-term success, VoIP is going to become a mainstream service—very likely attaining primary line service status in homes and businesses across the country (unlike the early promise of cellular service). So, creating a proposition that is focused on all the market 'hot buttons' holds the most promise of continued appeal to the market.

The choice of distribution channels is critical from the perspective of rapid customer acquisition and, since retail distribution can account for more than 20 percent of your operating expense, it's also critical from a financial viewpoint. Direct sales is the highest dollar-cost approach; however, indirect channels (agents, for instance) can be difficult to manage in a pre-mass-adoption market—these channels will need to learn how to sell VoIP services in the most time-effective way. Regardless of your chosen channel, you will need a well-constructed set of marketing support tools to help the sales effort, including high-impact communications, benefit/cost demos, product functionality demos, easy-to-use post-sales materials, and so forth. Providing educational information to your target market (despite the increased information flow from the large local exchange

carriers [LECs]) is an important part of the communication process. Once prospects understand the power of VoIP services in addressing their needs, the excitement level rises—the challenge lies in making that power readily and quickly visible.

*What Else Can I Do to Accelerate ROI?*
Service providers can "buy" market share through "low ball" pricing—but as tempting as this might be in the short term, it is not a sound financial strategy and will not maximize return on investment. The market is likely to respond very positively to price-positioning of 10 percent to 30 percent below pricing of comparable legacy (TDM) service, so that should be the target range for price-positioning. In fact, 72 percent of SMBs express some level of interest in VoIP services with just a 15 percent cost savings.

And, as the above chart indicates, 69 percent of service providers agree on a target range for cost-savings of 10 to 30 percent. Cost reductions afforded by IP technology make this price-positioning profitable and, we believe, sustainable.

**FIGURE 3**

| | Market Requirements |
|---|---|
| **RESIDENTIAL** | > Bundled local / long-distance with features, but cheaper<br>> Advanced features for power users<br>> Mobile and / or PC integration |
| **SOHO** | > Bundled local / long-distance with features, but cheaper<br>> Premium service package<br>> Mobile and PC integration |
| **SMB** | > Cheaper for same-day service as today<br>> Dial tone & Class 5 / Centrex feature set, tele-worker / remote worker features<br>> Big company features, without capital outlay<br>> Avoid dedicated, expensive staff |
| **ENTERPRISE** | > Uniform feature set across offices (voice mail, dialing plans, etc.)<br>> Outsourced network management<br>> Tele-worker / remote worker features<br>> Collaboration and business-grade IM&P |

The ability for VoIP services to attract new customers and to retain existing ones is strengthened tremendously through product bundling—for instance, with Internet access—to create a "one-stop" offer that consumers and businesses perceive as being much more interesting and compelling than buying VoIP service on its own.

As with all sound product strategies, creating real, positive financials comes down to ensuring you are delivering the value that customers want and will pay for. VoIP offers a wide range of exciting functionality that can be positioned against real consumer and business needs and will be the key to maximizing long-term margins and return on investment (ROI) for service providers.

## Engineering Focus

### How Can I Be Sure That a Solution Will Actually Work?
IP technology has been deployed for several years as a means of achieving compressed transmission in long-haul backbone carrier networks. But VoIP as a front-line voice service is relatively new, as are the plethora of products being deployed to enable VoIP services. So, first and foremost, a wise step is to request a demo account that permits a thorough trial of any product or service you might be contemplating. This will help you determine whether it works and will allow you to understand capabilities and limitations.

In addition, consider all the proactive tactics a network vendor or platform/software vendor can use to ensure reliability of their offering and to give you a degree of comfort. These can range from physical network redundancy and active network testing procedures to rock-solid service-level agreements (SLAs) and network operations center (NOC) coverage. Confirm that processes and procedures

**FIGURE 4**

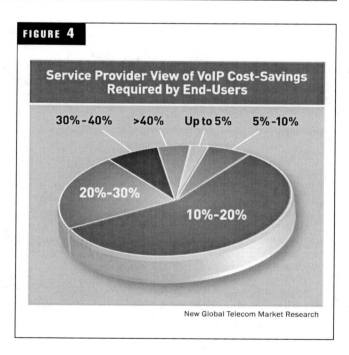

Service Provider View of VoIP Cost-Savings Required by End-Users

New Global Telecom Market Research

are tailored to the unique challenges of an IP network consisting of disparate, "decoupled" elements. Investigate what is (and what is not) standard procedure to be sure you are covered.

Further, make sure you understand plans for testing and extending the line of supported equipment, including media gateways and customer premise equipment (CPE)—the marketplace will continuously try out and adopt new equipment going forward.

**FIGURE 5**

| | Physical IP Network Redundancy | Active Network Testing | Call Quality Monitoring | VoIP Traffic Monitoring Metrics | Element Monitoring | SLA's | SIP Message Management | 24 x 7 NOC |
|---|---|---|---|---|---|---|---|---|
| **NETWORK VENDORS** | ✓ | ✓ | ✓ | ✓ | ✓ | ✓ | ✓ | ✓ |
| **PLATFORM/ SOFTWARE VENDORS** | | | ✓ | ✓ | ✓ | ✓ | ✓ | |

Finally, now that E911 service for VoIP has been mandated by the FCC, it is crucial that any such service is reputable and proven. Verify who the E911 service provider is. Since there is a limited number of these established providers, it should be easy to get information about their compliance with the order, geographic range of coverage, ability to offer mobile service, level of technical support, plans for improvements/upgrades and overall service performance. It is also important to know what the agreement between the wholesaler and the E911 service provider entails. Ensure what the level of service is, the coverage area that has been agreed to and what backup plans are in place in case of system failure.

### What Will the Biggest Challenges Be in Moving from Testing to Live Implementation?

This is a big leap, requiring sufficient internal expertise and vendor support. IP networks, by their very nature, are unlike consolidated TDM networks—IP networks are essentially "decoupled," requiring critical voice networking knowledge. Some important considerations include the following:

- Ensuring that your production systems are redundant and fully tested
- Testing media servers, POP3 servers, DNS setup and network connectivity (note that testing environments often do not include important ancillary systems)
- Ensuring that security policies are in place
- Ensuring that provisioning processes are in place and have been tested, including backup plans for each step along the way
- Ensuring that network documentation is in place, accurate, and sufficiently detailed
- Ensuring that procedures for carefully taking end users through planned changes are thorough and tested
- User acceptance testing to ensure that all features and functionality are working as planned—this should include troubleshooting checklists to systematically identify any minor problems

### What Level of Voice Expertise Will I Need to Acquire?

Delivering VoIP services does require a sufficient level of experience in the voice telecom world—this is not a data service, but it's not a traditional voice service, either. Engineering issues can be successfully managed with expertise inside your organization or available through vendors/partners in the following areas:

- Local number portability (LNP)
- 911 and e911
- Directory assistance and directory listings
- Operator services
- Inter–LATA and Intra–LATA dialing plans and billing
- 800 services
- Equal access
- Call flows (and feature use)
- TDM-based troubleshooting
- Corporate competitive local-exchange carrier (CLEC) status
- Dealing effectively with LEC and other CLEC procedures, processes and people

In addition, as the chart indicates, a formalized training program will be extremely valuable—for your staff and for business and residential end users. Particularly given the relatively "new" state of VoIP, access to relevant information is critical to ensure a level of comfort with your staff and your customers.

## Operations Focus

### What Ongoing Technical and Service Support Will I Need?

You will need to effectively address quality of service (QoS), which means bringing a focus to the major contributing factors to QoS, including the following:

- *Latency*: This is the time needed for a packet to traverse the network.
- *Packet Loss*: This is the percentage of packets lost while transiting the network.
- *Jitter*: This is a measure of the variation in arrival delay for a series of packets.

You must establish first- and second-tier customer support procedures and hierarchy, with clear delineation between your company's responsibilities and those of your vendors. Responsibilities to carefully consider include problem troubleshooting, maintenance of system integrity, and hardware and applications support. And, if you are providing VoIP services to the business market, technical knowledge for customers' LAN/wide area network (WAN) assessments and interworking will be needed.

Also, consider that success in VoIP will include short product development cycle times—quickly getting new features and capabilities from testing into production (to maintain high product value for your chosen target market). So, prod-

**FIGURE 6**

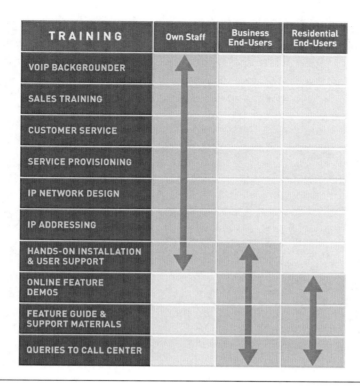

| TRAINING | Own Staff | Business End-Users | Residential End-Users |
|---|---|---|---|
| VOIP BACKGROUNDER | | | |
| SALES TRAINING | | | |
| CUSTOMER SERVICE | | | |
| SERVICE PROVISIONING | | | |
| IP NETWORK DESIGN | | | |
| IP ADDRESSING | | | |
| HANDS-ON INSTALLATION & USER SUPPORT | | | |
| ONLINE FEATURE DEMOS | | | |
| FEATURE GUIDE & SUPPORT MATERIALS | | | |
| QUERIES TO CALL CENTER | | | |

uct development and engineering functions need to be highly skilled and tightly integrated.

### How Will I Ensure Scalability of Back-Office and Delivery Functions?

As your sales volume grows, this will drive the need for automation in operational areas such as order input, account setup, feature assignment, billing, CPE fulfillment, on-site network assessment and installation (business market), and customer relationship management (CRM). This is a tremendous challenge that requires considerable attention to automation and integration of operational support systems (OSSes) and business support systems (BSSes) in the early stages of service delivery. And, in addition to dealing with growing volumes, OSS/BSS systems must also provide increased accuracy and audit capabilities through the entire service or product life cycle to ensure seamless delivery to customers.

### Once I Gain Traction, How Can I Replicate Service to Expand My Footprint?

Initial selection of geographic markets is a complex decision, considering your preferred sales approach, existing network, level of anticipated competition, and so forth.

**FIGURE 7**

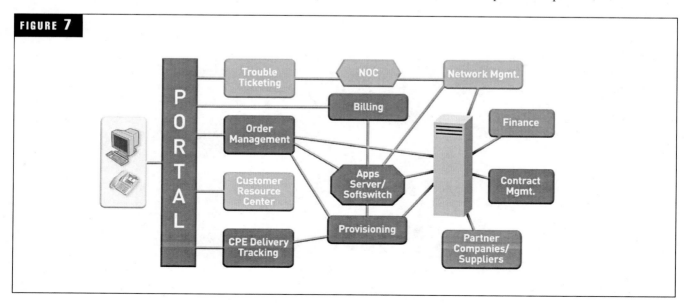

Appropriate phasing of growth is important to maintain smooth operations and sound financial results. An orderly expansion will be facilitated by learning from initial market experiences, but it will nonetheless require detailed cost analysis that takes significant variations in circumstances into account, including conditions such as LEC interconnection for public switched telephone network (PSTN) access, collocation costs and local support, as well as intrastate and interstate toll arrangements.

# QoS for Multiservice IP Networks: Challenges and Prospects

## Gerhard Heide
*Senior Engineer*
U4EA Technologies, Ltd.

## Annie Snelson
*Product Marketing Manager*
U4EA Technologies, Ltd.

## Peter Thompson
*Chief Scientist*
U4EA Technologies, Ltd.

## Abstract

Quality of service (QoS) in Internet protocol (IP) networks is a crucial tool in ensuring the reliability of a whole range of emerging new services such as voice over Internet protocol (VoIP) and IPTV. This paper examines QoS technologies and approaches at contention points and assesses their suitability to deliver quality for these new services. It establishes that most QoS mechanisms, developed nearly a decade ago and still widely deployed, have significant shortcomings for this purpose. An exploration of the underlying theory shows why these mechanisms are deficient and what needs to be addressed to resolve this for multiservice networks. One recently developed mechanism illustrates how this can be done, but it is still surprising that this problem has received little attention in recent years, given the rapid rise of IP convergence.

## Introduction

The range of applications supported by the Internet has increased tremendously in the past 10 years, first expanding from file transfer and e-mail to encompass the World Wide Web and, more recently, real-time applications such as Skype, with a user base that has grown to millions in a very short time. This has accelerated the trend to replace multiple special-purpose networks with general-purpose packet-switched networks, driven by the attractive prospect of substantially reduced capital and operational costs combined with increased flexibility. Combining different networks in this way is not without difficulty, however, particularly with regard to security and reliability, each of which is a well-known Achilles' heel of the Internet.

The focus of this paper is on delivering reliable performance, in particular solving the problem of delivering streams of packets in such a way that the applications depending on them behave satisfactorily. In a residential triple-play scenario, these applications might be telephony (audio and, possibly, video), television (perhaps video on demand [VoD]), and Internet access for Web browsing, etc., all of which are accompanied by some sort of performance requirement or at least expectation on the part of the consumer. These expectations can be compromised by mutual interaction between the different applications and/or interference from further sources of traffic such as peer-to-peer file sharing. In a business context, television would be substituted by business-critical applications such as remote file access and transaction processing, with even more stringent constraints on performance.

Although IP is the converged network protocol of choice, the best-effort philosophy of the Internet is inadequate to guarantee the level of reliable performance that paying customers expect. As the services carried on IP networks become more diverse and sophisticated (real-time, interactive, mission-critical, financial, etc.), IP networks must respond to ensure service performance. Hence, an effective form of QoS is required for this, especially since this trend is forecast to continue in an even stronger fashion with the advent of IPTV.

### QoS in Converged Packet Networks

Meeting performance constraints for multiple applications simultaneously is far from trivial. Packet-switched networks base their economies (even more than circuit-switched ones do) on statistical multiplexing, the assumption that applications do not fully use the network all the time. Dynamically allocating resources to applications on demand makes the network flexible and efficient but also makes performance unpredictable unless steps are taken to control the process. In core networks, this unpredictability is typically reduced by packet route control (for example, using multiprotocol label switching [MPLS]), combined with substantial over-provisioning (made possible by technologies such as dense wavelength division multiplexing [DWDM]) and admission control.

Moving toward the edges of the network makes such techniques less effective, because both the choice of routes and the capacity of individual links diminish—all the way to the final, unique, and generally low-bandwidth link to the customer. This creates a need for more fine-grained methods to control the allocation of network resources when demand exceeds supply, as it surely will, if only for short periods. These methods are the subject of this paper. In particular, we consider what methods are available and assess them in respect to their ability to support the sort of application mix discussed above, how scalable they are, and what fairness they provide.

## QoS Overview

Any device in a network that forwards packets is a shared resource. If the load offered to it is less than the output it can deliver (a combination of its processing power, memory, and the bit rates of the input and output media), then it will introduce a little delay but no loss. However, if the demands on any of these resources exceed their limits, then congestion may occur. Starvation may also occur; even when the offered load is below the potential output, i.e., in the form of competition between different flows, a more aggressive flow might continually outperform a less aggressive one.

The main problem caused by congestion is that some flows do not get the resources they aspire to or require. This can manifest itself as unacceptable performance of a particular flow or a general degradation for all flows. Since congestion is unavoidable, it is unrealistic to expect perfect service for every flow. The forwarding device will be presented with choices, which it must resolve according to the service specifications for the flows being transported, their compliance to agreed traffic specifications measured over a certain time interval and the aggregate state of the device measured over a range of timescales.

Another challenge in resolving these problems is to ensure some degree of fairness between customers. This means mostly a minimum level of access and equal treatment compared to similar customers, assuming some sort of classification. Important customers should not be made to suffer by less important ones. Ideally, every customer should get the treatment he or she requires.

This section gives an overview of the most common mechanisms that are being used to manage congestion in today's IP networks. To lay the groundwork, we start with short reviews of queuing theory, as the mathematical discipline that provides the underlying concepts, and fairness, since it is a vital performance criterion, whose meaning can often be unclear in a particular instance.

### Queuing Theory

On a very basic level, any packet-forwarding device such as a router or switch operates according to the same principles. Packets arrive and are then sent on or discarded according to some previously specified rules. If packets are forwarded, then usually some delay is also incurred. Therefore, any packet-forwarding device can be regarded as a queue, which allows us to use queuing theory (e.g., [ALL90], [ALB00], [KLE75], [TAK93]) to analyze its performance. As a mathematical research area, it gained prominence in the first half of the 20th century and is now well established as the theory of choice for investigating packet networks.

In general terms, any queue or waiting line is a facility that involves customers (items, packets, etc.) arriving then waiting to receive service they seek, so in its simplest form, a queue is a "waiting room" and a server. This basic model can be extended in many ways, as the following example questions indicate:

- How many servers are there?

- Do customers leave the queue on their own account (called reneging) if they have not received service after a time? If they do, do they return after a while? Similarly, do customers who are turned away because the queue is full try again some time later (i.e., in some kind of feedback loop)? Is the waiting room finite? (Obviously, this always holds in reality, since by definition there are only finitely many resources available. Nonetheless, a lot of queuing theory assumes infinite buffers; quite often, realistic scenarios can be approximated with the infinite case. Interest in finite queues has only increased significantly in the past three decades.)

- In which order are customers serviced? Common examples are as follows:

  ○ First in first out (FIFO), i.e., the customer who has waited longest in the queue is serviced next. Sometimes this is also called first come, first served (FCFS).
  ○ Last in first out (LIFO), i.e., the customer who arrived last is serviced first.
  ○ Random order of service (ROOS).
  ○ Priority systems, which may also include a preemptive mechanism (i.e., if a customer with higher priority than the one currently in service arrives, the service is interrupted and either resumed later or started anew).
  ○ Some form of service sharing, i.e., services can be shared equally between customers, or they can be interrupted and then resumed, e.g., by some form of round-robin.

Queuing theory is about describing such a facility in terms of various parameters, predicting its behavior and also optimizing it with respect to some quantities, thereby enabling a designer to build a system that maximizes/minimizes some variable (usually minimizing the overall time spent in the queue) without excessive costs.

The most commonly used parameters are the queue length (the average number of customers present in the system, which usually includes customers currently receiving service), loss and delay averages (for different priority classes as well as overall), etc. In most cases, at least where networks are concerned, only the averages and sometimes the variances of these parameters are computed (especially the variance of the delay times, which gives a measure of the expected jitter)—higher moments are rarely considered.

### Little's Law

In this context, it is worth mentioning a celebrated result that holds true not only for queues, but also for nearly every system where objects arrive, remain for some time, and then depart ([LIT61]). It is intuitively reasonable to assume that the average number of such objects in the system (i.e., the average queue length $L$) is equal to the average number arriving multiplied by the mean time spent in the system. Little's Law asserts that this is indeed the case, the only condition being that these quantities (which are formally defined as limits) exist. So if $\lambda$ is the number of packets that arrive (and are accepted) and $W$ the expected delay per customer, then $L = \lambda W$.

Little's Law illustrates a fundamental relationship between loss, delay, and the service rate of a queue, namely that fixing any one of them creates a proportional relationship between the other two. In the formula above, each of the three variables is influenced by one of the parameters ($L$ by the service rate, $\lambda$ by the loss and $W$ by the delay; note that in networking terms the service rate is closely related to the throughput). So if, for example, the loss is fixed, a faster service rate would cause $L$ to be smaller and, by the formula, also reduce the average waiting time in the system. In other words, this three-parameter system has two degrees of freedom. This means that a queue-controlling mechanism aiming to regulate loss, delay, and service rate can only move in a two-dimensional surface through the three-dimensional space created by loss, delay, and throughput. It is truly surprising that most commonly available mechanisms do not make use of this two-dimensional area but restrict themselves to a one-dimensional approach. We return to this later.

### Fairness

In its simplest and most intuitive form, "fairness" in networking terms refers to the property of assigning bandwidth "justly" among flows. Most users have an instinctive ability to assess if a given allocation is fair. However, there is no single technical definition of fairness accepted to be correct. Rather, a range of criteria have been standardized to evaluate the fairness of a network. This section reviews four fairness principles which are most commonly used in a networking context.

The fairness referred to in these definitions is that which is provided to streams within a class of service. In other words, if a stream is in a different priority class to another stream, the priority scheme employed should specify the treatment each stream receives.

### Min-Max Fairness

The min-max fairness principle is one of the most common definitions of fairness in networking terms. According to this principle, fair systems allocate as much resources as possible to smaller, less aggressive users while not unnecessarily wasting resources.

If there is contention, a system exhibiting min-max fairness will penalize the streams with the highest bandwidth and favor the smallest streams.

### Proportional Fairness

According to the proportional fairness principle, each flow gets a proportion of link rate inversely proportional to the length of the flow. This principle of allocating bandwidth in proportion to some quantity is commonly used to describe fairness between several concatenated links but can be extended to other quantities, the most obvious being the proportion of arrival rates of the streams. That is, one form of proportional fairness is to allocate to each stream a proportion of the link rate corresponding to the proportion of a stream's arrival rate to the total arrival rate.

### Jain's Fairness Index

Jain's fairness index allows a simple evaluation of the fairness of a system as it is represented by a single number. Here, a fair system is defined as one in which all connections get "equal access to bandwidth." The average throughputs of each flow are combined to compute a normalized number between 0 and 1; 0 denotes the maximum unfairness and 1 denotes the maximum fairness according to this principle.

### Coefficient of Variance

The coefficient of variance is the standard deviation divided by the mean. It can be used to indicate a system's fairness when applied to the egress rates of flows for a specific time scale. A low value implies that the sending rates of the flows are more closely clustered around the average, from which one could deduce that the bandwidth sharing is relatively fair. As with Jain's fairness index, the benefit of this measure is that a system's fairness is expressed in a single number. However, the meaning of the value is reversed from Jain's index—a coefficient of variance equal to zero is the most fair, and there is no upper limit.

### A Single Fairness Model

All these ideas indicate that fairness is a very subjective concept indeed. One way to combine these varied ideas is the following definition, which reflects the current state of the art:

To call a convergence of flows fair, the following need to be ensured:

- No connection consumes more than its fair share, which is defined by one of the criteria listed above (e.g., min-max fairness, proportional fairness, or Jain's fairness index)
- No connection is starved.

### QoS Mechanisms

We now describe the operation and basic properties of old and well-known queuing mechanisms, plus one that is newer (and hence less well-known).

### First In First Out

The packet-forwarding element consists of just a single queue and, as the name implies, packets leave in the order of their arrival. This essentially describes the best-effort model of the Internet.

The main advantage is that it is simple to implement. This model usually works acceptably at low loads and if there is no emphasis on service differentiation. Its disadvantage is that there is absolutely no control over any aspect of quality.

Performance depends on traffic patterns and conditions and the best-effort model becomes highly unfair toward less aggressive streams at high loads. This can lead to total starvation of some flows.

*PS, GPS, and Similar*

These intimately related models are mainly concerned with dividing service equally between any number of classes. A natural and intuitive method for achieving this is the processor sharing (PS) model, originally proposed by Kleinrock (KLE67). Every active user receives an equal amount of service or processor time in round-robin fashion. So if there are $n$ customers in the system, the server spends $1/n$ time units with each. If one of the customers departs, the service is adjusted immediately. The processor-sharing model is a limiting case where the service quantum can approach zero and originates from an idealized fluid model.

Generalized processor sharing (GPS) is an extension of this model where the service rate is still equal for each of the customers, but is adjusted by some function that depends on the number of customers in the system. For example, the server rate could increase if there are more flows active. It is easy to further extend PS and GPS by introducing the concept of weights. In this case, each incoming flow has a weight attached to it that adjusts the time the server spends on it accordingly. In the fluid model, these weights are analogous to the speed of the different fluids (or different widths of their respective pipes).

The flow-oriented nature of GPS makes it ideal for modeling IntServ (BEC01). Furthermore, it is widely recognized to provide (min-max) fairness between any number of streams. One important point to make is that since GPS requires arbitrarily small time increments, it can only ever be approximated in a network environment. The smallest time increment in a network is usually determined by a packet service time, as a packet has to be served in its entirety before it can be forwarded (unless some form of pre-emption is used). There are various algorithms to approximate GPS in real-life scenarios such as fair queuing (FQ) and weighted fair queuing (WFQ), which are covered in the following.

In addition to GPS, there are other methods of sharing the service between various jobs in the system. Of those we mention general head of the line proportional processor sharing (GHLPPS), shortest remaining processor time (SRPT), and foreground background (FB). It should be noted, however, that these methods are not usually implemented in IP networks as they are generally concerned with customers whose service times range over several orders of magnitude. In most IP networks, the service time a packet receives usually depends on its size. The typical range of packet sizes—though having profound effects on the queuing performance—is not large enough for disciplines such as SRPT and FB to be much different from GPS (these mechanisms are sometimes used in Web servers, where the downloading of a file is a single job whose size depends on the file size).

*FQ, WFQ, and CBWFQ*

FQ is an algorithm to implement PS (DEM90). All the incoming flows are served in round-robin fashion. WFQ is an extension of FQ, which associates a weight with each stream and allocates the service dynamically across flows based on these weights. In most implementations, these weights are dependent on the number of packets present in the system at any one time (actually, this is similar to GHLPPS). Class-based WFQ (CBWFQ) extends this again by letting the user set the weight given to each class (for ease of configuration, the typical way to implement this is by letting the user specify the desired bandwidth per class and then calculate the weights internally based on these bandwidth settings).

It has been shown [GRE92] that FQ approximates PS very closely with a difference of only one packet service time. Since this difference is the closest any algorithm in a non-pre-emptive network can approximate PS, it follows that there is no better approximation.

WFQ guarantees a certain minimum bandwidth per priority class and has the advantage that any unused bandwidth is shared out equally between the active flows. It also offers min-max fairness between the incoming flows. Its disadvantages are that the overall delay for each stream, even for the

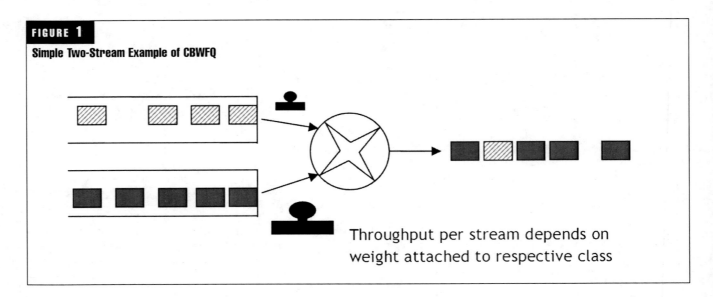

**FIGURE 1**

**Simple Two-Stream Example of CBWFQ**

Throughput per stream depends on weight attached to respective class highest priority, is based on the total number of active classes. It is also not possible to predict the amount of loss per stream and two streams within the same class can compete for the same bandwidth with the more aggressive one penalizing the other.

### PBS, HOL, PQ, RED, and TCP Rate Shaping

Most of the methods in the previous sections described different ways of sharing the service between the customers. Instead of sharing the service, it is also possible to allocate places in the queue in different ways. This can range from various quality classes having access to different buffers to not allowing customers to enter the queue at all.

Head of the line (HOL) is a priority-based buffer-sharing system. The queue is set up such that newly arriving customers of priority $l$ join the queue behind all packets of priority $l$ or better and in front of all packets of priority $l+1$ or worse. Customarily, lower priority packets in service are not pre-empted by arriving higher priority ones, but sometimes this is done as well. In any other respect the queue operates as a FIFO queue. Under HOL highest-priority packets will have at most one additional extra service time delay, namely if a packet of low priority is in service. The advantage of HOL is reduced delay for the highest-priority packets, but without any other mechanism, high-priority packets can potentially starve lower priority ones.

Priority queuing (PQ) is very similar to HOL. Here, there are as many queues as priority classes and the queues are serviced in order of priority. Customers join the queue corresponding to their priority. Note that this is functionally equivalent to a single-queue HOL system, which only ever allows a specific number of packets per priority class into the queue. So while PQ and HOL have quite similar effects, the main distinction is that PQ manages the priority classes using service and HOL using arrivals.

Partial buffer sharing (PBS) is also a priority-based buffer-sharing system, but set up in a different way. With PBS, one or more thresholds are set in the queue, each associated with a priority class. Once the number of packets in the queue exceeds a particular threshold, only packets of priority higher than the one associated with the threshold are accepted into the queue, while arriving lower-priority packets are discarded. PBS can range from a complete sharing model (which is equivalent to a normal queue) to complete partitioning, where there is a threshold associated with each buffer.

The main benefit of PBS is to ensure that a high load in low-priority streams will not cause excessive loss for more important flows. However, PBS on its own is not very effective at controlling the delay of high-priority streams. In general, the more buffers there are in a queue, the higher the average delay per packet. In fact, if only PBS is employed, one can expect the delay for higher-priority packets to be larger than for lower-priority ones.

Random early detection (RED) [FLO93] is a system to control the incoming load and thus tries to actually avoid congestion. If the network load increases beyond a certain level, then packets are dropped at random. Normally, RED is used in conjunction with TCP traffic, i.e., it exploits the TCP congestion control mechanism; if TCP packets get dropped,

then the source will realize this and react by reducing the transmission speed until the problem is resolved. The rate at which packets are discarded is a function of the queue length and can be weighted based on traffic characteristics, thus giving preferential treatment to certain flows. This variant is called weighted RED (WRED).

There is another method making use of the TCP information exchange mechanism (i.e., the handshake) that aims to ensure that all TCP flows transmit with a pre-configured rate. TCP rate shaping modifies the window size for a TCP traffic flow. Delaying acknowledgment packets effectively causes the rate to slow down and, therefore, aids in avoiding congestion. TCP rate shaping is very good at guaranteeing bandwidth but is hard to scale because it needs to keep track of all active TCP connections (contrary to RED).

While RED and TCP rate shaping are able to improve TCP performance, they are powerless with regards to other protocols—UDP in particular—used by most of the new breed of real-time applications such as VoIP.

### Token Bucket and Leaky Bucket

A token bucket is a facility with a numerical counter that determines the amount of traffic processed. It is parameterized by two numbers, $r$ and $b$. Here, $r$ is the rate at which tokens accumulate in the bucket, and $b$ is its capacity, i.e., $b$ determines the maximum number of tokens that can be stored. Tokens arriving when the bucket is full will be discarded. The token bucket regulates the output of the queue by requiring that each forwarded packet has to remove a number of tokens corresponding to its size in bytes (or possibly some other parameter). If there are fewer tokens present than a particular packet requires, then that packet is usually discarded, though there are variants which buffer the packet until enough tokens accumulate (of course, since the data buffer is limited, only a certain number of packets can be buffered).

In case of the former option, the primary function of the token bucket is to control the rate of the incoming traffic (policing). In the latter case, the token bucket also acts as a shaper by changing the inter-packet gaps of the flow. The advantages of the token bucket include a very effective control of the average throughput as well as reducing traffic bursts (since the length of a burst is limited by the parameter $b$). However, the throughput can nonetheless become very large on very short time scales. Token buckets are most commonly used as policers in connection with other mechanisms. For example, the token bucket is specified for IP services ([RFC2212]) and widely used, with modifications, in routing products.

The leaky bucket mechanism is similar to the token bucket in that it also deploys some "container" to regulate the traffic flow. Contrary to the token bucket, the leaky bucket collects the arriving packets in the bucket (instead of tokens). If the bucket is full, any further arriving packets are discarded. Furthermore, the bucket has a "hole" through which the packets "leak" away at a fixed rate. The method used for departure is generally FIFO, but other mechanisms can of course also be used. The main difference to a normal queue is that the departure times are fixed and do not (up to a point) depend on arrivals.

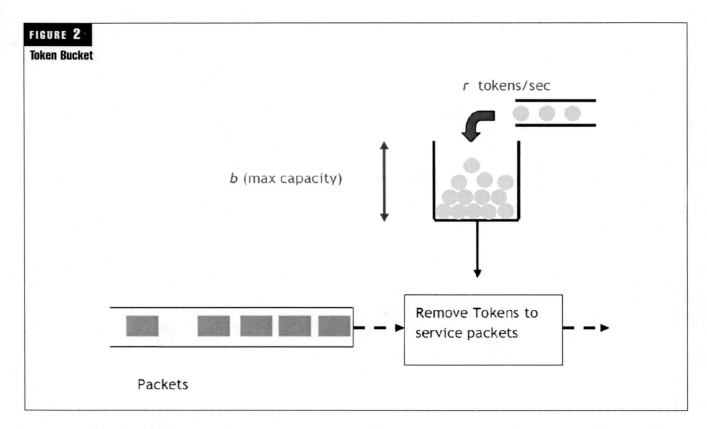

**FIGURE 2**
**Token Bucket**

Sometimes, leaky and token buckets are used in combination to form policies such as the dual-leaky bucket. For example, if a token bucket forwards packets to a leaky bucket, then the token bucket could be used as a policer, while the leaky bucket acts a shaper. The main disadvantage of both "bucket policies" is the inability to distinguish between types of traffic.

*LLQ and Other Combinations*
Many QoS implementations in IP networks consist of a combination of several of the above mechanisms. A typical example is low-latency queuing (LLQ), which is a combination of CBWFQ with a single-priority queue. The priority queue provides a low-delay service, which is normally intended for VoIP traffic, while all other flows are handled by the CBWFQ part. While this is an improvement on the delay control of CBWFQ, it is insufficient with regards to multiple low-delay priority services due to the limitations of a single-priority queue.

Other common combinations include WFQ and WRED (e.g., as an implementation of AF classes defined by differentiated services [DiffServ]) or WFQ with multiple token buckets to try to limit the delay given to individual streams. This is feasible [PAR92], but the resulting delay bounds are generally very large and still not suitable for real-time traffic.

*LDQ*
As mentioned earlier, loss, delay, and throughput form a system with two degrees of freedom. Reviewing many of the commonly used IP QoS mechanisms and combinations of mechanisms, it becomes quickly apparent that even though all of them provide an improvement in service, quite often, the method in question only improves a particular aspect of quality and possesses no flexibility with regards to

the other two. For example, a token bucket is only able to regulate loss and has no influence on the delay or throughput. Similarly, a WFQ discipline allows differentiated throughput control but is not able to affect the loss and delay more than a simple FIFO queue (so the combination of WFQ with the token bucket just mentioned will also not be very efficient at controlling delay).

Since the main strengths of the mechanisms reviewed so far are spread across the spectrum of parameters, it is surprising that there are hardly any combined implementations that play to these different strengths.

One such mechanism is loss-delay queuing, also known as guarantee of service (GoS), which is essentially a combination of HOL and PBS, plus some randomization and policing elements that overcome the danger of starvation inherent to HOL. Additionally, the underlying mathematical algorithm is based on a finite buffer model, contrary to those for several of the other methods. This ensures greater applicability to the real world and allows firm statistical predictions ([HOL00], [VOW01]).

In pure HOL, the priority assigned to a class determines its delay; however, there is no control of the loss experienced by that class. Furthermore, the delay is only ensured for the highest-priority class, and lower-priority classes can be squeezed out by more urgent ones (up to the point of starvation). In PBS, on the other hand, the classification of a stream will decide the loss percentage for that stream.

LDQ uses PBS to control packet arrivals and hence control and differentiate loss. It uses a priority service mechanism to control packet departures, thereby differentiating their delay.

Combining HOL and PBS allows for a more flexible approach where the priority class determines both the loss and the delay. Hence, the result is a shift away from the one-dimensional natures of HOL and PBS to form a two-dimensional system, where any loss priority can be associated with any delay priority. So, quality classes can be represented by a two-dimensional grid as shown in *Figure 3* for a 3x3 example:

Here, the capital letters denote the loss priorities and the numerals the delay priorities. So, for example, traffic assigned to priority class A2 will on average be delayed longer but will have a lower loss percentage than traffic in priority class C1. To guard against starvation and to ensure fairness, flows are policed and randomized. The most common cause of such unfairness is bursty traffic, or various phase effects between flows. Randomizing the gaps between packets alleviates phase effects and traffic bursts and guarantees that no class will be penalized more than its classification dictates.

## QoS Mechanisms Comparison

This section examines several important requirements for a QoS mechanism in a converged IP network and assesses the above approaches in this regard. The key requirements under consideration are as follows:

- *Multiservice capability*: Can the mechanism provide differentiated quality to multiple applications with different needs?

- *Scalability*: Is the mechanism efficient and effective when the number of users/applications increases?

- *Fairness*: Does the mechanism fulfill the fairness criteria defined earlier?

### Multiservice Capability

This is arguably the most important requirement for any QoS mechanism in a multiservice network. As we have seen above, most commonly used mechanisms have difficulty in providing quality to multiple applications with specific constraints on loss and delay. This is not surprising, since these mechanisms do not fully exploit the two-dimensional nature of the control space. To illustrate this further consider, the following example, which compares the use of the control space between CBWFQ and LDQ, has been provided:

Assume that there are just two arriving flows. Denote them by A and B and their respective average delays by $D_A$ and $D_B$. Suppose that each of the streams has the same arrival rate. To make the CBWFQ and LDQ systems comparable, we arrange them in the same way as much as possible by setting the speed of the outgoing link and the total number of buffers to be the same for both. Further presume that if a stream belongs to a specific class in CBWFQ, the bandwidth assigned to this class is determined by the arrival rate of the stream. Similarly, if a stream is placed into a particular LDQ class, the bandwidth allocated to that class is derived in the same way. Since this setup essentially fixes throughput, we concentrate on the possible performance ranges of the two models regarding delay and loss per flow. We are only interested in the relative values of these measurements, i.e., is $D_A$, greater, equal to, or smaller than $D_B$? (The absolute values depend of course on the number of buffers made available and the speed of the service.)

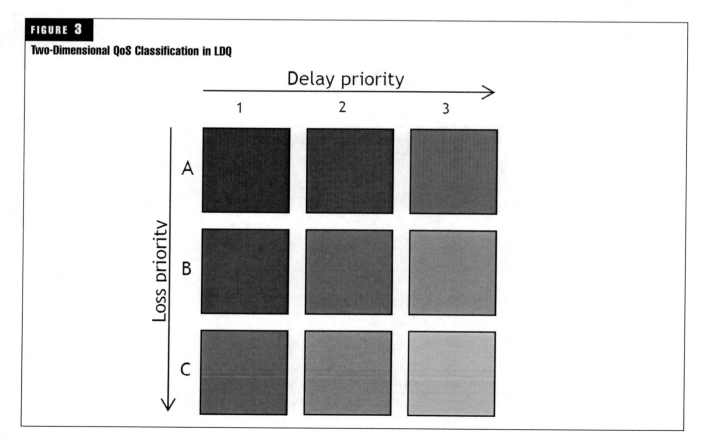

**FIGURE 3**

**Two-Dimensional QoS Classification in LDQ**

The relative performance of A and B in CBWFQ is determined by the weights assigned to them. Call these weights $a$ and $b$. These weights describe the time the server spends servicing each flow. If $\mu$ denotes the average service rate, then we have $b = \mu - a$, i.e., after choosing one of the weights, the other one is determined.

Suppose without loss of generality that $a > b$. This translates to an increase in service rate for flow A compared to B. Thus by Little's Law, packets of type A will on average be serviced faster and are also less likely to be lost than packets of type B. Hence, $D_A < D_B$ and, simultaneously, $L_A < L_B$, where $L_A$ and $L_B$ denote the average loss for A and B, respectively. Note, that it is not possible to have $D_A > D_B$ and simultaneously $L_A < L_B$. Thus in CBWFQ, the performance of one stream compared to another will always consist of either improved or deteriorated loss *and* delay, and never a mixture. Essentially, CBWFQ only increases or decreases the bandwidth for a particular stream.

On the other hand, in LDQ, we can place any stream in any of the grid classes. The relationship between any two streams is determined by their placement in the grid. If A is in A1, for example, and B in B2, then we also have $D_A < D_B$ and $L_A < L_B$ just as in the CBWFQ case. But if we place A in B1 and B in A2, relations such as $D_A > D_B$ and $L_A < L_B$ are also achievable, allowing far greater flexibility than in CBWFQ. *Figure 4* illustrates this:

In LDQ, A and B can be placed in any of the boxes, allowing for all possible loss/delay combinations. CBWFQ would correspond solely to the blue line, whereas A and B could be placed anywhere on the line. For further comparison, the green and red line mark the two incorporated components of LDQ, HOL, and PBS, and the yellow line indicates LLQ.

Since multiservice networks run many applications with many loss and delay requirements (and combinations thereof), it is clear that many older QoS mechanisms lack the flexibility to accommodate the respective needs of these applications.

### Scalability

Another important requirement is the ability of a QoS mechanism to deal with continuing convergence. For example, if it is not able to handle an ever-increasing number of users, it is unlikely to be a good candidate for a multiservice network, as it will probably have to be upgraded or replaced in the near future (or the network will have to be run continuously at very low efficiency levels).

Implementations that contain state-based mechanisms and need to keep track of every flow such as TCP rate shaping or WFQ can be quite problematic in this respect, whereas state-free mechanisms such as RED scale more easily.

### Fairness

The fairness provided to different flows sharing the same quality class is another important consideration. Quite often, these flows represent different users of the same application, and it is crucial that the QoS mechanism provides a reasonable degree of fairness to ensure that aggressive over-contract users do not excessively penalize the others. CBWFQ and TCP rate shaping have been specifically designed to deliver min-max fairness. However, for most mechanisms the users in a class are essentially competing in

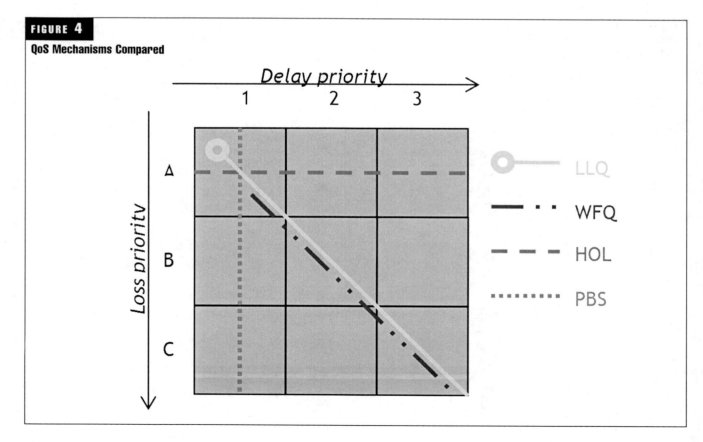

**FIGURE 4**

**QoS Mechanisms Compared**

a FIFO queue which, under ideal conditions, provides proportional fairness but has absolutely no protection against traffic patterns and phase effects.

For LDQ, a randomization element greatly reduces the impact of traffic bursts, resulting in much improved proportional fairness.

*Figure 5*, obtained from real-world testing, illustrates the fairness provided to 30 users under different schemes.

It shows a comparison of the average throughput between FIFO, LDQ, FQ, and RED (in clockwise order starting from the top left). Each area of a circle represents the throughput for a particular user and shows its proportion to the total area. The "fair share" is given by the equal bandwidth allocation, so the average percentage for each circle sector in a system with perfect fairness would be 3.33 percent. Since perfect fairness is unattainable in practice (in any work-conversing discipline), an acceptable fairness threshold might be for each stream to receive at least half its fair share (which translates to 1.67 percent of the total circle area).

If the users are managed using a 32-length FIFO queue, only 20 receive a fair bandwidth allocation. RED fares much worse, only effectively sharing the bandwidth between 7 streams as shown. With FQ, 26 users are observed to have at least their fair throughput. In a single LDQ class, 25 of the users consume at least their fair bandwidth allocation. Note how the proportional fairness achieved by the LDQ randomization is nearly as good as the FQ min-max fairness without requiring a state-based system.

## Conclusions

This paper presented a state-of-the-art overview of currently used QoS mechanisms in congested IP networks and assessed their capability in the face of growing convergence. Three main criteria (multiservice capability, fairness, and scalability) were identified as primary requirements for any qualified mechanism.

We have found that most of the mechanisms surveyed (with the exception of loss-delay queuing) are not ideally suited to provide QoS in such a multiservice network, falling short on at least one of the criteria.

It is surprising that this issue has not been better addressed in the past 10 years. Many researchers have predicted the rise of convergence, but their focus has been on improving quality mechanisms in the core such as route-finding protocols, while providing QoS at a contention point seems to have gone out of fashion. Most of the mechanisms surveyed have been designed without modern multiservice networks in mind, and little effort has been spent in improving them.

However, as discussed earlier, these are the points where an effective mechanism for managing contention so as to guarantee QoS is essential, if the ability of the network to deliver QoS end-to-end is not to be compromised.

Real-time or interactive entertainment and gaming, live broadcasts, etc., over efficiently utilized IP networks is not a feasible prospect until the hurdles of QoS for multiple services are overcome.

## FIGURE 5
### Fairness Comparison

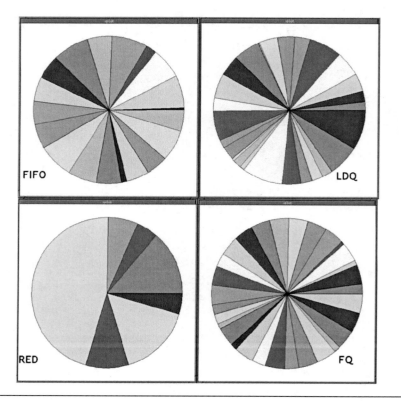

However, signs are emerging that more interest is being paid in this field. Multiservice QoS has become a non-negotiable requirement and we are therefore poised to see a lot more activity in the next decade, as network and service providers start to understand and resolve the problems associated with reliable and cost-effective multiservice networks.

## References

(1) Allen A., "Probability, Statistics, and Queueing Theory with Computer Science Applications, Second Edition," Academic Press Inc., 1990.

(2) Bansal, N. and Harchol-Balter, M., *"Analysis of SRPT Scheduling: Investigating Unfairness,"* School of Computer Science Carnegie Mellon University.

(3) Beckers, J., Hendrawan, I., Koi, R., and Van der Mei, R., *"Generalized Processor Sharing Performance Models for Internet Access Lines,"* 9th IFIP Conference on Performance Modeling and Evaluation of ATM & IP Networks 2001.

(4) Bernet, Y., Ford, P., Yakatkar, R., Baker, F., Zhang, L., Speer, M., Braden, R., Davie, B., Wroclawski, J., Felstaine, E., *"A Framework for Integrated Services Operation over Diffserv Networks,"* RFC 2998, 2000.

(5) Blake, S., Black, D., Carlson, M., Davies, E., Wang, Z., and Weiss, W., *"An Architecture for Differentiated Services,"* RFC 2475, 1998.

(6) Bonald, T., May, M., and Bolot, J., *Analytic Evaluation of RED Performance,* IEEE Infocom 2000.

(7) Braden, R., Clark, D., and Shenker, S., *"Integrated Services in the Internet Architecture: an Overview,"* RFC 1633, 1994.

(8) Braden, R., Zhang, L., Berson, S., Herzog, S., and Jamin, S., *"Resource Reservation Protocol (RSVP) Version 1 Functional Specification,"* RFC 2205, 1997.

(9) Chuang, L., Wanming, L., Baoping, Y., and Chanson, S., *"A Dynamic Partial Buffer Sharing Scheme for Packet Loss Control in Congested Networks,"* Department of Computer Science, Tsinghua University, Beijing.

(10) Crovella, M. and Bestavros, A., *"Self-Similarity in World Wide Web Traffic: Evidence and Possible Causes,"* Proc. of the 1996 ACM SIGMETRICS International Conference on Measurement and Modelling of Computer Systems, pp 160–169, 1996.

(11) Demers, A., Keshav, S., and Shenker, S., *"Analysis and Simulation of a Fair Queueing Algorithm,"* Internetworking: Research and Experience 1, pp 3–26, 1990.

(12) Feng, W., Kandlur, D., Saha, D., and Shin, K., *"A Self-Configuring RED Gateway,"* IEEE 1999.

(13) Floyd, S. and Jacobson, V., *"Random Early Detection Gateways for Congestion Avoidance,"* IEEE/ACM Transactions on Networking, Vol. 1, pp 397–413, 1993.

(14) Floyd, S. and Jacobson, V., *"Link-sharing and Resource Management Models for Packet Networks,"* IEEE/ACM Transactions on Networking, Vol. 3 No. 4, pp 365–386, 1995.

(15) Greenberg, A., Madras, N., *"How Fair is Fair Queueing?"* Journal of the Association for Computing Machinery 39, 3, pp 568–598.

(16) Hogan, D., *"Hierarchical Fair Queueing,"* Basser Department of Computer Science, University of Sydney, 1996.

(17) Holyer, J., *"A Queueing Theory Model for Real Data Networks,"* UK Performance and Engineering Workshop, 2000.

(18) Jacobson, V., Nichols, K., and Zhang, L., *"A Two-Bit Differentiated Services Architecture for the Internet,"* RFC 2638, 1999.

(19) Kleinrock, L., *"Time-Shared Systems: A Theoretical Treatment,"* Journal of the Association for Computing Machinery 14, 2, pp 242–261, 1967.

(20) Kleinrock, L., "Queueing Systems, Volume 1," John Wiley & Sons Inc, 1975.
[LIT61] LITTLE, J.D.C. *"A Proof of the Queueing Formula L=λW,?* Operations Research, 9, 383?387 (1961).

(21) Matsuo, T., Hasegawa, G., Murata, M., and Miyahara, H., *?Comparisons of Packet Scheduling Algorithms for Fair Service among Connections,?* Department of Information and Computer Sciences, Osaka University.

(22) Matsufuru, N., and Aibara, R., *"Efficient Fair Queueing for ATM Networks Using Uniform Round Robin,"* IEEE 1999.

(23) May, M., Bolot, J., Jean-Marie, A., and Diot, C., *"Simple Performance Models of Differentiated Services Schemes for the Internet,"* IEEE 1999.

(24) Morris, R., *"Variance of Aggregated Web Traffic,"* IEEE Infocom 2000.

(25) Parekh, A., "A generalized processor sharing approach to flow control in integrated services networks," Ph.D. thesis, Massachusetts Institute of Technology, February 1992.

(26) Pekergin, N., *"Stochastic Bounds on Delays of Fair Queueing Algorithms,"* IEEE 1999.

(27) Rom, R., and Sidi, M., "Multiple Access Protocols, Performance and Analysis," Springer Verlag, 1990.

(28) Rosen, E., Viswanathan, A., and Callon, R., *"Multiprotocol Label Switching Architecture,"* RFC3031.

(29) Takagi, H., "Queueing Analysis, Volume 1-2," Elsevier Science Publishers B.V., 1993.

(30) Tang, P., and Tai, C., *"Network Traffic Characterization Using Token Bucket Model,"* IEEE 1999.

(31) Vowden, C., and LaFave, L., *"Guaranteeing Quality: An Analytical Framework for Network Resource Sharing,"* UK Performance and Engineering Workshop 2001.

# Acronym Guide

| | |
|---|---|
| 2B1Q | two binary, one quaternary |
| 2B1Q | two binary, one quaternary |
| 2G | second generation |
| 3DES | triple data encryption standard |
| 3G | third generation |
| 3GPP | third-generation partnership project |
| 3R | regeneration, reshaping, and retraining |
| 4B3T | four binary, three ternary |
| 4F/BDPR | four-fiber bidirectional dedicated protection ring |
| 4F/BSPR | four-fiber bidirectional shared protection ring |
| 4G | fourth generation |
| AAA | authentication, authorization, and accounting |
| AAL–[x] | ATM adaptation layer–x |
| ABC | activity-based costing |
| ABR | available bit rate |
| AC | alternating current OR authentication code |
| ACD | automatic call distributor |
| ACF | admission confirmation |
| ACH | automated clearinghouse |
| ACL | access control list |
| ACLEP | adaptive code excited linear prediction |
| ACM | address complete message |
| ACR | alternate carrier routing OR anonymous call rejection |
| ADM | add/drop multiplexer OR asymmetric digital multiplexer |
| ADPCM | adaptive differential pulse code modulation |
| ADS | add/drop switch |
| ADSI | analog display services interface |
| ADSL | asymmetric digital subscriber line |
| AES | advanced encryption standard |
| AFE | analog front end |
| AGW | agent gateway |
| AIM | advanced intelligent messaging |
| AIN | advanced intelligent network |
| ALI | automatic location identification |
| AM | amplitude modulation |
| AMA | automatic messaging account |
| AMI | alternate mark inversion |
| AMPS | advanced mobile phone service |
| AN | access network |
| ANI | automatic number identification |
| ANM | answer message |
| ANSI | American National Standards Institute |
| AOL | America Online |
| AON | all-optical network |
| AP | access point OR access provider |
| APC | automatic power control |
| API | application programming interface |
| APON | ATM passive optical network |
| APS | automatic protection switching |
| ARCNET | attached resource computer network |
| ARI | assist request instruction |
| ARM | asynchronous response mode |
| ARMS | authentication, rating, mediation, and settlement |
| ARP | address resolution protocol |
| ARPANET | Advanced Research Projects Agency Network |
| ARPU | average revenue per customer |
| A-Rx | analog receiver |
| AS | application server OR autonomous system |
| ASAM | ATM subscriber access multiplexer |
| ASC | Accredited Standards Committee |
| ASCII | American Standard Code for Information Interchange |
| ASE | amplified spontaneous emission |
| ASIC | application-specific integrated circuit |
| ASIP | application-specific instruction processor |
| ASON | automatically switched optical network |
| ASP | application service provider |
| ASR | access service request OR answer-seizure rate OR automatic service request OR automatic speech recognition |
| ASSP | application-specific standard part |
| ASTN | automatically switched transport network OR analog switched telephone network |
| ATC | automatic temperature control |
| ATIS | Alliance for Telecommunications Industry Solutions |
| ATM | asynchronous transfer mode OR automated teller machine |
| ATMF | ATM Forum |
| ATP | analog twisted pair |
| ATU-C | ADSL transmission unit-CO |
| ATU-R | ADSL transmission unit-remote |
| A-Tx | analog transceiver |
| AUI | attachment unit interface |
| AVI | audio video interleaved |
| AWG | American Wire Gauge OR arrayed waveguide grating |
| AYUTOS | as-yet-unthought-of services |
| B2B | business-to-business |
| B2C | business-to-consumer |
| BCSM | basic call state model |
| BDCS | broadband digital cross-connect system |
| BDPR | bidirectional dedicated protection ring |
| BE | border element |
| BER | bit-error rate |
| BERT | bit error–rate test |
| BGP | border gateway protocol |
| BH | busy hour |
| BHCA | busy hour call attempt |
| BI | bit rate independent |
| BICC | bearer independent call control |
| BID | bit rate identification |
| BIP | bit interactive parity |
| B–ISDN | broadband ISDN |
| BLEC | broadband local-exchange carrier OR building local-exchange carrier |

| | | | |
|---|---|---|---|
| BLES | broadband loop emulation services | CDMS | configuration and data management server |
| BLSR | bidirectional line-switched ring | CDN | control directory number |
| BML | business management layer | CDPD | cellular digital packet data |
| BOC | Bell operating company | CDR | call detail record OR clock and data recovery |
| BOF | business operations framework | | |
| BOND | back-office network development | CD–ROM | compact disc–read-only memory |
| BOSS | broadband operating system software | CWDM | coarse wavelength division multiplexing |
| BPON | broadband passive optical network | CE | customer edge |
| BPSK | binary phase shift keying | CEI | comparable efficient interface |
| B–RAS | broadband–remote access server | CEO | chief executive officer |
| BRI | basic rate interface | CER | customer edge router |
| BSA | business services architecture | CERT | computer emergency response team |
| BSPR | bidirectional shared protection ring | CES | circuit emulation service |
| BSS | base-station system OR business support system | CES | circuit emulation service |
| | | CESID | caller emergency service identification |
| BTS | base transceiver station | CEV | controlled environment vault |
| BVR | best-value routing | CFB/NA | call forward busy/not available |
| BW | bandwidth | CFO | chief financial officer |
| CA | call agent | CGI | common gateway interface |
| CAC | call admission control OR carrier access code OR connection admission control | CHN | centralized hierarchical network |
| | | C–HTML | compressed HTML |
| CAD | computer-aided design | CIC | circuit identification code |
| CAGR | compound annual growth rate | CID | caller identification |
| CALEA | Communications Assistance for Law Enforcement Act | CIM | common information model |
| | | CIMD2 | computer interface message distribution 2 |
| CAM | computer-aided manufacture | CIO | chief information officer |
| CAMEL | customized application of mobile enhanced logic | CIP | classical IP over ATM |
| | | CIR | committed information rate |
| CAP | competitive access provider OR carrierless amplitude and phase modulation OR CAMEL application part | CIT | computer integrated telephone |
| | | CLASS | custom local-area signaling services |
| | | CLE | customer-located equipment |
| CAPEX | capital expenditures/expenses | CLEC | competitive local-exchange carrier |
| CAR | committed access rate | CLI | command-line interface OR call-line identifier |
| CARE | customer account record exchange | | |
| CAS | channel-associated signaling OR communications applications specification | CLID | calling-line identification |
| | | CLLI | common language location identifier |
| | | CLR | circuit layout record |
| CAT | conditional access table OR computer-aided telephony | CM | cable modem |
| | | CM&B | customer management and billing |
| CATV | cable television | CMIP | common management information protocol |
| C-band | conventional band | | |
| CBDS | connectionless broadband data service | CMISE | common management information service element |
| CBR | constant bit rate | | |
| CBT | core-based tree | CMOS | complementary metal oxide semiconductor |
| CC | control component | | |
| CCB | customer care and billing | CMRS | commercial mobile radio service |
| CCF | call-control function | CMTS | cable modem termination system |
| CCI | call clarity index | CNAM | calling name (also defined as "caller identification with name" and simply "caller identification") |
| CCITT | Consultative Committee on International Telegraphy and Telephony | | |
| | | CNAP | CNAM presentation |
| CCK | complementary code keying | CNS | customer negotiation system |
| CCR | call-completion ratio | CO | central office |
| CCS | common channel signaling | CODEC | coder-decoder OR compression/decompression |
| CD | chromatic dispersion OR compact disc | | |
| cDCF | conventional dispersion compensation fiber | COI | community of interest |
| | | COO | chief operations officer |
| CDD | content delivery and distribution | COPS | common open policy service |
| CDDI | copper-distributed data interface | CORBA | common object request broker architecture |
| CDMA | code division multiple access | | |
| CDMP | cellular digital messaging protocol | CoS | class of service |

| | | | |
|---|---|---|---|
| COT | central office terminal | DBS | direct broadcast satellite |
| COTS | commercial off-the-shelf | DC | direct current |
| COW | cell site on wheels | DCC | data communications channel |
| CP | connection point | DCF | discounted cash flow OR dispersion compensation fiber |
| CPAS | cellular priority access service | | |
| CPC | calling-party category (also calling-party control OR calling-party connected) | DCLEC | data competitive local-exchange carrier |
| | | DCM | dispersion compensation module |
| CPE | customer-premises equipment | DCN | data communications network |
| CPI | continual process improvement | DCOM | distributed component object model |
| CPL | call-processing language | DCS | digital cross-connect system OR distributed call signaling |
| CPLD | complex programmable logic device | | |
| CPN | calling-party number | DCT | discrete cosine transform |
| CPU | central processing unit | DDN | defense data network |
| CR | constraint-based routing | DDS | dataphone digital service |
| CRC | cyclic redundancy check OR cyclic redundancy code | DECT | Digital European Cordless Telecommunication |
| CRIS | customer records information system | demarc | demarcation point |
| CR–LDP | constraint-based routed–label distribution protocol | DEMS | digital electronic messaging service |
| | | DES | data encryption standard |
| CRM | customer-relationship management | DFB | distributed feedback |
| CRTP | compressed real-time transport protocol | DFC | dedicated fiber/coax |
| CRV | call reference value | DGD | differentiated group delay |
| CS | client signal | DGFF | dynamic gain flattening filter |
| CS–[x] | capability set [x] | DHCP | dynamic host configuration protocol |
| CSA | carrier serving area | DiffServ | differentiated services |
| CSCE | converged service-creation and execution | DIN | digital information network |
| CSCF | call-state control function | DIS | distributed interactive simulation |
| CSE | CAMEL service environment | DITF | Disaster Information Task Force |
| CS–IWF | control signal interworking function | DLC | digital loop carrier |
| CSM | customer-service manager | DLCI | data-link connection identifier |
| CSMA/CA | carrier sense multiple access with collision avoidance | DLE | digital loop electronics |
| | | DLEC | data local-exchange carrier |
| CSMA/CD | carrier sense multiple access with collision detection | DLR | design layout report |
| | | DM | dense mode |
| CSN | circuit-switched network | DMD | dispersion management device |
| CSP | communications service provider OR content service provider | DMS | digital multiplex system |
| | | DMT | discrete multitone |
| CSR | customer-service representative | DN | distinguished name |
| CSU | channel service unit | DNS | domain name server OR domain naming system |
| CSV | circuit-switched voice | | |
| CT | computer telephony | DOC | department of communications |
| CT–2 | cordless telephony generation 2 | DOCSIS | data over cable service interface specifications |
| CTI | computer telephony integration | | |
| CTIA | Cellular Telecommunications & Internet Association | DOD | Department of Defense |
| | | DOJ | Department of Justice |
| CTO | chief technology officer | DoS | denial of service |
| CWD | centralized wavelength distribution | DOS | disk operating system |
| CWDM | coarse wavelength division multiplexing | DOSA | distributed open signaling architecture |
| | | DOT | Department of Transportation |
| CWIX | cable and wireless Internet exchange | DP | detection point |
| DAC | digital access carrier | DPC | destination point code |
| DACS | digital access cross-connect system | DPE | distributed processing environment |
| DAM | DECT authentication module | DPT | dial pulse terminate |
| DAMA | demand assigned multiple access | DQoS | dynamic quality of service |
| DAML | digital added main line | D-Rx | digital receiver |
| DARPA | Defense Advanced Research Projects Agency | DS–[x] | digital signal [level x] |
| | | DSAA | DECT standard authentication algorithm |
| DAVIC | Digital Audio Video Council | DSC | DECT standard cipher |
| DB | database | DSCP | DiffServ code point |
| dB | decibel(s) | DSF | dispersion-shifted fiber |
| DBMS | database management system | DSL | digital subscriber line [also xDSL] |
| dBrn | decibels above reference noise | DSLAM | digital subscriber line access multiplexer |

| | |
|---|---|
| DSLAS | DSL–ATM switch |
| DSP | digital signal processor OR digital service provider |
| DSS | decision support system |
| DSSS | direct sequence spread spectrum |
| DSU | data service unit OR digital service unit |
| DTH | direct-to-home |
| DTMF | dual-tone multifrequency |
| DTV | digital television |
| D-Tx | digital transceiver |
| DVB | digital video broadcast |
| DVC | dynamic virtual circuit |
| DVD | digital video disc |
| DVMRP | distance vector multicast routing protocol |
| DVoD | digital video on demand |
| DVR | digital video recording |
| DWDM | dense wavelength division multiplexing |
| DXC | digital cross-connect |
| E911 | enhanced 911 |
| EAI | enterprise application integration |
| EAP | extensible authentication protocol |
| EBITDA | earnings before interest, taxes, depreciation, and amortization |
| EC | electronic commerce |
| ECD | echo-cancelled full-duplex |
| ECRM | echo canceller resource module |
| ECTF | Enterprise Computer Telephony Forum |
| EDA | electronic design automation |
| EDF | electronic distribution frame OR erbium-doped fiber |
| EDFA | erbium-doped fiber amplifier |
| EDGE | enhanced data rates for GSM evolution |
| EDI | electronic data interchange |
| EDSX | electronic digital signal cross-connect |
| EFM | Ethernet in the first mile |
| EFT | electronic funds transfer |
| EJB | enterprise Java beans |
| ELAN | emulated local-area network |
| ELEC | enterprise local-exchange carrier |
| EM | element manager |
| EMI | electromagnetic interference |
| EML | element-management layer |
| EMS | element-management system OR enterprise messaging server |
| E–NNI | external network-to-network interface |
| ENUM | telephone number mapping |
| E–O | electrical-to-optical |
| EO | end office |
| EoA | Ethernet over ATM |
| EOC | embedded operations channel |
| EoVDSL | Ethernet over VDSL |
| EPD | early packet discard |
| EPON | Ethernet PON |
| EPROM | erasable programmable read-only memory |
| ERP | enterprise resource planning |
| ESCON | enterprise systems connectivity |
| ESS | electronic switching system |
| ETC | establish temporary connection |
| EtherLEC | Ethernet local-exchange carrier |
| ETL | extraction, transformation, and load |
| eTOM | enhanced telecom operations map |

| | |
|---|---|
| ETSI | European Telecommunications Standards Institute |
| EU | European Union |
| EURESCOM | European Institute for Research and Strategic Studies in Telecommunications |
| EXC | electrical cross-connect |
| FAB | fulfillment, assurance, and billing |
| FAQ | frequently asked question |
| FBG | fiber Bragg grating |
| FCAPS | fault, configuration, accounting, performance, and security |
| FCC | Federal Communications Commission |
| FCI | furnish charging information |
| FCIF | flexible computer-information format |
| FDA | Food and Drug Administration |
| FDD | frequency division duplex |
| FDDI | fiber distributed data interface |
| FDF | fiber distribution frame |
| FDM | frequency division multiplexing |
| FDMA | frequency division multiple access |
| FDS–1 | fractional DS–1 |
| FE | extended framing |
| FEC | forward error correction |
| FEPS | facility and equipment planning system |
| FEXT | far-end crosstalk |
| FHSS | frequency hopping spread spectrum |
| FICON | fiber connection |
| FITL | fiber-in-the-loop |
| FM | fault management OR frequency modulation |
| FOC | firm order confirmation |
| FOT | fiber-optic terminal |
| FOTS | fiber-optic transmission system |
| FP | Fabry-Perot [laser] |
| FPB | flex parameter block |
| FPGA | field programmable gate array |
| FPLMTS | future public land mobile telephone system |
| FPP | fast-packet processor |
| FR | frame relay |
| FRAD | frame-relay access device |
| FSAN | full-service access network |
| FSC | framework services component |
| FSN | full-service network |
| FT | fixed-radio termination |
| FT1 | fractional T1 |
| FTC | Federal Trade Commission |
| FTE | full-time equivalent |
| FTP | file transfer protocol |
| FTP3 | file transfer protocol 3 |
| FTTB | fiber-to-the-building |
| FTTC | fiber to the curb |
| FTTCab | fiber-to-the-cabinet |
| FTTEx | fiber-to-the-exchange |
| FTTH | fiber-to-the-home |
| FTTN | fiber-to-the-neighborhood |
| FTTS | fiber-to-the-subscriber |
| FTTx | fiber-to-the-x |
| FWM | four-wave mixing |
| FX | foreign exchange |
| GA | genetic algorithm |

| | |
|---|---|
| Gb | gigabit |
| GbE | gigabit Ethernet [also GE] |
| GBIC | gigabit interface converter |
| Gbps | gigabits per second |
| GCRA | generic cell rate algorithm |
| GDIN | global disaster information network |
| GDMO | guidelines for the definition of managed objects |
| GE | [see GbE] |
| GEO | geosynchronous Earth orbit |
| GETS | government emergency telecommunications service |
| GFF | gain flattening filter |
| GFR | guaranteed frame rate |
| Ghz | gigahertz |
| GIF | graphics interface format |
| GIS | geographic information services |
| GKMP | group key management protocol |
| GMII | gigabit media independent interface |
| GMLC | gateway mobile location center |
| GMPCS | global mobile personal communications services |
| GMPLS | generalized MPLS |
| GNP | gross national product |
| GOCC | ground operations control center |
| GPIB | general-purpose interface bus |
| GPRS | general packet radio service |
| GPS | global positioning system |
| GR | generic requirement |
| GRASP | greedy randomized adaptive search procedure |
| GSA | Global Mobile Suppliers Association |
| GSM | Global System for Mobile Communications |
| GSMP | generic switch management protocol |
| GSR | gigabit switch router |
| GTT | global title translation |
| GUI | graphical user interface |
| GVD | group velocity dispersion |
| GW | gateway |
| HCC | host call control |
| HD | home domain |
| HDLC | high-level data-link control |
| HDML | handheld device markup language |
| HDSL | high-bit-rate DSL |
| HDT | host digital terminal |
| HDTV | high-definition television |
| HDVMRP | hierarchical distance vector multicast routing protocol |
| HEC | head error control OR header error check |
| HEPA | high-efficiency particulate arresting |
| HFC | hybrid fiber/coax |
| HIDS | host intrusion detection system |
| HLR | home location register |
| HN | home network |
| HOM | high-order mode |
| HomePNA | Home Phoneline Networking Alliance [also HomePNA2] |
| HomeRF | Home Radio Frequency Working Group |
| HQ | headquarters |
| HSCSD | high-speed circuit-switched data |
| HSD | high-speed data |

| | |
|---|---|
| HSIA | high-speed Internet access |
| HSP | hosting service provider |
| HTML | hypertext markup language |
| HTTP | hypertext transfer protocol |
| HVAC | heating, ventilating, and air-conditioning |
| HW | hardware |
| IAD | integrated access device |
| IAM | initial address message |
| IAS | integrated access service OR Internet access server |
| IAST | integrated access, switching, and transport |
| IAT | inter-arrival time |
| IBC | integrated broadband communications |
| IC | integrated circuit |
| ICD | Internet call diversion |
| ICDR | Internet call detail record |
| ICL | intercell linking |
| ICMP | Internet control message protocol |
| ICP | integrated communications provider OR intelligent communications platform |
| ICS | integrated communications system |
| ICW | Internet call waiting |
| IDC | Internet data center OR International Data Corporation |
| IDE | integrated development environment |
| IDES | Internet data exchange system |
| IDF | intermediate distribution frame |
| IDL | interface definition language |
| IDLC | integrated digital loop carrier |
| IDS | intrusion detection system |
| IDSL | integrated services digital network DSL |
| IEC | International Electrotechnical Commission OR International Engineering Consortium |
| IEEE | Institute of Electrical and Electronics Engineers |
| I-ERP | integrated enterprise resource planning |
| IETF | Internet Engineering Task Force |
| IFITL | integrated [services over] fiber-in-the-loop |
| IFMA | International Facility Managers Association |
| IFMP | Ipsilon flow management protocol |
| IGMP | Internet group management protocol |
| IGP | interior gateway protocol |
| IGRP | interior gateway routing protocol |
| IGSP | independent gateway service provider |
| IHL | Internet header length |
| IIOP | Internet inter–ORB protocol |
| IIS | Internet Information Server |
| IKE | Internet key exchange |
| ILA | in-line amplifier |
| ILEC | incumbent local-exchange carrier |
| ILMI | interim link management interface |
| IM | instant messaging |
| IMA | inverse multiplexing over ATM |
| IMAP | Internet message access protocol |
| IMRP | Internet multicast routing protocol |
| IMSI | International Mobile Subscriber Identification |
| IMT | intermachine trunk OR International Mobile Telecommunications |

| | | | |
|---|---|---|---|
| IMTC | International Multimedia Teleconferencing Consortium | ITU–T | ITU–Telecommunication Standardization Sector |
| IN | intelligent network | ITV | Internet television |
| INAP AU | INAP adaptation unit | IVR | interactive voice response |
| INAP | intelligent network application part | IVRU | interactive voice-response unit |
| INE | intelligent network element | IWF | interworking function |
| InfoCom | information communication | IWG | interworking gateway |
| INM | integrated network management | IWU | interworking unit |
| INMD | in-service, nonintrusive measurement device | IXC | interexchange carrier |
| I–NNI | internal network-to-network interface | J2EE | Java Enterprise Edition |
| INT | [point-to-point] interrupt | J2ME | Java Micro Edition |
| InterNIC | Internet Network Information Center | J2SE | Java Standard Edition |
| IntServ | integrated services | JAIN | Java APIs for integrated networks |
| IOF | interoffice facility | JCAT | Java coordination and transactions |
| IOS | intelligent optical switch | JCC | JAIN call control |
| IP | Internet protocol | JDBC | Java database connectivity |
| IPBX | Internet protocol private branch exchange | JDMK | Java dynamic management kit |
| IPcoms | Internet protocol communications | JMAPI | Java management application programming interface |
| IPDC | Internet protocol device control | | |
| IPDR | Internet protocol data record | JMX | Java management extension |
| IPe | intelligent peripheral | JPEG | Joint Photographic Experts Group |
| IPG | intelligent premises gateway | JSCE | JAIN service-creation environment |
| IPO | initial public offering OR Internet protocol over optical | JSIP | Java session initiation protocol |
| | | JSLEE | JAIN service logic execution environment |
| IPoA | Internet protocol over ATM | JTAPI | Java telephony application programming interface |
| IPQoS | Internet protocol quality of service | | |
| IPSec | Internet protocol security | JVM | Java virtual machine |
| IPTel | IP telephony | kbps | kilobits per second |
| IPv6 | Internet protocol version 6 | kHz | kilohertz |
| IPX | Internet package exchange | km | kilometer |
| IR | infrared | L2F | Layer-2 forwarding |
| IRU | indefeasible right to user | L2TP | Layer-2 tunneling protocol |
| IS | information service OR interim standard | LAC | L2TP access concentrator |
| IS-IS | intermediate system to intermediate system | LAI | location-area identity |
| | | LAN | local-area network |
| ISA | industry standard architecture | LANE | local-area network emulation |
| ISAPI | Internet server application programmer interface | LATA | local access and transport area |
| | | LB311 | location-based 311 |
| ISC | integrated service carrier OR International Softswitch Consortium | L-band | long band |
| | | LBS | location-based services |
| ISDF | integrated service development framework | LC | local convergence |
| | | LCD | liquid crystal display |
| ISDN | integrated services digital network | LCP | link control protocol |
| ISDN–BA | ISDN basic access | LD | laser diode OR long distance |
| ISDN–PRA | ISDN primary rate access | LDAP | lightweight directory access protocol |
| ISEP | intelligent signaling endpoint | LD–CELP | low delay–code excited linear prediction |
| ISM | industrial, scientific, and medical OR integrated service manager | LDP | label distribution protocol |
| | | LDS | local digital service |
| ISO | International Organization for Standardization | LE | line equipment OR local exchange |
| | | LEAF® | large-effective-area fiber |
| ISOS | integrated software on silicon | LEC | local-exchange carrier |
| ISP | Internet service provider | LED | light-emitting diode |
| ISUP | ISDN user part | LEO | low Earth orbit |
| ISV | independent software vendor | LEOS | low Earth-orbiting satellite |
| IT | information technology OR Internet telephony | LER | label edge router |
| | | LES | loop emulation service |
| ITSP | Internet telephony service provider | LIDB | line information database |
| ITTP | information technology infrastructure library | LL | long line |
| | | LLC | logical link control |
| ITU | International Telecommunication Union | LMDS | local multipoint distribution system |
| | | LMN | local network management |

| | | | |
|---|---|---|---|
| LMOS | loop maintenance operation system | MII | media independent interface |
| LMP | link management protocol | MIME | multipurpose Internet mail extensions |
| LMS | loop-management system OR loop-monitoring system OR link-monitoring system | MIMO | multiple inputs, multiple outputs |
| | | MIN | mobile identification number |
| | | MIPS | millions of instructions per second |
| LNNI | LANE network-to-network interface | MIS | management information system |
| LNP | local number portability | MITI | Ministry of International Trade and Industry (in Japan) |
| LNS | L2TP network server | | |
| LOL | loss of lock | MLT | mechanized loop testing |
| LOS | line of sight OR loss of signal | MM | mobility management |
| LPF | low-pass filter | MMDS | multichannel multipoint distribution system |
| LQ | listening quality | | |
| LRN | local routing number | MMPP | Markov-Modulated Poisson Process |
| LRQ | location request | MMS | multimedia message service |
| LSA | label switch assignment OR link state advertisement | MMUSIC | Multiparty Multimedia Session Control [working group] |
| LSB | location-sensitive billing | MNC | mobile network code |
| LSMS | local service management system | MOM | message-oriented middleware |
| LSO | local service office | MON | metropolitan optical network |
| LSP | label-switched path | MOP | method of procedure |
| LSR | label-switched router OR leaf setup request OR local service request | MOS | mean opinion score |
| | | MOSFP | multicast open shortest path first |
| LT | line terminator OR logical terminal | MOU | minutes of use OR memorandum of understanding |
| LTE | lite terminating equipment | | |
| LUNI | LANE user network interface | MPC | mobile positioning center |
| LX | local exchange | MPEG | Moving Pictures Experts Group |
| M2PA | message transfer protocol 2 peer-to-peer adaptation | MPI | message passing interface |
| | | MPLambdaS | multiprotocol lambda switching |
| M2UA | message transfer protocol 2–user adaptation layer | MPLS | multiprotocol label switching |
| | | MPOA | multiprotocol over ATM |
| M3UA | message transfer protocol 3–user adaptation layer | MPoE | multiple point of entry |
| | | MPoP | metropolitan point of presence |
| MAC | media access control | MPP | massively parallel processor |
| MADU | multiwave add/drop unit | MPx | MPEG–Layer x |
| MAN | metropolitan-area network | MRC | monthly recurring charge |
| MAP | mobile applications part | MRS | menu routing system |
| MAS | multiple-application selection | MRSP | mobile radio service provider |
| Mb | megabit | ms | millisecond |
| MB | megabyte | MSC | mobile switching center |
| MBAC | measurement-based admission control | MSF | Multiservice Switch Forum |
| MBGP | multicast border gateway protocol | MSIN | mobile station identification number |
| MBone | multicast backbone | MSNAP | multiple services network access point |
| Mbps | megabits per second | MSO | multiple-system operator |
| MC | multipoint controller | MSP | management service provider |
| MCC | mobile country code | MSPP | multiservice provisioning platform |
| MCU | multipoint control unit | MSS | multiple-services switching system |
| MDF | main distribution frame | MSSP | mobile satellite service provider |
| MDSL | multiple DSL | MTA | message transfer agent |
| MDTP | media device transport protocol | MTBF | mean time between failures |
| MDU | multiple-dwelling unit | MTP [x] | message transfer part [x] |
| MEGACO | media gateway control | MTTR | mean time to repair |
| MEMS | micro-electromechanical system | MTU | multiple-tenant unit |
| MExE | mobile execution environment | MVL | multiple virtual line |
| MF | multifrequency | MWIF | Mobile Wireless Internet Forum |
| MFJ | modified final judgment | MZI | Mach-Zender Interferometer |
| MG | media gateway | N11 | (refers to FCC–managed dialable service codes such as 311, 411, and 911) |
| MGC | media gateway controller | | |
| MGCF | media gateway control function | NA | network adapter |
| MGCP | media gateway control protocol | NAFTA | North America Free Trade Agreement |
| MHz | megahertz | NANC | North American Numbering Council |
| MIB | management information base | NANP | North American Numbering Plan |

| | | | |
|---|---|---|---|
| NAP | network access point | NSAP | network service access point |
| NARUC | National Association of Regulatory Utility Commissioners | NSAPI | Netscape server application programming interface |
| NAS | network access server | NSCC | network surveillance and control center |
| NASA | National Aeronautics and Space Administration | NSDB | network and services database |
| | | NSP | network service provider OR network and service performance |
| NAT | network address translation | | |
| NATA | North American Telecommunications Association | NSTAC | National Security Telecommunications Advisory Committee |
| NBN | node-based network | NT | network termination OR new technology |
| NCP | network control protocol | NTN | network terminal number |
| NCS | national communications system OR network connected server | NTSC | National Television Standards Committee |
| | | NVP | network voice protocol |
| NDA | national directory assistance | NZ–DSF | nonzero dispersion-shifted fiber |
| NDM–U | network data management–usage | O&M | operations and maintenance |
| NDSF | non-dispersion-shifted fiber | OA&M | operations, administration, and maintenance |
| NE | network element | | |
| NEAP | non-emergency answering point | OADM | optical add/drop multiplexer |
| NEBS | network-equipment building standards | OAM&P | operations, administration, maintenance, and provisioning |
| NEL | network-element layer | | |
| NEXT | near-end crosstalk | OBF | Ordering and Billing Forum |
| NFS | network file system | OBLSR | optical bidirectional line-switched ring |
| NG | next generation | OC–[x] | optical carrier–[level x] |
| NGCN | next-generation converged network | OCBT | ordered core-based protocol |
| NGDLC | next-generation digital loop carrier | OCD | optical concentration device |
| NGF | next-generation fiber | OCh | optical channel |
| NGN | next-generation network | OCR | optical character recognition |
| NGOSS | next-generation operations system and software OR next-generation OSS | OCS | original call screening |
| | | OCU | office channel unit |
| NHRP | next-hop resolution protocol | OCX | open compact exchange |
| NI | network interface | OD | origin-destination |
| NIC | network interface card | ODBC | open database connectivity |
| NID | network interface device | ODSI | optical domain services interface |
| NIDS | network intrusion detection system | O–E | optical-to-electrical |
| NIIF | Network Interconnection Interoperability Forum | O–EC | optical–electrical converter |
| | | OECD | Organization for Economic Cooperation and Development |
| NIS | network information service | | |
| NIU | network interface unit | OEM | original equipment manufacturer |
| nm | nanometer | O–E–O | optical-to-electrical-to-optical |
| NML | network-management layer | OEXC | opto-electrical cross-connect |
| NMS | network-management system | OFDM | orthogonal frequency division multiplexing |
| NND | name and number delivery | | |
| NNI | network-to-network interface | OIF | Optical Internetworking Forum |
| NNTP | network news transport protocol | OLA | optical line amplifier |
| NOC | network operations center | OLAP | on-line analytical processing |
| NOMAD | national ownership, mobile access, and disaster communications | OLI | optical link interface |
| | | OLT | optical line termination OR optical line terminal |
| NP | number portability | | |
| NPA | numbering plan area | OLTP | on-line transaction processing |
| NPAC | Number Portability Administration Center | OMC | Operations and Maintenance Center |
| | | OMG | Object Management Group |
| NPN | new public network | OMS SW | optical multiplex section switch |
| NP–REQ | number-portable request query | OMS | optical multiplex section |
| NPV | net present value | OMSSPRING | optical multiplex section shared protection ring |
| NRC | Network Reliability Council OR nonrecurring charge | | |
| | | ONA | open network architecture |
| NRIC | Network Reliability and Interoperability Council | ONE | optical network element |
| | | ONI | optical network interface |
| NRSC | Network Reliability Steering Committee | ONMS | optical network-management system |
| NRZ | non–return to zero | ONT | optical network termination |
| NS/EP | national security and emergency preparedness | ONTAS | optical network test access system |
| | | ONU | optical network unit |

| | | | | |
|---|---|---|---|---|
| OP | optical path | PE | provider edge |
| OPEX | operational expenditures/expenses | PER | packed encoding rules |
| OPS | operator provisioning station | PERL | practical extraction and report language |
| OPTIS | overlapped PAM transmission with interlocking spectra | PESQ | perceptual evolution of speech quality |
| | | PFD | phase-frequency detector |
| OPXC | optical path cross-connect | PHB | per-hop behavior |
| ORB | object request broker | PHY | physical layer |
| ORT | operational readiness test | PIC | point-in-call OR predesignated interexchange carrier OR primary interexchange carrier |
| OS | operating system | | |
| OSA | open service architecture | | |
| OSC | optical supervisory panel | PICS | plug-in inventory control system |
| OSD | on-screen display | PIM | personal information manager OR protocol-independent multicast |
| OSGI | open services gateway initiative | | |
| OSI | open systems interconnection | PIN | personal identification number |
| OSMINE | operations systems modification of intelligent network elements | PINT | PSTN and Internet Networking [IETF working group] |
| OSN | optical-service network | PINTG | PINT gateway |
| OSNR | optical signal-to-noise ratio | PKI | public key infrastructure |
| OSP | outside plant OR open settlement protocol | PLA | performance-level agreement |
| | | PLC | planar lightwave circuit OR product life cycle |
| OSPF | open shortest path first | | |
| OSS | operations support system | | |
| OSS/J | OSS through Java | PLCP | physical layer convergence protocol |
| OSU | optical subscriber unit | PLL | phase locked loop |
| OTM | optical terminal multiplexer | PLMN | public land mobile network |
| OTN | optical transport network | PLOA | protocol layers over ATM |
| OUI | optical user interface | PM | performance monitoring |
| O-UNI | optical user-to-network interface | PMD | physical-medium dependent OR polarization mode dispersion |
| OUSP | optical utility services platform | | |
| OVPN | optical virtual petabits network OR optical virtual private network | PMDC | polarization mode dispersion compensator |
| OWSR | optical wavelength switching router | PMO | present method of operation |
| OXC | optical cross-connect | PMP | point-to-multipoint |
| P&L | profit and loss | PN | personal number |
| PABX | private automatic branch exchange | PNNI | private network-to-network interface |
| PACA | priority access channel assignment | PnP | plug and play |
| PACS | picture archiving communications system | PO | purchase order |
| PAL | phase alternate line | PODP | public office dialing plan |
| PAM | Presence and Availability Management [Forum] OR pulse amplitude modulation | POET | partially overlapped echo-cancelled transmission |
| | | POF | plastic optic fiber |
| PAMS | perceptual analysis measurement system | POH | path overhead |
| PAN | personal access network | POIS | packet optical interworking system |
| PBCC | packet binary convolutional codes | PON | passive optical network |
| PBN | point-to-point–based network OR policy-based networking | PoP | point of presence |
| | | POP3 | post office protocol 3 |
| PBX | private branch exchange | POS | packet over SONET OR point of service |
| PC | personal computer | PosReq | position request |
| PCF | physical control field | POT | point of termination |
| PCI | peripheral component interconnect | POTS | plain old telephone service |
| PCM | pulse code modulation | PP | point-to-point |
| PCN | personal communications network | PPD | partial packet discard |
| PCR | peak cell rate | PPP | point-to-point protocol |
| PCS | personal communications service | PPPoA | point-to-point protocol over ATM |
| PDA | personal digital assistant | PPPoE | point-to-point protocol over Ethernet |
| PDC | personal digital cellular | PPTP | point-to-point tunneling protocol |
| PDD | post-dial delay | PP–WDM | point-to-point–wavelength division multiplexing |
| PDE | position determination equipment | | |
| PDH | plesiochronous digital hierarchy | PQ | priority queuing |
| PDN | public data network | PRI | primary rate interface |
| PDP | policy decision point | ps | picosecond |
| PDSN | packet data serving node | PSAP | public safety answering point |
| PDU | protocol data unit | PSC | Public Service Commission |

| | | | |
|---|---|---|---|
| PSD | power spectral density | RPR | resilient packet ring |
| PSDN | public switched data network | RPRA | Resilient Packet Ring Alliance |
| PSID | private system identifier | RPT | resilient packet transport |
| PSN | public switched network | RQMS | requirements and quality measurement system |
| PSPDN | packet-switched public data network | | |
| PSQM | perceptual speech quality measure | RRQ | round-robin queuing or registration request |
| PSTN | public switched telephone network | | |
| PTE | path terminating equipment | RSU | remote service unit |
| PTN | personal telecommunications number service | RSVP | resource reservation protocol |
| | | RSVP–TE | resource reservation protocol–traffic engineering |
| PTP | point-to-point | | |
| PTT | Post Telephone and Telegraph Administration | RT | remote terminal |
| | | RTCP | real-time conferencing protocol |
| PUC | public utility commission | RTOS | real-time operating system |
| PVC | permanent virtual circuit | RTP | real-time transport protocol |
| PVM | parallel virtual machine | RTSP | real-time streaming protocol |
| PVN | private virtual network | RTU | remote test unit |
| PWS | planning workstation | RxTx | receiver/transmitter |
| PXC | photonic cross-connect | RZ | return to zero |
| QAM | quadrature amplitude modulation | SAM | service access multiplexer |
| QoE | quality of experience | SAN | storage-area network |
| QoS | quality of service | SAP | service access point OR session announcement protocol |
| QPSK | quaternary phase shift keying | | |
| QSDG | QoS Development Group | SAR | segmentation and reassembly |
| RAD | rapid application development | S-band | short band |
| RADIUS | remote authentication dial-in user service | SBS | stimulated Brillouin scattering |
| RADSL | rate-adaptive DSL | SCAN | switched-circuit automatic network |
| RAM | remote access multiplexer | SCCP | signaling connection control part |
| RAN | regional-area network | SCCS | switching control center system |
| RAP | resource allocation protocol | SCE | service-creation environment |
| RAS | remote access server | SCF | service control function |
| RBOC | regional Bell operating company | SCL | service control language |
| RCP | remote call procedure | SCM | service combination manager OR station class mark OR subscriber carrier mark |
| RCU | remote control unit | | |
| RDBMS | relational database management system | | |
| RDC | regional distribution center | SCN | service circuit node OR switched-circuit network |
| RDSLAM | remote DSLAM | | |
| REL | release | SCP | service control point |
| RF | radio frequency | SCR | sustainable cell rate |
| RFC | request for comment | SCSI | small computer system interface |
| RFI | request for information | SCSP | server cache synchronization protocol |
| RFP | request for proposal | SCTP | simple computer telephony protocol OR simple control transport protocol OR stream control transmission protocol |
| RFPON | radio frequency optical network | | |
| RFQ | request for quotation | | |
| RGU | revenue-generating unit | SD | selective discard |
| RGW | residential gateway | SD&O | service development and operations |
| RHC | regional holding company | SDA | separate data affiliate |
| RIAC | remote instrumentation and control | SDB | service design bureau |
| RIP | routing information protocol | SDC | service design center |
| RISC | reduced instruction set computing | SDF | service data function |
| RJ | registered jack | SDH | synchronous digital hierarchy |
| RLL | radio in the loop | SDM | service-delivery management OR shared data model |
| RM | resource management | | |
| RMA | request for manual assistance | SDN | software-defined network |
| RMI | remote method invocation | SDP | session description protocol |
| RMON | remote monitoring | SDRP | source demand routing protocol |
| ROADM | reconfigurable optical add/drop multiplexer | SDSL | symmetric DSL |
| | | SDTV | synchronous digital hierarchy |
| ROBO | remote office/branch office | SDV | switched digital video |
| ROI | return on investment | SE | service element |
| RPC | remote procedure call | SEC | Securities and Exchange Commission |
| RPF | reverse path forwarding | SEE | service-execution environment |

| | | | |
|---|---|---|---|
| SEP | signaling endpoint | SOE | standard operating environment |
| ServReq | service request | SOHO | small office/home office |
| SET | secure electronic transaction | SON | service order number |
| SFA | sales force automation | SONET | synchronous optical network |
| SFD | start frame delimiter | SOP | service order processor |
| SFF | small form-factor | SP | service provider OR signaling point |
| SFGF | supplier-funded generic element | SPC | stored program control |
| SG | signaling gateway | SPE | synchronous payload envelope |
| SG&A | selling, goods, and administration OR sales, goods, and administration | SPF | shortest path first |
| SGCP | simple gateway control protocol | SPIRITS | Service in the PSTN/IN Requesting Internet Service [working group] |
| SGSN | serving GPRS support node | SPIRITSG | SPIRITS gateway |
| SHDSL | single-pair high-bit-rate DSL | SPM | self-phase modulation OR subscriber private meter |
| SHLR | standalone home location register | | |
| SHV | shareholder value | SPoP | service point of presence |
| SI | systems integrator | SPX | sequence packet exchange |
| SIBB | service-independent building block | SQL | structured query language |
| SIC | service initiation charge | SQM | service quality management |
| SICL | standard interface control library | SRF | special resource function |
| SID | silence indicator description | SRP | source routing protocol |
| SIF | SONET Interoperability Forum | SRS | stimulated Raman scattering |
| sigtran | Signaling Transport [working group] | srTCM | single-rate tri-color marker |
| SIM | subscriber identity module OR service interaction manager | SS | softswitch |
| | | SS7 | signaling system 7 |
| SIP CPL | SIP call processing language | SSE | service subscriber element |
| SIP | session initiation protocol | SSF | service switching function |
| SIP–T | session initiation protocol for telephony | SSG | service selection gateway |
| SISO | single input, single output | SSL | secure sockets layer |
| SIU | service interface unit | SSM | service and sales management |
| SIVR | speaker-independent voice recognition | SSMF | standard single-mode fiber |
| SKU | stock-keeping unit | SSP | service switching point |
| SL | service logic | STE | section terminating equipment |
| SLA | service-level agreement | STM | synchronous transfer mode |
| SLC | subscriber line carrier | STN | service transport node |
| SLEE | service logic execution environment | STP | shielded twisted pair OR signal transfer point OR spanning tree protocol |
| SLIC | subscriber line interface circuit | | |
| SLO | service-level objective | STR | signal-to-resource |
| SM | sparse mode | STS | synchronous transport signal |
| SMC | service management center | SUA | SCCP user adaptation |
| SMDI | simplified message desk interface | SVC | switched virtual circuit |
| SMDS | switched multimegabit data service | SW | software |
| SME | small-to-medium enterprise | SWAN | storage wide-area network |
| SMF | single-mode fiber | SWAP | shared wireless access protocol |
| SML | service management layer | SWOT | strengths, weaknesses, opportunities, and threats |
| SMP | service management point | | |
| SMPP | short message peer-to-peer protocol | SYN | IN synchronous transmission |
| SMS | service-management system OR short message service | TALI | transport adapter layer interface |
| | | TAPI | telephony application programming interface |
| SMSC | short messaging service center | | |
| SMTP | simple mail transfer protocol | TAT | terminating access trigger OR termination attempt trigger OR transatlantic telephone cable |
| SN | service node | | |
| SNA | service node architecture OR service network architecture | | |
| | | Tb | terabit |
| SNAP | subnetwork access protocol | TBD | to be determined |
| SNMP | simple network-management protocol | Tbps | terabits per second |
| SNPP | simple network paging protocol | TC | tandem connect |
| SNR | signal-to-noise ratio | TCAP | transactional capabilities application part |
| SO | service objective | TCB | transfer control block |
| SOA | service order activation | TCIF | Telecommunications Industry Forum |
| SOAC | service order analysis and control | TCL | tool command language |
| SOAP | simple object access protocol | TCM | time compression multiplexing |
| SOCC | satellite operations control center | TCO | total cost of ownership |

| | | | |
|---|---|---|---|
| TCP | transmission control protocol | TV | television |
| TCP/IP | transmission control protocol/Internet protocol | UA | user agent |
| | | UADSL | universal ADSL |
| TC–PAM | trellis coded–pulse amplitude modulation | UAK | user-authentication key |
| TDD | time division duplex | UAWG | Universal ADSL Working Group |
| TDM | time division multiplex | UBR | unspecified bit rate |
| TDMA | time division multiple access | UBT | ubiquitous bus technology |
| TDMDSL | time division multiplex digital subscriber line | UCP | universal computer protocol |
| | | UCS | uniform communication standard |
| TDR | time domain reflectometer OR transaction detail record | UDDI | universal description, discovery, and integration |
| TE | traffic engineering | UDP | user datagram protocol |
| TEAM | transport element activation manager | UDR | usage detail record |
| TED | traffic engineering database | UI | user interface |
| TEM | telecommunications equipment manufacturer | ULH | ultra-long-haul |
| | | UM | unified messaging |
| TFD | toll-free dialing | UML | unified modeling language |
| THz | terahertz | UMTS | Universal Mobile Telecommunications System |
| TIA | Telecommunications Industry Association | | |
| TIMS | transmission impairment measurement set | UN | United Nations |
| TINA | Telecommunications Information Networking Architecture | UNE | unbundled network element |
| | | UNI | user network interface |
| TINA-C | Telecommunications Information Networking Architecture Consortium | UOL | unbundled optical loop |
| | | UPC | usage parameter control |
| TIPHON | Telecommunications and Internet Protocol Harmonization over Networks | UPI | user personal identification |
| | | UPS | uninterruptible power supply |
| TIWF | trunk interworking function | UPSR | unidirectional path-switched ring |
| TKIP | temporal key integrity protocol | URI | uniform resource identifier |
| TL1 | transaction language 1 | URL | universal resource locator |
| TLDN | temporary local directory number | USB | universal serial bus |
| TLS | transparent LAN service OR transport-layer security | USTA | United States Telecom Association |
| | | UTOPIA | Universal Test and Operations Interface for ATM |
| TLV | tag length value | | |
| TMF | TeleManagement Forum | UTS | universal telephone service |
| TMN | telecommunications management network | UWB | ultra wideband |
| TMO | trans-metro optical | UWDM | ultra-dense WDM |
| TN | telephone number | V&H | vertical and horizontal |
| TNO | telecommunications network operator | VAD | voice activity detection |
| TO&E | table of organization and equipment | VAN | value-added network |
| TOM | telecom operations map | VAR | value-added reseller |
| ToS | type of service | VAS | value-added service |
| TP | twisted pair | VASP | value-added service provider |
| TPM | transaction processing monitor | VBNS | very–high-speed backbone network service |
| TPS–TC | transmission control specific–transmission convergence | | |
| | | VBR | variable bit rate |
| TR | technical requirement OR tip and ring | VBR–nrt | variable bit rate–non–real-time |
| TRA | technology readiness assessment | VBR–rt | variable bit rate–real time |
| TRIP | telephony routing over Internet protocol | VC | virtual circuit OR virtual channel |
| trTCM | two-rate tri-color marker | VCC | virtual channel connection |
| TSB | telecommunication system bulletin | VCI | virtual channel identifier |
| TSC | terminating call screening | VCLEC | voice CLEC |
| TSI | time slot interchange | VCO | voltage-controlled oscillator |
| TSP | telecommunications service provider | VCR | videocassette recorder |
| TSS | Telecommunications Standardization Section | VCSEL | vertical cavity surface emitting laser |
| | | VD | visited domain |
| TTC | Telecommunications Technology Committee | VDM | value delivery model |
| | | VDSL | very-high–data-rate DSL |
| TTCP | test TCP | VeDSL | voice-enabled DSL |
| TTL | transistor-transistor logic | VGW | voice gateway |
| TTS | text-to-speech OR TIRKS® table system | VHE | virtual home environment |
| | | VHS | video home system |
| TUI | telephone user interface | VITA | virtual integrated transport and access |
| TUP | telephone user part | | |